The publisher gratefully acknowledges the generous support of the General Endowment Fund of the University of California Press Foundation.

BARNUM BROWN

BARNUM BROWN

The Man Who Discovered
Tyrannosaurus rex

LOWELL DINGUS
AND MARK A. NORELL

UNIVERSITY OF CALIFORNIA PRESS
BERKELEY   LOS ANGELES   LONDON

University of California Press, one of the most distinguished university presses in the United States, enriches lives around the world by advancing scholarship in the humanities, social sciences, and natural sciences. Its activities are supported by the UC Press Foundation and by philanthropic contributions from individuals and institutions. For more information, visit www.ucpress.edu.

University of California Press
Berkeley and Los Angeles, California

University of California Press, Ltd.
London, England

© 2010 by Lowell Dingus and Mark A. Norell

Library of Congress Cataloging-in-Publication Data

Dingus, Lowell.
    Barnum Brown : the man who discovered Tyrannosaurus rex / Lowell Dingus and Mark A. Norell.
       p.   cm.
    Includes bibliographical references and index.
    ISBN 978-0-520-25264-6 (cloth : alk. paper)
    1. Brown, Barnum. 2. Paleontologists—United States—Biography. 3. Tyrannosaurus rex. I. Norell, Mark. II. Title.
    QE707.B77D56   2010
    560.92—dc22   2009040528

Manufactured in the United States of America

19 18 17 16 15 14 13 12 11 10
10 9 8 7 6 5 4 3 2 1

This book is printed on Cascades Enviro 100, a 100% post consumer waste, recycled, de-inked fiber. FSC recycled certified and processed chlorine free. It is acid free, Ecologo certified, and manufactured by BioGas energy.

CONTENTS

List of Illustrations — VII
Prologue: The Mindset of Barnum Brown — XI

1. Child of the Frontier (1873–1889) — 1
2. Student... of Sorts (1889–1896) — 19
3. Apprentice Extraordinaire (1896–1898) — 38
4. To Land's End: Patagonia (1898–1900) — 59
5. To the Depths of Hell Creek (1900–1903) — 79
6. Love (1903–1906) — 97
7. Loss (1906–1910) — 111
8. The Canadian Dinosaur Bone Rush (1910–1916) — 128
9. Cuba, Abyssinia, and Other Intrigues (1916–1921) — 153
10. Jewels from the Orient: Raj India (1921–1923) — 174
11. Perils and Pearls Up the Irrawaddy: Burma (1923) — 198
12. Samos: Isle of Intrigue (1923–1925) — 208
13. Ancient Americans Hunting Bison? Birds as Dinosaurs? (1925–1931) — 227
14. Digging—and Flying—for Dinosaurs: Howe Quarry and the Aerial Survey of Western Fossil Beds (1931–1935) — 246
15. Toward the Golden Years: The Mystery Track-Maker and the Glen Rose Trackway (1935–1942) — 264
16. Brown as a Spy, Movie Consultant, and Showman at the World's Fair (1942–1963) — 285
Epilogue — 296

Appendix 1. List of Major Specimens Collected
by Barnum Brown on Display in the AMNH Fossil Halls   304

Appendix 2. Memoirs of Barnum Brown: Discovery,
Excavation, and Preparation of the Type Specimen
Tyrannosaurus rex   309

Appendix 3. Summary of Fossil Collections by
Barnum Brown and His AMNH Crews   312

Notes   315

Bibliography   343

Acknowledgments   351

Index   353

ILLUSTRATIONS

MAPS

1. Western North America, showing Brown's fossil sites — 51
2. Southern Patagonia, 1899 — 64
3. Red Deer River in Alberta, Canada, 1910–15 — 130
4. Southern Cienfuegos Province, Cuba — 156
5. Route of the Dudley Expedition to Abyssinia, 1920 — 164
6. Brown's route through Raj India during the 1921–23 expedition — 178
7. Brown's fossil site near Mogaung, Burma, 1923 — 201
8. Greek isle of Samos, showing the quarries where Brown collected in 1923–24 — 212

FIGURES

1. Portraits of Barnum Brown and his family taken around 1880 — 6
2. Class portrait of Barnum Brown as it appeared in the 1897 yearbook of the University of Kansas — 20
3. Williston's field crew for the 1895 KU expedition to southeastern Wyoming — 36
4. Riggs and Brown collecting a specimen of *Coryphodon* in Wyoming, 1896 — 46
5. Wagon at the base of the Wasatch badlands, Wyoming, 1896 — 47
6. Barnum Brown and Henry Fairfield Osborn at Como Bluff during the AMNH expedition of 1897 — 54

7. *Diplodocus* quarry at Como Bluff 55
8. Miocene exposures at Lake Pueyrredon, Patagonia, 1898 66
9. Hatcher's image of the team and wagon that Brown drove for him, 1899 68
10. Brown collecting skeleton of the Miocene horse *Hypohippus,* Colorado, 1901 84
11. Brown's crew using horse team and scraper to remove overburden from Quarry no. 1, Montana, 1902 92
12. Brown collecting specimen of *Triceratops* at Quarry no. 2, Montana, 1902 93
13. Brown's first wife, Marion, washing matrix for Pleistocene fossils, Arkansas, 1904 98
14. Crow Indians at Pryor Gap, Montana, 1904 101
15. Hauling a 4,100-pound block containing the pelvis of type *Tyrannosaurus rex* out of Quarry no. 1, Montana, 1905 108
16. Portrait of Frances Brown, only child of Barnum and Marion Brown, 1929 115
17. Blasting overburden covering the *Tyrannosaurus rex* bone layer, Big Dry Creek, Montana, 1908 116
18. Celebrants at the Fourth of July party held at the Twitchell ranch in 1908 118
19. Skull and vertebral column of the *Tyrannosaurus* skeleton, 1908 120
20. Brown's crew using block and tackle to crate the cast containing the pelvis of the 1908 *Tyrannosaurus rex* specimen 122
21. The AMNH field crew on a flatboat on the Red Deer River, Alberta, 1912 132
22. Brown's crew perched on the bluffs above the Red Deer River, Alberta, 1912 138
23. Brown excavating the skeleton of "Prosaurolophus" (*Corythosaurus*), Alberta, 1912 144
24. Brown's crew hauling their collection of fossils out of Red Deer Canyon, 1914 149
25. Brown and an assistant draining the Chapapote Spring at Baños de Ciego Montero, Cuba, 1918 157
26. Criminals hanged in an Abyssinian town, 1920 165

27. Group of Issa warriors, Abyssinia, 1920–21 — 168
28. Outcrop of Upper Siwalik strata exposed near Siswan, India, 1921–23 — 181
29. Brown and assistants with skull of "Platelephas," Siswan, India, 1921–23 — 183
30. Barnum and Lilian Brown, Punjab, India, 1921–23 — 192
31. Lilian Brown dressed in a sari, ca. 1922 — 196
32. Workers preparing to retrieve oil from a well by hand, Yenangyaung, Burma, 1923 — 200
33. Brown's crew excavating Quarry no. 1, Samos, Greece, 1923 — 214
34. Brown and Carl Schwachheim next to an in situ Folsom point, New Mexico, 1927 — 235
35. One of Brown's crew members photographing the Howe Quarry, Wyoming, bone bed from overhead in a barrel, 1934 — 254
36. Bone bed at Howe Quarry, Wyoming, 1934 — 255
37. Brown's crew at Howe Quarry, Wyoming, watching the plane in which Brown conducted the aerial survey, 1934 — 258
38. R. T. Bird drilling out a hadrosaur track in the roof of the States Mine, Colorado, 1937 — 270
39. R. T. Bird and his WPA crew excavating sauropod trackways, Texas, 1940 — 278
40. Brown, Schlaikjer, and Bird with the largely reconstructed skull of *Deinosuchus*, 1940 — 279
41. Brown in his famous beaverskin coat, Montana, ca. 1916 — 286
42. Brown measuring the femur of *Tyrannosaurus rex* in the Cretaceous Hall of the American Museum of Natural History, ca. 1938 — 292
43. Headstones of Marion Brown, Barnum Brown, and Lilian Brown in Oxford, New York, 2008 — 294

Prologue
The Mindset of Barnum Brown

THE GLITTER OF BARNUM BROWN'S celebrity star has dimmed considerably during the forty-five years since his death and the century since he resurrected *Tyrannosaurus rex* from the daunting badlands of the Hell Creek Formation in Montana. In the mid-twentieth century, however, Brown—known by his adoring admirers as "Mr. Bones"—was one of the most famous scientists in the world. People would flock to the train station when he arrived in the field to collect dinosaur fossils, vying for the right to drive him to his hotel or camp.[1] Thousands of other aficionados would gather around the radio to listen live to his tales from the field about the latest exploits and adventures on his decades-long quest for dinosaurs. In a sense, he became every bit as legendary as the dinosaurs he discovered.

Then as today, Brown was recognized as the greatest dinosaur collector of all time. There is no question that he was well built for the task, with a sturdy body that rose to over six feet, crowned as an adult with piercing eyes and a prominent bald pate.[2] Finding dinosaurs came easily to this man raised on a farm and trained at a university on the Kansas frontier. This background led him to pursue the life of a relative loner in the field, and there is little doubt that it was a lifestyle he preferred. It freed Brown to focus on his passion for fossil collecting and allowed him to roam the world in search of both dinosaurs and oil, as well as to gather intelligence for the U.S. government.

Despite his preference for fieldwork, Brown left a significant body of scientific publications that still inform paleontological research. His study involving the growth patterns of the early horned dinosaur *Protoceratops*, for example, was the first to consider the importance of ontogeny in dinosaurs, a field of research that remains vibrant to this day. His analysis of the rock layers in Montana that produced both *Tyrannosaurus* and the minute mammals that succeeded the large dinosaurs set the stage for later research regarding the demise of nonavian dinosaurs 65 million years ago. Although Brown published fewer scientific studies than many of his contemporaries, to state simply that he was the greatest dinosaur collector of all time seriously underestimates his impact on the discipline of paleontology, in terms of both research and collecting. For in addition to dinosaurs, Brown also discovered a marvelous menagerie of other fossil vertebrates, including the largest crocodile ever found (until recently) and a prehistoric bison with the Paleolithic spear point that killed it still lying by its skeleton.

Given all that he accomplished, the fact that no full biography of Brown has ever been written is a mystery to us. Now, essentially a century after his stunning discovery of *Tyrannosaurus*, seems an especially fitting moment to tell his tale. Our account relies heavily on Brown's field correspondence and expedition reports preserved in the archives of the American Museum of Natural History (AMNH). It has been both a privilege and a pleasure to explore the life of this man who so greatly influenced our profession and our careers. Although the intervening decades have shrouded some aspects of his life in a frustrating fog, one element of Brown's behavior is very clear: he was always willing to drop whatever he was doing and assume any reasonable risk in order to venture into the field in search of fossils, particularly if his destination was some far-off corner of the globe.

So it was that, on December 7, 1898, the twenty-five-year-old Barnum Brown awoke and prepared to trudge through a snowy New York City to his job at the American Museum of Natural History, completely unaware that he would end the day sailing away from New York. As he later related,

> The morning... started like any other winter day—snow piled high along the streets—and I arrived at the museum at 9 o'clock as usual, but before I had taken my hat off, Professor Osborn called me into his office.
> "Brown, I want you to go to Patagonia today with the Princeton expedition.... The boat leaves at eleven; will you go?"
> "This is short notice, Professor Osborn, but I'll be on that boat..."[3]

It had been a year and a half since Brown made a pitch to participate in an international expedition on behalf of the museum, and now his boss, Henry Fairfield Osborn, had finally given him the nod. It would be no short sojourn, either; as Osborn advised the young paleontologist, he might well be gone another year and a half.[4] Undaunted, Brown, who had never been out of the country, instantly set about gathering his equipment with the help of museum staff. Although it's not clear from existing documents whether Osborn had already given Brown an inkling that something was afoot, Brown does reveal that he "had anticipated another expedition and had brought back from my field trip in western Kansas my boots, saddle, bridle, winchester rifle, blankets, and heavy tarpaulin cover."[5]

Brown was the sole AMNH representative on the expedition, which also included two highly experienced field men, the expedition leader, Princeton's John Bell Hatcher, and his assistant, Olaf A. Peterson, both of whom had participated in previous fossil-collecting expeditions to Patagonia.

On a moment's notice, Brown found himself boarding the Grace Line freighter *Capac* for a nonstop, month-long journey to the small but bustling port of Punta Arenas on the Chilean shore of the Straits of Magellan, near the southern tip of South America. The long cruise proved trying for Brown, for as he later admitted, he "had never been to sea before and soon was a victim of seasickness, first hoping I would die, then afraid I wouldn't."[6]

To pass the time and ease his discomfort, Brown played cards with his colleagues. Unbeknownst to him, however, Hatcher was also a "master at poker." In fact, he had funded his three expeditions to Patagonia by tutoring ranch hands in how to play.[7] Hatcher, whom colleagues described as cantankerous, was according to Brown "a man of average height, lean, about 35 years of age, with an uncommon knowledge of human nature." At the poker table, his face was "inscrutable, and you never knew whether he had a bob-tail flush, or a full house." Allegedly, Hatcher on occasion lost "his entire year's salary" in a poker session, then vowed to win it all back in the next.

Early on in the voyage, before Brown had a chance to gauge his opponent:

> Peterson came up and asked if I would like to join in a poker game. Being fond of poker, I agreed but neglected to ascertain the values of the ... chips. When this game ended ... I was the sixteen dollar loser.
>
> From then on during the entire voyage we played poker night and day, everyone losing to Hatcher. "Brown," he would say, "I hate to take your money as I know your salary is only fifty dollars a month." But in the last game ... when Petersen and the captain had dropped out leaving only Hatcher and myself betting, we kept raising the bets until finally Hatcher called, showing three tens to my three queens.... I regained almost all I had lost during the voyage.[8]

Soon after arriving in Patagonia, Brown found himself on his own after Hatcher and Peterson, having completed their collecting work, returned to the United States. Undaunted by the challenge of coping alone in a foreign land, he pressed on for well over a year in his pursuit of fossil mammals for Osborn and the museum.

Brown's willingness to drop whatever he was doing for an opportunity to traverse a new field area searching for fossils is emblematic of both his personality and his professional dedication. His more or less solitary jaunt through the Patagonian wilderness set many precedents that would play out over the rest of his life and career.

Nevertheless, Brown's preparations for such a spur-of-the-moment undertaking as the Patagonian project had not begun in New York. In reality, he had long honed his skills in very different surroundings from those of the big city, for he hailed from the American West.

ONE

Child of the Frontier
(1873–1889)

AS HE NEARED HIS NINETIETH YEAR, Barnum Brown determinedly struggled to compile his notes for an autobiography. After decades of racing around the planet in pursuit of paleontological plunder, this project afforded him a rare opportunity to reflect on a life filled with adventure and intrigue. Even his very first memory involved a solitary encounter with a universe that he would later come to revere and explore on his own terms with energy and determination. As Brown described the moment: "My earliest recollection is of lying in a clothes basket under a tree as I looked up and saw the leaves moving overhead. I was probably less than two years old at the time."[1]

What could be more prescient for the most successful field paleontologist in history than to root his perception of consciousness in the wonder of the natural world? Fortunately for Brown, an eye for nature was not unique to him within his family. His parents nurtured his instincts for the outdoors throughout his childhood on the raucous frontier of the Midwest's rolling hills and grassy plains.

Almost twenty years before Barnum's birth, his father took a calculated gamble and renounced the relatively settled landscape of the East to strike out for the fabled expanses of the American West. Brown, who in his notes anoints his father as "My Most Unforgettable Character," admired his father's sense of adventure, but he also had profound respect for the sense of responsibility that tempered it. As Brown explained,

> My parents both came from old pioneer stock: William Brown, my father, was born in Virginia in 1833. He had a deep abiding love for horses and other livestock, for the soil, and for his country. There were those among the pioneers who were merely drifters, fiddle-footed and restless, that wandered westward either to escape an unpleasant situation in the east, or in the hope of getting something for nothing in the west.... Father's pioneering was

purposeful: he was hard working, with a good head for business; he sought and found promising opportunities worthy of the heavy investment of thought, time and labor that he poured into them.

Cognizant of the burgeoning wave of westward migration, and catalyzed by events following the Mexican War and California gold rush in the late 1840s, the twenty-one-year-old William Brown hitched his oxen to his covered wagon and headed west in 1854, the same year that Stephen A. Douglas proposed the Kansas-Nebraska Act establishing the Territory of Kansas. With his eye trained on Kansas, Brown first traveled to Wisconsin by way of a bustling Chicago, with its thirty thousand residents, in order to assess "how the pioneers of the generation before his own had dealt with the problems of *their* frontier." The rich fields of Wisconsin were already studded with prosperous farms, one of which belonged to Charles Silver, a former army officer who had "fought as a private against Tecumseh at Tippecanoe in 1811." Silver attained the rank of captain during the War of 1812 and participated in the Battle of Bad Axe in 1832, where he fought against the legendary chief of the Sacs and Foxes, Black Hawk. In the years between his tours in the military, Silver developed a large dairy farm in Green County, where he owned his own cheese house. In 1855, Brown, while accumulating more capital and livestock in preparation for his foray into Kansas, met Silver's fifteen-year-old daughter, Clara. They married the same year. Four years later, the young couple—now with a daughter, Melissa—

> loaded such of their possessions as were not on the hoof into [their] ox-drawn covered wagon, and headed westward.... They averaged 10 miles a day on the way to Kansas Territory where, near Lickskillet in Osage County, my second sister, Alice Elizabeth, was born on the 4th of January, 1860. Father went from place to place in the Osage County area, sizing up the opportunities.... Finally, he picked the spot for his future home on one of the rolling hills.... Seams of coal cropped out around its slopes, so the pioneers named it Carbon Hill.... Father set to work with a will; they lived in the wagon for the short time it took him to build a log house. The windows at first were greased paper; the tables and chairs were made of boxes and barrels that contained supplies. And they were home!

Although this pioneer family's home seems initially to have been a happy one, the same cannot be said for the greater region into which they had immigrated. Tremendous tensions were building in the Kansas Territory, especially regarding the question of slavery. The Kansas-Nebraska Act had repealed the Missouri Compromise, which stipulated that any new states admitted to the union north of latitude 36° 30'N must be "free" states. Stephen Douglas, who was a railway promoter, believed that the citizens of a territory being considered for statehood should have the right to vote, through "popular" or "squatter" referendums, on whether their territory would become a free or slave state. He was also pressing for the first transcontinental railway to run through Chicago. In order to overcome the opposition of southern legislators to his preferred railroad route (they wanted the transcontinental railway to run from New Orleans to southern California), Douglas sponsored the Kansas-Nebraska Act, which southerners favored because it would essentially overturn the Missouri Compromise by letting residents of the Kansas and Nebraska Territories vote on whether to become free or slave states. The population of Nebraska was dominated by free-staters, so little question existed about the anti-slavery orientation of that territory.

But Kansas was a different matter. In his biographical notes, Brown wrote that waves of settlers from both northern and southern states rushed into the region to establish new farms, with many of the "Yankees" being spurred on by "abolitionist promoters . . . specifically to provide an anti-slavery majority." Violent raids, such as the sacking of Lawrence in 1856 by pro-slavery militias, were quickly followed by retaliatory attacks, such as the one by John Brown in which five pro-slavery settlers were killed along the Potawatamie River.

All these tumultuous events, often referred to as "Bleeding Kansas," helped bring about the Lincoln-Douglas debates in the Illinois senatorial campaign of 1858, a central issue of which was the question of slavery in new states. Douglas continued to advocate his position of "popular sovereignty," whereas Lincoln opposed the expansion of slavery into new territories and states. Their widely watched debates, noteworthy for their eloquence, fueled Lincoln's ascendancy to the presidency in 1860. Shortly thereafter, Southern forces fired on Fort Sumter—the first shots of the Civil War. The effects of these epic events in U.S. history were directly felt on Carbon Hill. Barnum relates:

> Kansas was admitted as a Free State in 1861. Raiders of all persuasions, lawless guerrillas and partisans, stole or destroyed much of the crops and livestock. Father saw that he would never get ahead at that rate, so he put adversity to good use. He was a good man . . . with wide experience in wagon-train freighting: he kept close watch on his animals, got good, strong covered freight wagons and a government contract. During the war years

he was home at the cabin during the winter; spring, summer and autumn he hauled supplies from the railheads at Fort Leavenworth and thence by a roughly triangular route to the frontier posts in western Kansas, Nebraska, Wyoming, Colorado, Santa Fe, and back to Osage County. He hauled anything for which the Army gave orders . . . [including] staples such as corn, flour, sugar, and coffee. His train consisted of five enormous covered wagons, each with a capacity of six tons of cargo, and each with its reliable driver of the three yokes of oxen that pulled the loads.

Although the family prospered, William's absence for much of the war years presented numerous challenges for Clara and her two young daughters. In essence, the hostilities in Kansas and Missouri during the Civil War represented an extension of the prewar tensions and violence of the late 1850s. In Osage County, combatants included not only the troops of the Union and Confederacy, but also guerrilla fighters and militias not directly under the command of either formal army. The pro-Union guerrillas were called jayhawkers, a term that lives on as the nickname for the sports teams at the University of Kansas. Their pro-Confederate counterparts were called bushwackers, in reference to their most common tactic, the ambush of opposing individuals or families in rural regions—like Carbon Hill. As Barnum recounts: "Mother used to say that during the War it was not uncommon for Federals to stop for food in the morning, Rebels at noon, and Bushwackers at night. She never dared to say where their sympathy lay for fear of retaliation by shooting the family, burning down their house, or destroying their property."

After the Civil War, changes transformed the region as newly constructed transcontinental and regional railroad lines sliced across the frontier, following the old covered wagon trails. In his biographical notes, Brown explained that the burgeoning rail lines

little by little spelled the end of profitable wagon freighting in the years after the war: the Union Pacific joined the Central Pacific to complete the first transcontinental line in 1869, and by 1880 the old Atchison and Topeka Railroad, chartered in Kansas in 1859, had reached Santa Fe. . . . Father's last freighting trip ended with the home stretch from Santa Fe, along the trail of that name, to Carbondale, close to whose cemetery traces of the old trail could still be seen a few years ago. Gone are the oxen and the men who drove them, but in some unbroken pasture land, in the summer, you can still see the remnants of the historic trail outlined by a golden blaze of Mexican thistles, the seeds of which had been transplanted by the feet of the oxen.

With peace in the country more or less restored and more time to spend around his homestead, William set about increasing the acreage of his claim and diversifying his farm's productivity. The oxen that had once pulled the freight wagons were now yoked to plows and scrapers in order to clear and tend fields for crops as well as to expose the coal seams for fuel, which the family used themselves and sold to outside buyers. William also sought to enhance the comfort of frontier life for Clara and his two daughters by constructing a more "modern house ... of brick and clapboards, 'salt-box' fashion, with two big rooms downstairs, bedrooms upstairs under a gable roof, and a large cellar." In time, William came to control 640 productive acres around Carbon Hill, a full square mile that supported "500 head of cattle, numerous hogs, and a fine well of ever-cold water." In an account published by the Carbondale Centennial Association in 1972, Mary Snell and Rosalind Metzler reveal that the Brown home was acclaimed as "the best residence in this section of the country," based on a historical document compiled in 1883. (Unfortunately, as Snell and Metzler go on to recount, the house burned down in 1971, "as a blow torch was being used to remove paint in the process of restoring the old house.")[2]

This phase of domestic expansion coincided with a population boom in the family. William and Clara's first son, Frank, arrived on the scene in 1867. Six years later, Frank, along with the rest of the family, eagerly awaited the arrival of a third sibling. However, another impending event was competing for Frank's attention, and it would come to leave an indelible mark on the life of the new child. As Barnum's daughter, Frances, explained in the short biography she penned about her father in 1987,

> In the early 1870s, the fame of P. T. Barnum had already spread from Brooklyn to the midcontinent. "Barnum's Great Traveling World's Fair" ... was traveling far and wide.... The master showman heralded his approach with posters plastered on barns, trees, town buildings, everywhere a likely spot could be found to titillate public attention.... The barns and buildings of Topeka were no exception. Small wonder that the bright eyes of a six-year-old lad [Frank] ... spied the gaudy pictures. He could dream of no greater delight than to be taken to see the real thing....
>
> An event of greater delight to his parents was the safe arrival of their second son on February 12, 1873.... Frank was pleased that he was going to have a

FIGURE 1. Portraits of Barnum Brown and his family taken around 1880, when Barnum was about four and they were living on their homestead in Carbondale, Kansas. The photo of the family is labeled 1881; however, the photo of Barnum, who is wearing the same clothes as in the family portrait, is labeled "? Age 4," which he would have been in 1877. Front row (from left): Clara Silver Brown, Barnum Brown, Frank Brown, William Brown; back row (from left) Alice Elizabeth Brown and Clara Melissa Brown (AMNH Vertebrate Paleontology Archive 2:5 B1 F1)

younger brother to boss around, a task he felt well qualified to assume after being on the receiving end so long from his older sisters. However, nothing could long hold his attention away from the approaching P. T. Barnum Show. Therefore, when the new baby remained unnamed for several days because the family could not agree on what he should be called, it was Frank who burst in on one of the arguments with "Let's call him Barnum."[3]

The name stuck, and almost ninety years later, as the boy with the showman's name looked back on his life, Brown could only conclude: "There must be something in a name, for I have always been in the show business of running a fossil menagerie."

Although Barnum was a child of the frontier, his early years were anything but an exercise in loneliness. In addition to his family, a seemingly enormous crew of thirty-one men helped William run the farm, and a local girl was hired to help Clara keep everyone fed, a monumental chore performed daily with spectacular aplomb, especially considering that all thirty-eight dined in two separate sittings in the large kitchen. Summoned by a large bell, the family and crew feasted on

> eggs and bacon, pancakes with sorghum syrup, and coffee for breakfast; dinner at noon when the table groaned under steaks, potatoes, green vegetables fresh from the garden or canned, apple butter, jam, pickles, kraut, beans, thick slabs of bread, piles of hot biscuits smothered in honey and butter, apple, peach, blackberry, raspberry or pumpkin pie fresh from the oven—all homemade. Supper at six was another banquet of fricasseed or fried chicken, fried potatoes, stacks of hot biscuits with honey and butter, dessert that was sometimes hot shortcake topped with wild strawberries we had gathered earlier in the day and thick whipped cream. Gallon pitchers of milk stood on the long table. No one ever got up hungry from the Brown table.

Nevertheless, Brown freely admits that life was not all "rosy" on the farm, as livestock losses from "disease, drought, flood, and famine" bedeviled his family's operation. Brown became acutely aware that "all farmers are gamblers at heart and in practice, whether they know or admit it or not. They gamble on what the coming season will produce, they gamble on what the prices will be for what they can produce ... but they have nothing to say about the prices of the things they must buy. Lo, the poor farmer!"

Hen cholera and other poultry diseases took their toll, as did an invisible blight on corn in drought years that resulted in black leg, a bovine disease triggered by the ingestion of bacteria living in the soil that caused cows to abort their calves or, at worst, ended in death. According to Barnum's recollections, drought years were somewhat cyclical, with four wet years "of plenty" followed on average by three years of drought, during which "there would be so little rain that the corn wouldn't even sprout." As the wells dried up, "We would watch the primer which checked the amount of water left in the cistern. . . . We frequently had to haul water from Wakarusa, five miles north of us, for our livestock, sometimes making two or three trips a day."

In wet years, climatic conditions could veer toward the biblical. "One spring," Barnum recalled, "it rained off and on for forty days and forty nights. Then Berry Creek became a surging river, and much of Carbondale built in its valley was destroyed. Up on our hill, we escaped the flood.... The cattle ... were so covered with mud that you couldn't tell a cow from a steer, or a Poland China hog from a Cheshire White.... Nevertheless, these abnormally heavy rains replenished the ground water and improved the yield of crops next season."

As soon as he was old enough, Barnum was enlisted to pitch in on the daily and seasonal chores. Presumably due to his relative youth and lack of seniority, his first assignments fell under the supervision of his mother. In his notes, Barnum is unabashedly proud of this early supporting role, even though it involved what might have been viewed as "women's work": "[My mother] always said I was the best 'girl' she had, for Melissa was romancing and Alice, who resented being a girl and loved men's work, was raising cattle on her own, helped by brother Frank.... So little brother got a lot of housework."

Through his mother's tutelage, Barnum not only learned how to feed a large crew of workers but also acquired a wealth of information about the world of the frontier. In his notes, he rhapsodizes about his mother's love for the animals and plants that enlivened their rugged homestead existence:

> Mother liked a happy, cheerful home, and kept native songbirds. Redbirds are notorious fighters, especially with other redbirds that are caged; mother caught a great many by opening a window and placing a captured, caged bird close to it. A wild bird would fly in to ... attack; mother would close the window, capture it as it flew around the room, and cage it. Redbirds are beautiful whistlers and mother loved their cheery notes.... The first notes of the robins and bob-whites mingled with the scent of apple and peach blossoms with the coming of spring; then the meadow larks sang in the pastures while the bluebirds hovered around the house. Later, the woodpeckers could be heard hammering away [on] the telegraph poles.

Part of Barnum's chores as a young boy involved helping to tend his mother's garden:

> Mother and I loved flowers. Honeysuckle ran along the white picket fence around the house and yard. She had a large bed of petunias, phlox, red and white roses. Along the lane were oleanders and snowballs. We had pine and cedar trees in front of the house. In back there was a large mulberry tree with delicious fruit that we enjoyed in the summer. Along the walk to the privy were lilacs on one side, and Concord grape vines on a trellis on the other....
>
> In the east corner of the yard was the kitchen garden, where we raised lettuce, radishes, beets, wax beans, onions and sweet corn, between the rows

of which were pumpkin and squash vines. Beyond the yard to the east we grew large quantities of Irish and sweet potatoes, peanuts, popcorn and cabbage from which I made a barrel of sauerkraut each year.

Flipping through the tattered pages of his lightly annotated field diaries confirms the fact that Barnum never lost his love of flowers. For although written notes describing his daily scientific activities are often frustratingly rare, it's not uncommon to encounter the dried petals and stems of pressed plants that captured his curiosity.

As he grew, more physically demanding chores were added to Barnum's repertoire, bringing him under the guidance of his father as well.

When I was about ten years old, I was able to help with the farm work. In the late summer, I sometimes milked 20 cows, morning and night, sitting on a one-legged stool, so if a cow kicked I would fall over without resistance. . . .

During the harvest season, I worked on a corn-cutting sledge; it was V-shaped with a knife blade on each side that cut off the stalks about eight inches above the ground—we didn't have modern harvesting machines. Two of us, one on each side, rode the sledge, which was pulled along between two rows of corn by a single horse.

Barnum's sidekick around the homestead was "Old Bruno, my big curly-shaggy Newfoundland dog." Weighing in at 128 pounds, twice as heavy as Barnum, Old Bruno was energetic and playful, although his energies were often expended on flights of fancy unrelated to chores on the farm, such as scaring up jackrabbits and pursuing them until late in the evening, when, to Barnum's great amusement, he would limp home with sore feet, "all tuckered out." Another family stalwart was the pet cat, Old Maltese, who, though a skilled mouser, suffered from a substance abuse problem: "Old Maltese loved the smell of camphor, and used to get drunk from whiffing it, after which she would stagger around like a drunken man."

Eventually, Frank inadvertently ran the cat over with the mowing machine, cutting off most of one leg, and although he "held the stump . . . to stop the bleeding, and carried her back to the house to get a bandage on it," she could no longer hunt and took to killing Clara's chicks instead. Thus, "poor Frank had the sad task of killing her with the shotgun. So ended Old Maltese, the mighty huntress."

Not surprisingly, injuries on the farm were not limited to the family pets, and Barnum suffered through his fair share. One year was especially disastrous:

> One day when I was about eight... Frank dropped an iron mowing machine wheel on my heel by accident, and laid bare the flesh to the tendons. The family doctor came and dressed the wound; I was laid up for three months. Old Bruno tried to lick the wound, and the doctor said it wouldn't do any harm but would have curative properties; at least his gentle tongue was soothing.... Before I had completely recovered from this accident, a nail in my shoe caught in the stair carpet one morning causing me to trip and fall down the last three steps, striking my elbow on the door jamb and breaking all three bones in my left arm. There were no X-rays in those days, but with the help of chloroform, old Doctor Wood set my arm as best he could, but it was far from perfect, and is crooked to this day.

Barnum's early education took place in the local two-room schoolhouse about a mile's walk from home, and along with his fellows, Barnum seems only too happy to have participated in the usual pranks: "Professor A. V. Sparhawk was the school principal, a fine man and a good teacher with only one weakness: he chewed tobacco. There was a big stove across the hall from our classroom; he would step across the hall while classes were going on to spit in the stove. We boys, knowing this, always tried to trap him by reciting so long and asking so many questions that he would have to swallow his tobacco juice to answer."

There was, however, at least one burning question that the good schoolmaster could not answer. It arose from an activity on the homestead that would form the foundation of Barnum's legendary career:

> Father had about two dozen yokes of big, strong draught oxen that he had used when he was a government freighter. He used these same oxen to pull the plows and scrapers that stripped the overburden of rock from the underlying coal seams that we worked. Sometimes as much as eighteen feet of overlying rock had to be stripped away before the coal seam was laid bare. There were many fossil sea shells and the remains of other invertebrates in all of this overburden. When I was a little shaver, aged five, I was old enough to notice and take interest in such things, and used to follow the strippers in order to pick up all the fossils that they turned up. I collected the fossils in boxes that I took home to put in my bureau drawers. I well remember proudly showing two of my first discoveries to visitors: one a horn- or cornucopia-shaped specimen about three inches long; the other looked like a piece of

honeycomb with the beeswax showing.... Later I learned that both were fossil corals that had lived in an ancient sea whose bottom sediments had been deposited upon the coal when the sea flooded in where before there had been dry land with coal-producing forests. Father, though untrained in geology, encouraged me in making these collections, for he thought that by so doing we could find out why sea shells could be entombed in a Kansas hilltop 650 miles from the nearest seacoast today, the Gulf of Mexico. In time, my collection became so varied and bulky that Mother made me take them from the overflowing bureau drawers and put them in the laundry house, which was a building separate from our farmhouse. This became my first museum, where I had my first experience as a showman regaling visitors with these treasures, together with Indian arrow points and scrapers I picked up while plowing our cornfields.

From modern-day geologic maps, it appears that the coal- and fossil-bearing strata on Brown's homestead fall within the boundaries of the 300-million-year-old, Pennsylvanian Wabaunsee Group and probably represent beds within the Severy Shale or Howard Limestone.[4] Surviving photographs of the coal-mining works around Carbondale indicate that they were extensive operations, not just deep but forming extensive pits. This makes sense, for the cost of shipping coal from the Appalachians into this remote region must have made the locally mined fuel a valuable commodity.

As Barnum progressed through adolescence and his collection of Paleozoic invertebrates burgeoned, the family realized that, if he were to fully answer the riddle of the seashells on their Kansas hill, Barnum would need to continue his education. However, educational opportunities in the local area were limited. The nearby town of Carbondale had a population of only around two thousand during Barnum's boyhood. Its main street boasted of a bank, two churches, some grocery stores, and "two meat markets where you could buy steak for ten cents a pound, with bones, liver and 'lights' (other internal organs) thrown in free for cat and dog food."

The town did sport its own newspaper, the *Astonisher and Paralizer*, "run by two cantankerous veterans of the War Between the States." At the end of his life, Brown still harbored a palpable fondness for the duo's irreverence. In particular, he recalled their entreaties directed at "delinquent subscribers": "Some folks knead bread with their gloves on; others knead bread with their clothes on; but if subscribers to this paper don't send in their dues, the Editors [are] going to need bread without a darned thing on."

The paper kept the locals up on livestock and crop prices and other goings-on—and of the latter there were plenty, given that the city "boasted eighteen saloons," which served whiskey and beer to "miners, but not minors." According to Barnum, attractions also included

> Shady Lane, . . . the town's red light district. One summer day, when business was slack, a well-known farmer, hitching his team to the post outside, made a call at one of these houses. While he was engrossed with the girl of his choice, one of her colleagues stripped off her clothes, unhitched and jumped in the farmer's wagon, and raced the horses down through the center of town, up onto the hill opposite . . . back again the length of Main Street and on to the house on Shady Lane, stark naked. There were mixed reactions to this enlivenment of an otherwise dull day: The Astonisher and Paralizer had gratuitous copy for several issues. . . . The chief and most lasting effect was a remarkable upsurge in business along Shady Lane; this proved that it pays to advertise.

Yet despite all the social institutions that the town supported, one key ingredient had yet to materialize: a high school. By 1889, Barnum had completed the final year of schooling that was available in Carbondale. His parents were determined to have him continue the next fall, "but first Father wanted me to see what was left of the Old West before it faded away, to show me some of the places he had been in his pioneer days, and to broaden the outlook of an adolescent farm boy who had never been more than twenty miles from home."

(The last point seems to be an exaggeration, since Brown also speaks of a trip in his youth to visit his maternal grandfather, the former captain, in Wisconsin. Barnum was clearly proud of his grandfather's achievements, but when he visited the family farm in Wisconsin, he was appalled to find his grandfather using "his Captain's sword as a cheese knife! Here, as a boy, I recovered the sword from its ignoble use and kept it with his epaulets, bayonet, and accoutrements.")

Barnum's four-month foray with his father into the remnants of the Wild West—designed also as a search for a good spot to set up a new cattle ranch—would be a coming-of-age adventure of epic proportions for the boy. It also provided a glimpse of the role his father had played in taming the frontier. The elder Brown had made his last freighting trip when Barnum was still a young boy: "Standing in one of the wagon beds and stretching up my arms," he later recalled, "I could just reach the top of the wagon box, but the bows and canvas seemed to tower overhead." Now he could experience the wonders of the road as a full-fledged partner.

In addition to the reasons that he gives in the story of his childhood, the trip may have been triggered by a legal case involving Barnum's father and sister. On April 25, 1889, Melissa (Brown) Taylor filed a formal complaint warrant with a justice of the peace in Osage County, Kansas, J. M. Pleasant, against her father, William Brown, alleging an act of incest committed eight months earlier.[5] At that time, Melissa was about to turn thirty-two, William was approximately fifty-five, and Barnum was fifteen.[6] The complaint reads that, on August 15, 1888, William Brown did

> unlawfully and feloniously[,] lewdly and lasciviously cohabit with and carnaly know Melissa Taylor[;] he the said William Brown then and there being the father of the whole blood of the said Melissa Taylor and the said William Brown and Melissa Taylor being then and there persons within the degree of consanguinity within which marriages are by law declared to be incestuous and void to wit being then and there father and daughter of the whole blood and the said William Brown did then and there unlawfully, incestuously, knowingly, feloniously, and willfully commit incest in and upon the person of the said Melissa Taylor by then and there unlawfully committing adultery with and having carnal knowledge of the said Melissa Taylor he the said William Brown then and there knowing that the said Melissa Taylor was his said daughter.[7]

An accompanying document of Criminal Action was executed, also on April 25, 1889, by J. M. Pleasant based on the sworn statement described in the warrant, which initiated the legal case (No. 11) involving the State of Kansas vs. William Brown.[8] The criminal action indicates that the warrant of April 25 was served by a sheriff's deputy on April 26; the deputy then arrested William Brown and brought him to appear before the justice of the peace. The criminal action goes on to document the subsequent proceedings associated with the case. First, on April 26, both William and Melissa appeared before Justice Pleasant after William's arrest. Both parties consented to a continuance of the case until May 8, and William was required to enter into an legal agreement, in the form of a $1,000 bond (equivalent to almost $21,000 in today's dollars), to guarantee his appearance at the stipulated date. The document further indicated that if William failed to provide the bond, he would be jailed until May 8. William filed the $1,000 with the appropriate officials, however, and on the appointed day for the next hearing both William and Melissa showed up with their lawyers; the case was continued until May 17, but Justice Pleasant demanded that William "file another recognizance in the sum of $1000 for his appearance at said time, which he forthwith files, with the same sureties as before, and the same is by me approved, and defendant discharged until said time."

When William failed to appear for his next hearing, the justice of the peace rendered the following statement:

> The plaintiff appeared... [with her lawyer] P. E. Gregory... and the defendant came not and for more than one hour after the time named in said recognizance for the hearing of this cause came not but wholly made default. Whereupon the plaintiff asked that said default be recorded, and the record thereof with said recognizance be certified to the District Court of said county, which is accordingly done and a transcription of the docket entries together with all papers in said cause and said recognizance duly certified is forthwith filed in the office of the Clerk of the District Court of said County. [signed] J. M. Pleasant, J. P.

A document recording the forfeiture of the recognizance was filed by the clerk of the district court in Osage County in the case of Kansas vs. William Brown on May 20, 1889.[9] We have no further documents regarding the case. Thus, at this point it appears that the case was closed upon the forfeiture of the $2,000 recognizance, which we presume was in some way petitioned for and later transferred to Melissa.[10] It appears that the forfeiture represents either an admission of guilt in the case or an attempt to settle the case without further action. No record of other fines or of incarceration is known to us.

Brown never mentions the incident in any of the documents preserved at AMNH, so we have no evidence to indicate how he felt about the alleged crime or the case. However, it is clear that he retained a high degree of respect for his father, as evidenced by his own story of his childhood. It seems circumstantially suggestive that shortly after William forfeited the bond, he and Barnum set out to roam the west in search of a new homestead to start a cattle ranch. Furthermore, although there is no connection between the two events, it is interesting to note that Barnum would similarly flee for the field when, in 1919, he got entangled in a less pernicious, yet still somewhat scandalous, legal case involving an apparent lover.

In any event, in the late spring or early summer of 1889 Barnum and William customized a "light spring wagon" and rigged it with a canvas covering supported by bows. A trunk bolted to the back of the bed contained extra clothes, and a "grub box" on top of that held supplies of staples, including "sugar, bacon, flour, meal, beans, raisins, coffee, and so on." Cooking was done with a sheet-iron stove that was lashed on top of the grub box, along with a "pan, kettle, coffee pot and skillet." The bed of their souped-up wagon was lined with hay and a canvas covering to form a sleeping mattress. Horses could be fed from a feed box bolted to the back of the wagon, and a canvas sling strung underneath served as storage space for dried cow chips that would be collected along the way and burned for fuel. As Brown proudly recalled more than seventy years later, father and son then "took leave of the family, hitched our team to the wagon, and set off on my first expedition."

They usually rose before sunup, and while his father fed and harnessed the team, Barnum rustled up breakfast, usually "coffee, bacon, and such eggs as we could buy or gather along the way." After washing the dishes, the boy, sheltering from the morning chill, would huddle under the blankets as his father grabbed the reins and took to the road. A brisk pace for the first few hours ensured that they could stop at midday when good land for grazing came into view. After lunch, they renewed their trek until they found good grass in the late afternoon; they would then unhitch the horses and Barnum would prepare a somewhat repetitive dinner of beans, bacon, and eggs. When they were in Indian country, they chained and padlocked their horses to the wagon wheel at night.

Traveling along the Middle Loup River in Nebraska, the Browns had to keep moving to avoid run-ins with the large cattle companies, which monopolized the range and did not appreciate travelers grazing on their grass. In the Sand Hills of Cherry County, "rattling gunfire like that of a skirmish line" at first aroused their concerns, but it turned out to be only "hunters shooting Prairie Chickens that lived in the tall grass bordering the waterways and lakes" for shipment to lucrative markets in the east.

Back when William had run freight for the government during the Civil War and immediately afterward, bison were still a common sight on the plains. "When his scouts used to tell Father of a great herd . . . coming toward his wagon train they would circle to make a corral out of the wagons, placing the oxen and horses inside so that they would not be carried away or destroyed by the surging herd. Father told of other times when he saw great herds . . . streaming across the Missouri River in such numbers as to stop the river boats."

But beginning in 1865 with the construction of the transcontinental and more regional railways, "companies hired men to kill off the bison. Together with the hide hunters and barbed wire, wholesale slaughter and separation of the great herds reduced the former countless millions to 541 in 1889.... But they had been exterminated already from the country we were to pass through.... We saw only their heads, preserved by the long shaggy hair and tough hide, strewn across the prairies like small barrels."

In northeastern Wyoming, the government stagecoach that ran between Deadwood, South Dakota, and Billings, Montana, blew by them, powered by a team of four sturdy horses as it transported "mail and passengers at the rate of sixteen cents a mile." Shortly, the Browns approached the site of the Battle of the Little Bighorn, which had stunned the nation on the eve of the nation's centennial when more than 250 soldiers of the Seventh Cavalry under the command of Major General George Custer were killed by some 2,000 Sioux and Cheyenne warriors led by legendary chiefs, including Sitting Bull, Crazy Horse, and Gall. Ironically, as Barnum and his father neared the battlefield just thirteen years later on the Fourth of July, an Indian guided them to the ford of the Bighorn just in time for the Crow's reenactment of the battle.

> Although this tribe had always been friendly to the whites, the Commandant was taking no chance, so he had two companies of the garrison under arms, and two Gatling guns trained on the battleground.
>
> I well remember the occasion: squaws with papooses on their backs or in their laps sat all around the edge of the battlefield....
>
> We camped there beside the river that night. Little did I dream lying there in our covered wagon, that I would be camped in this same region not many years in the future, searching for fossil dinosaurs for the American Museum of Natural History.

Their odyssey then continued westward along the Yellowstone River, past Billings and present-day Livingston, to what would become the north entrance of Yellowstone Park, just above Mammoth Hot Springs, which still steams and bubbles today as it did then, creating a popular destination for park visitors. Although a troop of U.S. cavalry was stationed there, the Browns had no luck in supplementing their diminishing supply of provisions, because regulations prohibited the quartermaster from selling goods to anyone. So they carried on, following the Yellowstone River up to its headwaters at the lake; along the way, Barnum succeeded in catching "great numbers of lake trout, averaging two pounds apiece, sometimes getting two with one cast." Fried in bacon grease, then salted and placed in a keg, the fish provided a welcome break from the pair's habitual diet of eggs, bacon, and beans. It also gave them some bartering power. As Barnum recounted,

This was wild country in those days, and [the troopers] were not even allowed to go hunting or fishing for recreation. On our return trip, the Commandant heard that we had some fish, and offered to exchange any provisions that we required for the fish.... The trout made a welcome change of food for the troopers, and we needed the fresh provisions that the Quartermaster couldn't sell to us. Where there's a will, there's a way! One of the things we got was a luxury: a big jar of pickles. We were starved for something sour, and I ate all my share at one sitting.

Reluctantly, after their idyllic sojourn beneath the snow-capped peaks of the Rockies, Barnum and his father began the return trip through Billings before heading south along a different trail toward eastern Kansas. "There were more new sights and every day was a fascinating adventure. And then we were home again with our loved ones. We had traveled about three thousand miles at an average rate of twenty-five miles a day when we were on the march. It took us a little more than four months from start, to finish. What an experience! This was Father's finest gift to me; it was of himself."

It doesn't require much magnification to read between Barnum's lines and see why he felt his first foray into the fields of the western frontier was so valuable. Brown's early life prepared him for his successes to follow. His sense of awe at "seeing new sights every day" inspired a wanderlust that he would seek to satiate for the rest of his life.

Even more directly, Barnum's childhood can be viewed through the same experiential prism. None of the individual events that he recalled in his unfinished autobiography, except perhaps the trek to Yellowstone, comprised anything but the normal activities of a typical frontier upbringing. Nonetheless, they paved the way for a remarkable life spent predominantly in a determined search for fossils across both familiar and foreign landscapes in almost every corner of the globe.

Serendipitously, his father's choice for a family homestead site was blessed with intriguing fossils and geologic strata. These sowed the seeds of curiosity that quickly sprouted in the young boy's fertile mind, nurtured by the encouragement of his father and by his mother's practical love of the natural world that surrounded them. The daily and seasonal chores of managing the farm, from caring for livestock to driving the wagons and other farm equipment, provided Barnum with an extended tutorial for living and traveling in rugged and remote regions. Helping his mother to cook and care for the large crew of field hands and strip-miners fostered Barnum's ability to feed, manage, and organize large crews of diverse characters in his later fossil-collecting operations. The business acumen that he absorbed from his father enabled him to carry out shrewd financial transactions during his expeditions for the museum and during reconnaissance for oil and mining companies. Last but certainly not least, his three-thousand-mile odyssey in a covered wagon, launched at the insistence of his father, must have indelibly tempered and reinforced Barnum's prodigious skills and instincts for taking reasonable risks in his later quests for elusive fossil quarry, hidden among inhospitable and uninviting badlands throughout the world.

TWO

Student... of Sorts
(1889–1896)

BARNUM BROWN'S FIRST STRIDES TOWARD adulthood were not as ambitious or auspicious as his recent odyssey through the western frontier might lead one to suspect. Upon returning from the wilds of the American West, he enrolled in high school in 1889 at the age of sixteen. Because there was no high school in Carbondale, his parents sent him to one in the academic mecca of the newly minted state, Lawrence, a bustling town built around the fledging University of Kansas (KU). For children of the frontier, this represented an unusual opportunity. Few youngsters on the plains had any chance to extend their education beyond the local one-room schoolhouse in their hometown; most families were desperately poor, and responsibilities for the seasonal tasks on the family farm almost invariably precluded intellectual indulgence. Yet despite the privilege that Brown's foray into higher learning represented, he makes sparse mention of his high school days in the notes for his autobiography.

Brown was not alone among his cohort from Carbondale in his high school endeavors, however. In fact, in Lawrence he shared an inexpensively rented house with two other young men from his hometown. Given the twenty-five miles that separated Carbondale and Lawrence, Brown likely spent most weekdays at the school, with trips home on the weekends. He graduated in 1893 and that fall entered KU as an engineering student.[1]

In contrast to the laconic account of his high school days, Brown seemed to take great pleasure in recounting his days as a Jayhawk at KU. In large part, his nostalgia can be traced to the fondness he held for his primary professor and academic advisor, Samuel Wendell Williston. As Brown later recalled, having set off for KU with his "precious" collection of fossils from the family homestead under his arm, he quickly "came under the tutelage of... Williston... from whom I soon learned that I wanted to be a Paleontologist

FIGURE 2. Class portrait of Barnum Brown as it appeared in the 1897 yearbook of the University of Kansas, where Brown studied under Samuel Wendell Williston (University Archives, University of Kansas Library, P/ Brown, Barnum)

and Geologist."[2] Theirs was a relationship that would forever alter the vision of this child from the frontier. To his dying day Brown would proudly identify himself as a student of Williston's, because he played the seminal role in helping Brown forge a career that transcended landscapes, distances, and historical perspectives even greater than those Barnum encountered on the epic journey with his father.

Despite being separated by a generation in age, Williston and Brown shared numerous credentials in their formative backgrounds. Williston's family moved to Kansas from Boston in 1857 as part of the migration of New England farmers seeking to establish the territory as a slave-free state, just as William and Clara did two years later. Life was difficult for the family of six, crammed into a one-room cabin and consumed with simply finding enough to eat. "Bookish compared to his brothers," Williston enrolled in the Kansas State Agricultural College when he was fifteen.[3] There, under the guidance of the state's most prominent geologist and paleontologist, Benjamin Franklin Mudge, Williston acquired a fondness for collecting insects and fossils that spurred his academic career.

Academically, Williston's and Brown's lives were rooted in the first real flowering of vertebrate paleontology in the United States. While Brown was growing up on the western frontier, two titans of paleontology—Othniel Charles Marsh of Yale and Edward Drinker Cope of the Academy of Natural Sciences in Philadelphia—were battling it out on the East Coast over who was the preeminent paleontologist in the country. The main front of their battle, however, was actually well to the west of Carbondale, in the lands that Brown and his father traversed on their trip.

In personal terms, Cope's life embodied a storyline of boom to bust, whereas Marsh's embodied the opposite.[4] Marsh was born in 1831 to a farming family in upstate New York. When he was three his mother died, whereupon his upbringing fell to his father, eventual stepmother, and an extended group of aunts and uncles. During his adolescence, his intellectual curiosity was sparked through his association with an engineer helping to manage the Erie Canal, with whom he collected invertebrate fossils. Marsh was also fortunate enough to be taken under the well-feathered wing of his uncle, George Peabody. Peabody had amassed a fortune, first in the United States through the overseas merchant trade, and later as a London banker. In 1852, Marsh inherited money from his mother's dowry, which Peabody had funded, and enrolled in the Phillips Academy. Though a late bloomer academically, Marsh had the determination to gain an education, and this meshed well with Peabody's philanthropic priorities, which included the establishment of the Peabody Education Fund as well as a university in Nashville. Following his stint at Phillips, Marsh sought Peabody's financial support to enroll at Yale in 1856, which his uncle willingly granted. Upon graduation, Marsh received further financial assistance for postgraduate studies in Europe, when the budding paleontologist attended universities in Berlin and Breslau and surveyed the fossil collections in numerous museums. The crowning achievement of their familial partnership was realized in 1865, when Marsh convinced his uncle to establish the Peabody Museum of Natural History at Yale, at which Marsh would serve as the country's first Professor of Vertebrate Paleontology and launch his ambitious plans to develop a world-class collection of fossils.

Cope, for his part, was born into a wealthy family of Philadelphia-based shipping merchants in 1840.[5] As with Marsh, Cope's mother died when he was only three, but his well-endowed family nurtured the boy's precocious curiosity in natural history from the start. Although his father's plans for Cope involved a career in agriculture, by the age of six the boy was already making notes and observations about the fossil skeletons of ichthyosaurs at the Philadelphia Academy of Sciences. He received his education at the Quakers' Westtown School and the University of Pennsylvania (although he did not receive a formal degree), and he authored his first scientific article at the age of eighteen. During the Civil War, Cope accommodated his Quaker beliefs by refusing to fight and instead traveled through England and Europe studying collections in museums and universities. During his swing through Germany, he even crossed paths with Marsh, and the two struck up an incipient personal and professional friendship, especially through occasional correspondence.[6] In 1864, Cope returned to the United States and assumed a teaching role at another Quaker institution, Haverford College; however, because he preferred a less structured environment with more time for research, he soon became associated with the Philadelphia Academy of Sciences.

From a convivial and collegial beginning in the early 1870s, Cope's and Marsh's quest for fossil dinosaurs and mammals devolved into a professional and personal spitting contest later in the 1870s and 1880s. One fuse for their feud ignited in 1877, when Arthur Lakes, a schoolteacher in Colorado, sent both men samples of immense Jurassic dinosaur bones that he had collected in the Front Range of the Rocky Mountains, outside the town of Morrison.[7] While Cope immediately began to prepare a scientific paper on the fossils, Marsh took a different course, presenting Lakes with a contract to hire him as a collector and secure his present and future collections from the site. Elated, Lakes wrote to Cope, explaining the new arrangement and requesting that Cope forward the fossils he had sent on to Marsh. Cope, though incensed, was powerless to stop the transaction.

Soon thereafter, however, and to his great delight, Cope received another shipment of Jurassic dinosaur bones from another Colorado schoolteacher, O. W. Lucas, collected from the same rock unit, exposed at Garden Park, outside Cañon City, about 125 miles southwest of Morrison. In fact, the sample Lucas sent contained larger and better-preserved bones, which infuriated Marsh. His recompense came in the form of yet another shipment of Jurassic dinosaur bones, sent by two railroad workers, W. H. Reed and W. E. Carlin, from Como Bluff, Wyoming, a site that still shines in the annals of vertebrate paleontology. Through the rest of the 1870s and into the 1880s, the field crews of Cope and Marsh boisterously jockeyed each other for dominance at Como Bluff and at many other fossil-rich localities in Wyoming, Colorado, and other territories in the West.

Although both spent time in the field, Cope more than Marsh, the antagonists were usually far removed from the field-fray in their East Coast ivory towers. They fueled their feud with vituperative claims and counterclaims in the press and halls of government as they competed for funding and dominance over the agencies that controlled the nation's geological resources. Politically, Marsh gained the upper hand by becoming the chief paleontologist for the United States Geological Survey in 1882, which allowed him to control government funding and field access for his operations throughout the next decade. Marsh also served as president of the prestigious National Academy of Sciences. Cope eventually served as president of the American Association for the Advancement of Sciences, but his skills at garnering power were less acute than those of Marsh.

On the plus side of their feud was a continuous stream of new dinosaur discoveries, with subsequent publications that, for the first time, began to illuminate the amazing menagerie of Jurassic and Cretaceous beasts that roamed North America between 150 and 65 million years ago. Their names are now indelibly etched in the collective consciousness of all dinosaur aficionados: *Stegosaurus*, *Allosaurus*, *Diplodocus*, *Triceratops*, and *Brontosaurus* (now called *Apatosaurus* because in the frenzy of the bone rush Marsh gave two generic names to different parts of the same kind of animal). The blitz of scientific publications on dinosaurs and numerous other topics based on their discoveries and battles boggles the mind. Marsh tallied 270, in large part about fossil dinosaurs and mammals, while Cope penned around 1,400, which addressed not only fossil dinosaurs and mammals, but also modern fishes, amphibians, and other reptiles. Comparatively, Cope, who is widely recognized as the greatest herpetologist of his day, eclipsed Marsh on intellectual fronts.[8] However, by the end of his life Cope had exhausted all of his financial resources, and his collection was sold to the American Museum of Natural History. Most of Marsh's treasures still remain in the Peabody Museum at Yale, the institution that he founded.

Into this fray wandered Samuel Williston, who, thanks in part to the recommendation of Benjamin Mudge, landed a position as the primary assistant in Marsh's lab at Yale. As part of his responsibilities, Williston also headed Marsh's field operations, along with his mentor, Mudge. But Williston's academic interests were in no way limited to paleontology. By 1890, when he left Yale, his accomplishments included a Ph.D. in entomology, a degree in medicine, and four years' experience teaching anatomy.[9]

Williston soon grew weary of the turmoil between his boss and Cope. He was especially irked that Marsh refused to put his assistants' names on the publications that they helped to research and write. Although at that time it was traditional, following European custom, for studies done by workers in a particular lab to be published solely under the name of the supervising professor, the practice not only denied researchers public recognition for their efforts but also deprived them of scientific publications, the currency for academic advancement. Instead of being intellectually supported and promoted by his mentor Marsh, Williston found that his career was stymied. This frustration was exploited by Cope, who encouraged Marsh's minions to air their grievances in public. For the most part, his biographers note, Williston demurred, "but in an unguarded moment [Williston] wrote Cope in 1886 that Marsh 'has never been known to tell the truth when a falsehood would serve the purpose as well.' Much to Williston's dismay, Cope printed his letter in a newspaper article. Clearly, Williston's future as a paleontologist would not be at Yale."[10]

Williston left Yale in 1890 to return to the relatively peaceful surroundings of Kansas. When the board of regents at KU sought to hire a new chancellor, they picked the university's professor of science and math, Francis Snow, who also happened to be an entomologist. In a stroke of administrative foresight, Snow agreed to take the position if the regents would allow him to hire Williston, essentially to succeed him. Thus, Williston wound up back in Lawrence, ready to set up his own paleontological expeditions without the interference of Marsh.[11] Shortly thereafter, Williston then set about recruiting students for his field trips to help transport gear, collect fossils, and keep the camps operational. Serendipitously, three years after his return to Kansas, into Williston's world walked an enthusiastic and willing young Barnum Brown.

In 1894, Williston was planning an expedition to the White River Badlands in South Dakota to obtain a collection of Tertiary mammals from the already famous exposures. These highly fossiliferous outcrops, composed predominantly of blindingly white, beige, and light gray sandstones and mudstones sculpted into treacherously steep ridges, rugged ravines, and fluted pinnacles, were deposited by ancient streams during the Oligocene between about 35 and 23 million years ago. The picturesque exposures now form the landscape of Badlands National Park, some seventy-five miles southeast of Rapid City. Still today, these badlands draw determined crews of paleontologists to conduct field research, especially from late spring to early fall. Summertime temperatures can soar over 110 degrees, and the days are often punctuated by tremendous thunderstorms, accompanied by lightning, hail, and violent winds. Although Brown was familiar with these conditions from his upbringing on the plains, his abilities to perform on a scientific expedition under such circumstances would be thoroughly tested.

Brown had not yet taken any classes from Williston, but somehow, foreshadowing his skills as a negotiator, he charmed his way onto the list of nine student crew members. At long last, Brown's chance to experience the life of a real paleontologist lay directly before him.

Researchers have often bemoaned Brown's aversion to keeping detailed field notes, but this trip, presumably at the direction of Williston, represents a fortuitous exception. On June 13, the crew left Lawrence by train for Kansas City, then continued up the Missouri River by boat to Lincoln, Nebraska. They seem to have dallied around the university there, for the next entry is on June 24, by which date they had arrived in the field: "Did washing and odd jobs. Rained in morning. Got dinner and went out collecting. Found four oreodon skulls and some rodent remains." Oreodonts were sheeplike, herbivorous mammals that roamed the middle of the continent in the Oligocene and Miocene, about 34 to 5 million years ago. They are extraordinarily common as fossils, and many species are known. The same entry reveals that Barnum cultivated an eye for more than just beautiful fossils, a propensity that would last the rest of his life: "Princeton girls did not come today which was a great disappointment after greasing our shoes and washing up. Saw a beautiful sunset. In the evening drowned out [a] Nebraska [crew, also in the field,] with 'Carmine' and other familiar college songs."[12]

The following day proved even more momentous, as Barnum began what would become a lifelong friendship with a fellow student who would also become a prominent paleontologist, Elmer S. Riggs. "Got breakfast, washed dishes, and Riggs and I took lunch and started south for an all day tramp. Found one fine oreodon skull.... Riggs and I climbed a crag at the risk of our lives. Saw some lovely scenes; to the north one could see from sixty to seventy-five miles. To the east rise needles and spires to a height of eight hundred feet."[13]

Even in the nineteenth century, paleontology was a competitive business. Brown's entry for June 26 shows that, despite Williston's move from Yale to KU, he still hadn't completely escaped spats in the field. The adversary on this trip was the crew from the University of Nebraska, led by thirty-eight-year-old Erwin Hinckly Barbour, another former Marsh assistant and vertebrate paleontologist who specialized in fossil mammals. On this day, which Brown lamented was "as hot as the lower regions are said to be," Williston had led his team northwest from camp to collect in exposures known as the *Titanotherium* beds. They had apparently prospected there before and were returning to begin excavating the specimens. In the meantime, however, Barbour and his crew had also prospected in the area, and Brown and his mates "found a great number of our bones tagged Neb." In a fit of pique, Brown reports, "we promptly tore them off."

From the outside, paleontological expeditions seem like adventurous, romantic treasure hunts. But this image doesn't take into account the hard work and often tedious daily routines and hardships of working in desolate regions, where blistering heat can make finding water a life-and-death concern, as one of Brown's later accounts points out: "While we were in these 'Bad Lands' it didn't rain and our water hole got lower and lower until finally we tested it with a stick. If the stick fell over, [we judged] the substance [to be] water and we could work a while longer; if the stick stood up, it was mud and we had to pull out for White River 40 miles distant."[14]

Despite the heat and dehydration, by June 27, just four days into the fieldwork, Brown had already assumed a critical role as Williston's field hand. When the professor found a fossil worth collecting, "He would start to work on a specimen and then turn it over to me and say, 'Brown, you take it out. I am so anxious to see the specimen when it is out that I am afraid I will injure it in excavation.'"[15] Thus, one key skill that Brown learned on this trip was how to excavate fossils. At that time, excavating a bone involved digging around the specimen with a pick, chiseling the matrix off near the bone, and covering the fossil with strips cut from gunny sacks soaked in flour paste, which hardened into a protective jacket. Brown was a natural at this task, and he took on the role of chief excavator for both Williston's specimens and his own. But beyond the excavation and jacketing, Brown, with his experience of moving large objects around his family's farm, was also adept at getting the delicate specimens down the steep, treacherous outcrops without damaging them—no easy task considering these jacketed specimens could often weigh upward of fifty to even several hundred pounds. Brown describes the task in his field journal:

> Went back to the same beds and cleaned up most of our bones. Boxed up the skull I found [presumably the "oreodon" skull] and let it down from the top of the ridge about 100 feet with a picket rope.
> Neb [i.e., Barbour's crew] got on her ear about us taking some specimens they had tagged but that we had found before.... The Dr. is well pleased with my skull. It is the greatest find so far.
> Going out to work this morning I tagged a prairie dog hole as one of Barbour's fossil holes.[16]

The "fossil hole" reference is to large, corkscrew-shaped bodies of cemented rock imbedded in the ancient sediments, called *Daemonelix* or "devil's corkscrews." At this time, there was a good deal of controversy about their origin, and Barbour was one of the key participants in the debate. In a series of papers written between 1892 and 1897, he laid out his observations and analyses of the *Daemonelix* structures, complete with microscopic imagery. At first, he was inclined to think that they might be sponges,[17] but he also entertained the idea that they might be roots or rhizomes. By 1895, those explanations had been challenged by Theodor Fuchs, who argued that they were burrows of Miocene rodents, possibly gophers. Fuchs even noted that one of the *Daemonelix* that Barbour described and illustrated in his 1982 paper contained the fossil skeleton of a jackrabbit-sized rodent. From Fuchs's perspective, therefore, the answer to the mystery was obvious from the start. But Barbour mounted a spirited defense, claiming that the structures were plant fossils and citing the improbability that any rodent, either living or dead, could have dug such perfectly shaped, spiraling burrows; he also presented microscopic evidence that the *Daemonelix* contained material with the clear cellular structure of vascular plants.[18]

Barbour's rebuttal was published in June 1895, the same month that the Kansas and Nebraska crews left for the field and when the intensity of the debate was at a high point. Thus, it may well have been a topic of spirited conversation between Williston and his students on their way out, especially since they gathered in Lincoln before heading on north. By jokingly marking a modern prairie dog hole as one of Barbour's *Daemonelix,* Brown demonstrated that he was more than willing to poke fun at his professional colleagues. His prank, along with a visit from a traveling religious group, enlivened the camp on the 28th.

Almost seventy years later, in 1962, Brown related the story, including his confession of guilt, in a letter to a relative of Williston's who was compiling the professor's biography. As Williston's wagon slowly crossed the field area, Brown explained, he and his buddy, Riggs, sitting together on the rear seat, would surreptitiously jump off when a prairie dog colony came into view and plant a label marked *"Daemonelix"* on one of the holes. Following in their wake, Barbour found the labels and became incensed at the impertinence of Williston's crew: "That night Dr. Barbour came striding into our camp, very angry, exclaiming: 'Dr. Williston, you and your men are not gentlemen, labeling dog holes *Daemonelix,* and I will have you recalled!' Dr. Williston replied, 'Why Dr. Barbour, I don't know what you are talking about!' Nor did he. Then Riggs and I confessed that we were the culprits and had labeled the dog holes in a spirit of fun. Thus explained, the matter passed off amicably."[19]

Things settled back down for the next week until another rite of passage for Americans in the field approached. On July 3, "Overton... came back from town, brought a case of beer for the Fourth. Neb. brought in an ice cream freezer full of cream. McCormick sent over a quarter of beef which is very fine." The next day, the field party celebrating the country's birthday became the first in a tradition that Brown would continue for his own field crews throughout his career.

> We ushered in the glorious fourth of July with twenty or thirty shots which made Neb think the Sioux were upon them.
> ... Tapped our beer this morning. Dr. was one of the boys.
> ... Neb invited us over to their camp. Everybody washed and changed socks that had any to change. They had a table with table cloth and boquets of sunflowers. First course was potato soup, 2nd roast beef, potatoes and beans, bread and coffee, 3rd raspberries and cheeries, 4th ice cream and three kinds of cake....
> After supper, three of Neb came over with us and we had a few bottles and a jolly time.[20]

Brown's entries end on July 5, but the trip's importance to Brown is manifest in the fact that he kept his notes until the end of his life, though he often disdained writing them later in his career. It was also the beginning of an apprenticeship under Williston that would form the foundation for the rest of his legendary career. Williston clearly provided a real-life role model for Brown to emulate—though he did exhibit some untidy examples, as Riggs points out regarding the professor's wardrobe in the field:

> When you get a fossil worked out and want to take it up and pack it the best procedure is to cut strips of gunny sacking and soak them in flour paste. Then you work the bandages all around the fossil and let them dry. Meanwhile, your hands get covered with the paste; the more particular collectors scrabble their hands in the dirt and then wipe them on the sage brush, but our chief, being a direct and practical-minded man, simply wiped his hands on his trousers with an efficient up and down motion and periodic changes from front to rear. After a few days the only part of his garment that would bend was the knees and when he went to bed he crawled out of [his pants] and leaned [them] up against the tent.[21]

In Brown's notes for his unrealized autobiography, he nostalgically concludes that his first foray into the field amounted to "glorious days for our teacher and his students."

Despite all the antics, the expedition was scientifically successful, as was documented in the University of Kansas *Students Journal* of midsummer 1894. The crew, the article dramatically observed, "took their lives in their hands and became martyrs on the field, in the interest of science"—apparently a reference to the extreme heat, thunderstorms, and occasional lack of good drinking water. Then the glorious results were enumerated:

> A complete skull of the Ilotherium; a nearly complete skeleton of a Sabertooth tiger [*Hoplophoneus*]—the only specimen of the kind that has ever been found...; and the larger part of the skeleton of a very peculiar animal—a ruminant with claws....
>
> Besides these specimens—which are of great value—many other finds were made, including the complete skull of a Mosasaur from the Cretaceous beds; four different kinds of reptiles from the lower Miocene beds; and twenty-four different genera of Mammals, ranging in size from the mouse to the elephant. There are enough of those specimens to fill entirely another case in the [KU] museum.

Elotherium was a large boarlike mammal that belonged to the group called entelodonts, and the clawed ruminant mentioned here was presumably a chalicothere, such as *Moropus*. Chalicotheres were strange beasts that looked something like a cross between a cow and a giant sloth.

With the heady days of fieldwork behind him, Barnum returned to another year of classes. In his photo for the Class of '97 yearbook, he displays a serious countenance, perhaps reflecting the demands of his coursework and professors. In addition to Williston, he recalls in his autobiographical notes, "I had other brilliant professors at K. U.—Dr. Edwin M. Hopkins in English, all of whose courses I took; Dr. Frederick Hodder in History; Dr. William H. Carruth in German; Dr. Edgar Henry Summerfield Bailey in Chemistry; Dr. William Chase Stevens in Botany stand out in memory."[22]

All was not completely serious on campus, of course. In the notes for his autobiography, Brown relates several anecdotes about his classes, including one taught by Lewis Lindsay Dyche, an irrepressible collector whose archives of regional organisms form the basis of the university's teaching and research collections. Today, the natural history building at KU is named after him. A renowned taxidermist, Dyche was famous for his mount of Comanche, Custer's horse that survived the Battle of Little Bighorn, which ended up at KU when the 7th Cavalry couldn't come up with the $700 to pay Dyche for the job. Brown writes in his notes about the general zoology class Dyche taught, a popular class because it was reputed to be a "snap."

> It was given three days a week and at the end of the hour five minutes were allowed for questions. Mr. Baker, one of the students, had been absent at the preceding lecture on bats and he had taken his notes from other students but had forgotten the terms oviporous and viviporous. When asked if there were any questions a hush fell over the students until Baker stood up and said "Professor Dyche do bats lay eggs or do they ah-ah?" "Mr. Baker the bats ah-ah," said Dyche. Whereupon the girls figuratively swallowed their handkerchiefs and Baker had to leave school for a week to escape ridicule.[23]

Brown's adolescent sense of humor was similarly tickled by the following event: "Among other subjects Dr. Williston was an authority on Diptera (flies) and other insects. Cora Becker a graduate student was working for her Master's degree and had taken several drawers of insects home for study. When she felt ready she came to Dr. Williston's door and knocked. Frank Snow, son of the Chancelor, went to the door and seeing Miss Becker with the drawers announced, 'Dr. Williston, Miss Becker is here with her drawers down, ready for the examination.'"[24]

Of course, Barnum was not averse to staging practical jokes of his own. One of them involved a sneak attack on a chamber pot. KU was a small school at the time, and there was only one dorm, with "girls on the 1st floor, boys on the 2nd; different doors for entrance; . . . no separate bathrooms."

> One year I with a roommate had an upper room heated by a stovepipe that came from the room below occupied by two girls. My mate and I went to bed early and the girls below bedeviled us every night by pounding on the stovepipe so that it was difficult to sleep.
>
> In the chemistry class we had just learned that metallic sodium would be fired [ignite] if water touched it so while the girls were at supper that night we sneaked into their room, carefully dried out the pot under the bed and laid a coil of metallic sodium in it. When they went to bed a scream came out of their room for one of the girls thought she was sitting over a sizzling snake. They weren't sure who had played this prank, but one of them, also a chemistry student, suspected it was the boys above them and thereafter we studied and slept in peace.[25]

As the year progressed, Williston began firming up plans for another summer in the field, this time in the wilderness of eastern Wyoming. Fossil mammals were not the focus of this summer campaign, however; he was now in search of larger and more spectacular prey: dinosaurs. His quest was fueled by several personal and professional objectives. Now free from the strictures of O. C. Marsh at Yale, Williston enthusiastically envisioned collecting dinosaurs under his own banner for KU. With numerous other states in the process of establishing natural history museums, including the American Museum of Natural History in New York and the Field Columbian Museum in Chicago, Williston was determined that Kansas not be left out of the parade. If Williston was seeking to one-up Marsh, one way to do so would be to lead an expedition to the Cretaceous exposures of Wyoming. As other chroniclers of the trip noted, "Williston knew the history surrounding the discovery of *Triceratops*—how a skull had been found [in 1888] by a cowboy on horseback who had seen it sticking out of a bank just out of reach, lassoed it, and dislodged the horn and part of the skull, which was sent to Marsh at Yale."[26] Previously, Marsh thought a similar horn sent to him from Colorado in 1887 belonged to an extinct bison, but in 1888 John Bell Hatcher, one of Marsh's collectors and a colleague of Williston's, happened through the

Late Cretaceous exposures near Lusk, Wyoming, where he found that the 1888 horn had been attached to a gigantic skull embedded in a sandstone concretion.[27] Hatcher sent the horn to Marsh, who realized that some of his assistants, including probably Williston, had been correct in suggesting that the horn was not from a bison but from a new group of dinosaurs. Marsh directed Hatcher to return to Lusk and collect the massive skull, which weighed more than a ton. Hatcher sent it to Yale, where Marsh described it as the type specimen of the famous *Triceratops*. Between 1889 and 1892, Hatcher went on to collect the skulls and skeletons of fifty horned dinosaurs for Marsh. Some of the skulls weighed as much as 3.5 tons, and they established Hatcher as a superior collector and Marsh as the preeminent expert on ceratopsians.[28] In fact, Hatcher was the true brains behind the ceratopsian work, and after leaving Marsh he was senior author on the definitive monograph on the Ceratopsia published posthumously in 1907.

Still, in 1895 no specimen of *Triceratops* was on public exhibit anywhere. "A skull," Williston's chroniclers write, "would be a splendid addition to the small museum at the university. Williston understood the public fascination with paleontology and he hoped to enhance his reputation as well as his bank account through articles in popular magazines. In particular, he hoped to publish an illustrated travelogue of the 1895 trip, and brought a large box camera with glass plates along for the purpose."[29] This use of a camera may have inspired Brown, who, although he left few field notes, was a prodigious photographer. In fact, the only records we have of many of his expeditions and discoveries are photographs and negatives in the museum's collections and detailed accounting summaries.

With experience from the previous summer's expedition under his belt, Barnum Brown was Williston's choice to spearhead the trip westward. Along with that expedition's teamster, Brown would guide the wagon to the field area, following much the same route he had traversed with his father six years earlier. Williston and the rest of the crew would travel by train to Denver, then north to Wyoming to meet Brown and the supply wagon. From mundane chores of daily routine to his tribulations with the weather and interactions with the locals, the embryonic paleontologist dutifully recorded his thoughts during the trip out in a tall, thin notebook, of which only a faded photocopy remains.

Although he was only twenty-two, Brown's confidence in his role, buttressed by the knowledge gleaned from the odyssey with his father, is apparent from the very first page: "Rogers and myself started from Washington

Wed. morn May 29 with a 1200# team of grays, a ten hundred wagon and about 600# of provisions.... We have a 12x14 tent and a 7x7 Sibley tent for use on the road. Our outfit is most complete have a little sewing table for a mess table and a good camp outfit of cooking utensils and a four hole sheet iron stove." Heading north across the state line into Nebraska, Brown bemoaned the state of the parched plains. Hence, when a rainstorm beset them on the first night out, he and his partner maintained a stiff upper lip and "let it rain for the good of the people and didn't grumble."[30]

As they traveled toward the Platte River on May 31, Brown again kept his keen eyes focused on more than just the drought-plagued plains. "We spied two women presumably but when we reached them they were girls. Asked them [to] ride and they got in the wagon. They were going to Nelson. We learned they were 14 yrs old and several other things. One was a corker. I had more fun than a barrel of monkeys. She wouldn't go along with us so we parted with them at Smyrna regretfully and camped for dinner."[31] This passage clearly reveals Brown's eye for the ladies. In addition to his status as the greatest fossil collector that ever lived, vertebrate paleontologists have also long celebrated his status as one of the discipline's most prodigious Don Juans. Rumors still abound regarding the legions of illegitimate children that were left in his wake as he traversed the globe in search of fossil treasures. Yet whether any concrete evidence exists to support Brown's reputation as such a truly rapacious rake remains in question.

Although some basics about Brown's relationships with women are clear, such as that he was married twice and fathered a daughter, Frances, with his first wife, several problems arise when attempting to compose a complete portrait. The main difficulty is that few of Brown's close confidants are still alive, and those who are remain loyal and refuse to divulge details about his personal life. Frances, moreover, has passed away without leaving any offspring to provide more details. Thus, stories of Brown's indiscretions are by and large secondhand accounts passed down from professors to students as colorful and highly amusing legends.

One story often told among vertebrate paleontologists at AMNH involves an incident that occurred upon Brown's death in 1963. According to Gene Gaffney, now retired curator of fossil reptiles and then a student at the museum, Gil Stucker, one of Brown's most trusted assistants, rushed into Barnum's office as soon as he heard the news and conducted a thorough search of the files, gathering up any potentially incriminating photos and correspondence. This purge is sometimes referred to as Stucker's attempt to "sanitize" Brown's legacy. Apparently, much of what Stucker gleaned from the office was sent to Brown's daughter, Frances, and a letter that Frances wrote to Stucker, which simply thanks Stucker and acknowledges receipt of the material, still resides in the archives of the Vertebrate Paleontology Department. We have no idea, however, where the documents that Stucker sent to Frances are now.

While Stucker's attempt may have been largely successful, some items either slipped through Stucker's fingers or were placed in the archives after his purge. Suggestive evidence includes an unsigned and undated love poem from a person with the initials JGS—clearly not either of his wives—and a few unlabeled photos of women, which Brown seemingly slipped under the cover flaps of his financial field diaries while he was married to his second wife. Although intriguing, these hardly represent concrete evidence of the man's alleged infidelity.

Brown's words about meeting the girls on the trail, though suggestive of a sexual encounter, are inconclusive. Nonetheless, perhaps fired by the prospect of further flirtatious adventures, Brown drove the wagon on into Nelson, where he "found no word from the Dr. nor saw neither of the girls but made a mash on another one"—that is, he expressed his attraction to her. The next day, Barnum continued his pursuit: "This afternoon we came through Hastings, saw some beautiful damsels and created a sensation with our Kansas Geological Expedition on the side of the wagon. People took us for a menagerie."[32]

Barnum's traveling geological circus continued northwest through Kearney and up toward the fork where the North and South Platte part ways. Funds were getting low by June 7, when the wagon reached Scouts Rest Ranch, where Brown confessed: "I am greatly disappointed at not receiving word and money from Dr. Williston. Rogers and I have only $.60 between us and are out of feed. Will stay here until morning and see if we receive anything."[33]

Apparently, Brown had sent a note of alarm to Williston, who wired four dollars (about $90 in today's currency) to the wagon crew; this allowed Barnum to buy corn and pay "$.20 for a feed" before the team set off up the North Platte for eastern Wyoming.[34] Brown's entries for the trip out end on June 13, but another account written by J. P. Sams, a KU trustee accompanying Williston, reveals that on June 16, Brown and Rogers reunited with the rest of the crew at Badger, Wyoming. All then set out for the *Triceratops* beds to the north near Lusk, arriving just in time for the annual festivities, which Brown recounts: "July 4th. Party arrived in Lusk yesterday evening. . . .

Pitched camp and agreed to work Sunday if they would let us stay and celebrate. This morning we went up and took in the town, which consists of five saloons, 2 confectionaries, a hardware, and a meat market. Did our trading at Baker Bros. and got some meat of Minnie's father. The natives..."[35] Unfortunately, a torn page truncates both the entry and Brown's journal at that point. So an account of the celebration and the identity of Minnie would be lost to posterity if not for the journal of Sams, who with his aversion to alcohol and "rowdying in the saloons" seemed rather out of place in this corner of "the wild and wooly west." On the eve of the Fourth, Sams relates, "As soon as dark we can hear the clatter of hoofs and know that from all directions the cowboys are coming to begin the celebration. Very soon a very large bonfire is started in the middle of main street and the sound of all kinds of fire works begins. The fellows running their horses through town, full tilt, fire their revolvers into the fire just to see it fly. They keep up this orgy nearly all night and we get but little sleep."[36]

Sams attended the next morning's more sober community program, at which "the Declaration of Independence is read by Minnie Schwartz, the Belle of the Town. She is fourteen, tall and not very good looking, dressed in white, short hair curled to a finish.... And still there is something about her I like." For Sams, what he admitted to liking turned out to be Minnie's name, which was also the name of Sams's wife. But for the students of the field crew, who bore no such prior burdens, a twirl with the Town Belle may well have been a motivation for "some of the boys [going] to the dance."[37]

The following week found the crew prospecting deep in the badlands. Sams's entry for July 8 relates:[38] "Today we all start out and at night we find that we have found two fossil heads away up on the highest bluffs embedded in the sandstone. The modern name given to this fellow is *Tri-sari-tops*."

From Sams's journal, it's not clear what role Brown played in either the discovery or collection of the skull, though one photo shows him in a pith helmet at the outcrop helping to load one of the blocks onto a cart to begin transporting it back to the university. What is clear is that Barnum left a favorable impression on the trustee: "Our boys have all been good faithful workers. Riggs, Brown and Gowell are young stout fellows, able to endure any amount of hardship, on the tramp or in camp; good agreeable boys well up in their studies in the University. Certainly I wish them well and if it is my good fortune in after years to assist these boys in gaining honorable places in their chosen professions... it will be a delight to do them any favor possible."[39]

FIGURE 3. Portrait of Williston's field crew for the 1895 KU expedition to southeastern Wyoming, where they discovered and collected a specimen of *Triceratops*. Brown sits atop the outcrop wearing a helmet; Elmer Riggs reclines on the far left; Williston sits second from the right (University Archives, University of Kansas Library, 33/0 1895)

The trip was again an unqualified success. Apparently keen to justify the expenditure of state funds to support the enterprise, the Kansas University *Weekly* trumpeted that "even in a business way for the market value of the fossils [the expedition] brought home amounts to several times the cost of the expeditions." The account focused especially on the immense *Triceratops* skull, "6 ft. long, 4 ft. broad and 3 ft. thick," but it went on to say as well that "some very valuable specimens were also found in western Kansas; among them two complete skeletons of extinct buffalo which was 2 ft. taller than those now living.... Also found was a skeleton of the bird with teeth, which is so rare and valuable—probably the best in the world. Altogether about five tons of fossils were brought in, and Dr. Williston is elated by his success."[40]

Upon their return from Wyoming, Williston felt comfortable enough with Barnum to let him take a room in his own house. In correspondence with her family, Williston's wife said: "Wendell wanted to help along Barnum Brown—who was in his class last year & collecting with him last

& this summer.... He took a great deal of the care & looked after things in camp. Things others wouldn't think of."[41]

Having spent two highly successful summer seasons in the field, the professor, feeling the need to catch up on his research, planned "to spend the [next] summer in the east or in Europe" visiting museums and studying in their collections.[42] In a fortunate twist of fate, Williston's decision to forgo fieldwork in 1896 would forever alter the trajectory of Brown's career.

THREE

Apprentice Extraordinaire
(1896–1898)

DURING THE 1894 EXPEDITION TO the badlands of South Dakota, Williston's crew encountered not only their collegial nemeses from the University of Nebraska but other crews as well. Among them was a crew from Princeton directed by John Bell Hatcher, who, like Williston, began his career as a collector and assistant for Marsh. More significantly, Williston also encountered a crew from New York's American Museum of Natural History, under the leadership of Jacob Wortman, a former collector for Cope.[1]

Like many nineteenth-century paleontologists, Wortman had been schooled in medicine before becoming intrigued with fossil vertebrates.[2] At the time, medical school was the most common form of postgraduate training for students interested in science, and such programs often included courses in many disciplines related to natural history. Describing him as "taciturn" behind his "rather saturnine appearance," Brown recalled Wortman being decidedly anti-Semitic and deeply resentful that his parents had chosen to name him Jacob Levi. Professionally, Wortman spent much of his paleontological career careening from one institution to another. He joined the staff of AMNH in 1891, but departed in 1898, after a falling out with Osborn, to join the Carnegie Museum in Pittsburgh.[3] In 1908, apparently fed up with fossils, he left paleontology altogether to start a drugstore in Brownsville, Texas, where he worked for the rest of his life.[4]

In 1896, Hatcher lured one of Wortman's principal assistants, Olaf Peterson, away from AMNH, so Wortman needed to find a replacement. He wrote Williston to see if any of the KU students he had met two years earlier might be up to the job. Apparently, Wortman had one particular student in mind, but Brown's professor unabashedly recommended a different candidate:

Brown has been with me on two expeditions, and is the best man in the field that I ever had. He is energetic, has great powers of endurance, walking thirty miles a day without fatigue, is very methodical in all his habits, and thoroughly honest. He has good ability as a student also and has been a student with me in anatomy, geology and paleontology. He practically relieved me of all care in my last expedition, looking after te'm [horses], provisions, outfit [wagon], etc. I cannot say enough in his praise. . . .

The man whom you remember in Dakota was probably Dickinson. He was a very good student in the University and a good fellow, but a complete failure in the field.[5]

Williston's recommendation carried great weight with Wortman, and Brown was offered a spot on an AMNH field crew the next summer. More than twenty years later, upon hearing of Williston's death, Brown would repay the compliments Williston had bestowed on him in a letter to Williston's wife:

It is difficult to find language at a time like this, but my heart simply compels some effort to let you know that I grieve with you, and I want you to know personally how much I appreciated the great man that we have all lost.

I was particularly privileged to have lived in his home and to have been with him in the classroom, and in camp as well as in his work and so feel that I have lost a friend, councillor and master, all in one. How many, many of us look back on our early association with him in College days and know that he, above all men, was our model, inspiration, ideal.

Many of us lose glamour by close contact, but association with Dr. Williston in daily life was ever an inspiration. He made one feel that life was real and earnest and worthwhile, and his remarkably varied achievements of solid worth stimulated men and women that he guided to go forth and try to emulate.[6]

In essence, that was exactly the task that Barnum now faced. It was as if a farm boy from eastern Kansas had been given a tryout with the New York Yankees of the paleontological world, and Brown was determined to make the best of the opportunity.

Or perhaps we should say the soon-to-be Yankees, because throughout the Cope-Marsh fray, the AMNH had been content to sit on the sidelines. Established in 1869, the museum spent the next two decades developing collections related to scientific disciplines other than paleontology. But it was becoming increasingly difficult to ignore the notoriety attending the incredible dinosaur skeletons from the American West.

Realizing it was a late entrant in this quest, in 1891 the AMNH launched a determined effort to catch up, hiring an ambitious young Princeton-educated scientist, Henry Fairfield Osborn, to found the Department of Vertebrate Paleontology.[7] Osborn came from solidly entrepreneurial stock.[8] His mother, Virginia Reed Sturges, was the daughter of a highly successful merchant in New York, Jonathan Sturges, while his father, William Henry Osborn, spent the first part of his prosperous career in the international shipping business; in the 1850s, however, William joined with Sturges to help finance and reorganize the Illinois Central Railroad, which became one of the nation's most successful.[9] Osborn's family, in short, possessed considerable wealth and extensive connections with the business community in New York, assets that Osborn would use to the museum's great advantage.

As part of the New York aristocracy, Osborn is often described as having been imperious and a difficult person to deal with. In his biography of Osborn, Ronald Rainger paints a particularly imposing portrait:

> In promoting himself and his department Osborn functioned as an aristocrat....
>
> Osborn often flaunted his status.... He offered criticisms on the personal lives of his assistants.... Osborn was a wealthy man who was fully convinced of his own self-importance and often treated others in the museum, including scientists, in a condescending manner....
>
> Osborn also became convinced of the truth and great value of his work.... Over the years... he lost all critical perspective on the character of his work.... He spoke of his own work in the same breath as the researches of Darwin, Huxley, or other towering figures in biology. He frequently referred to his... chance meeting with Darwin in Huxley's laboratory in 1879. Osborn came to view himself as the inheritor of their scientific mantle. He considered his own life exemplary, a model for making important scientific discoveries and achieving social and scientific prominence. His autobiographical sketches, notably *Fifty-Two Years of Research,* were didactic, egotistical works designed to portray his life as a lesson for aspiring young naturalists. In his own mind Osborn was a great man, and he did not hesitate to make that opinion known or to employ it to his advantage.[10]

Nonetheless, archival correspondence also shows that Osborn developed cordial and often warm relationships with many of his employees, including

Brown—even if his pompous and strong-willed manner grated on them at times. Rainger cites the loyalty and respect that several of the "less senior members" of Osborn's staff felt for their boss, including those who eventually moved on to work at other institutions, such as O. A. Peterson, whom Brown would later join on the Patagonian expedition. One source of this respect included the "limited independence" that Osborn granted them to develop their careers and carry on their work.[11]

Brian Regal offers a more nuanced summary of Osborn's personality:

> Osborn has been characterized as a pompous, self-important character from a politically incorrect past. While that is generally an accurate appraisal, he is far more complex than this one-dimensional portrayal. He was a man of contradictions: a man who loved Charles Darwin, but was not a Darwinist; a deeply religious Christian who vigorously opposed fundamentalism; a Nordicist and eugenicist who argued that all races had their geniuses who ought to be cultivated and promoted; a man who railed against the influx of immigrant children yet fought to improve the public school system to educate those very children; a man who believed that predestination ruled evolution and that personal motivation and will could alter the course of evolution; a snobbish patrician who promoted the careers of many of his underlings.[12]

Rainger and Regal both delve deeply into various aspects of Osborn's scientific research and social philosophy, including his participation in the eugenics movement and his quest to discover the paleontological roots of humanity, so there is no reason to dwell on those topics here. But how did these traits develop?

At Princeton, Osborn did not intend his studies to lead to a career in science.[13] His father, not unsurprisingly, hoped and expected that Henry would join the family business.[14] But Henry, along with a close friend and classmate, William Berryman Scott, became enthralled with geology and paleontology when they took Arnold Guyot's famous course in 1876 and traveled to Yale to study the collections there.[15] Marsh, however, was unwelcoming: not only did he deny them the opportunity to see new and unpublished specimens, but he surreptitiously followed them around the museum in his bed slippers to make sure that they stayed in areas that he had approved. Having developed a distinct distaste for Marsh, therefore,[16] they became disciples of his rival, Cope. In 1877, the same year the "bone rush" for dinosaurs began, both helped organize and conduct an expedition to Wyoming's Bridger Basin to collect fossil mammals.[17] After a year of graduate school at Princeton, Scott set sail for Cambridge, England, where he studied comparative anatomy under the guidance of Thomas Henry Huxley, the irrepressible proponent of Darwin's evolutionary theories, and the famous embryologist Francis Balfour.[18] Osborn soon followed to receive training from the same pair of eminent scholars. In 1881, Osborn finished his graduate degree at Princeton and accepted an assistant professorship there in the field of natural science.[19] Although Scott returned to Princeton upon his return from Europe and stayed there for the rest of his career, Osborn eventually left, first accepting a position at Columbia University, where he established the Department of Zoology and the graduate school in 1891. In the same year, he took on the challenge to establish the Department of Vertebrate Paleontology at the American Museum of Natural History.[20]

The reason for Osborn's interest in AMNH was multifaceted. Osborn's influential father feared that his son's opportunities at Princeton were limited and felt that if Henry was to be a scientist, he should strive to develop a career that would be both "socially useful and maintain the family prestige."[21] William Osborn's network of business associates in New York would allow his son to raise substantial funding for staffing an independent paleontological operation at the museum capable of taking on large-scale projects in the field and mounting major public exhibitions.[22]

Osborn's first initiative at AMNH involved the development of a program in paleomammalogy, his specialty, which would encompass collecting expeditions, related research, and public exhibitions.[23] Osborn quickly hired highly experienced and successful collectors like Wortman, who initially became his primary assistant, to develop a team that could scour the West for fossils. Other key hires included Adam Hermann, a gifted preparator who would help Osborn construct lifelike skeletal mounts, along with Rudolph Weber, Erwin Christman, and the freelancer Charles R. Knight, artists who began the museum's tradition of precedent-setting paleontological art that eventually included illustrations, murals, and sculptures of extinct animals.[24] One of Osborn's goals was to outdo Marsh both in developing superior collections and in fostering public literacy about evolutionary history through displays of fossil mounts and associated artwork, an approach that Marsh disdained.[25] By 1896, these enterprising efforts had led Osborn and Wortman to the doorstep of Brown.

Brown's tryout with AMNH led him to leave classes at KU before the term ended—a decision that, in a sense, set a precedent for the rest of his career. When faced with the choice of either staying home to complete his research or escaping the confines of the museum by heading to the field, he almost always chose to pursue his opportunities in the field. Williston's wife commented on Brown's early departure from KU in a note to her daughter: "Mr. Brown has gone home tonight and Sunday expects to start for Colorado where he will meet Dr. [Wortman] & party of the New York museum and go collecting 'doggie bone fossils' for them. It seems pretty cold to start off camping but he thinks they are going to Arizona or New Mexico."[26]

Major paleontological expeditions are expensive and often risky operations. Because they are so costly, it is essential to develop a carefully considered plan, which lays out requirements in terms of personnel, equipment, supplies, and duration. They are also designed and conducted to accomplish a specific set of goals. All these goals might not be realized, and unexpected discoveries are fairly common, but before an expedition even starts, a lot of research and planning goes into establishing an overarching agenda for the endeavor.

Osborn's agenda for the crew was to augment collections made by the first AMNH paleontologic expedition to the same region in 1892. Initially, Osborn directed Wortman to collect fossil mammals from extensive exposures of Paleocene and Eocene floodplain sediments in New Mexico's San Juan Basin. In general, these specimens would help Osborn shed light on the earlier phases of mammal evolution right after the extinction of nonavian dinosaurs. More specifically, as noted in the report written by Wortman and Osborn following the trip, "The particular object of this exploration was to secure if possible a skull and other missing parts of the skeleton of *Coryphodon* the immediate ancestor or forerunner of the later Uintatheres of the later Bridger epoch. While it is true that the collections of the Museum were unusually rich in the remains of these animals..., they are more or less fragmentary and too imperfect to be used in mounting a complete skeleton."[27] *Coryphodon,* a quadrupedal herbivore about the size of a cow, was one of the first mammals to attain a large body size after the Cretaceous extinction of large dinosaurs, thereby exploiting some of the ecological niches left open by the disappearance of the ceratopsians, such as *Triceratops,* and the hadrosaurs, or duckbills.

Joining Wortman and Brown was another of Osborn's "young guns," Walter Granger. He would eventually become Roy Chapman Andrews's chief assistant during the museum's Central Asiatic Expeditions in the 1920s, which garnered global fame for discovering the first well-documented eggs and nests of extinct dinosaurs in the Gobi Desert at Mongolia's Flaming Cliffs. Granger worked at the administrative and scientific core of the department for nearly fifty years.

From mid-April, when the weather was "extremely cold and unfavorable for rapid travel," through the end of June, under the scalding early-summer sun, the crew, with its mule-drawn wagon and pair of saddle horses, traversed the desolate drainages and divides etched across the basin.[28] Once in the badlands, it became difficult to obtain sufficient food for the animals, especially since there was little grazing and grain was exorbitantly expensive.

Near the turn of the twentieth century, the wagons and beasts that pulled them were the lifeline of any paleontological expedition. Even today, transportation issues are often the Achilles' heel of an expedition. A shortage of grass and feed for the horses and mules raised serious problems for getting in and out of the remote regions where the fossils could be found. The crew's tribulations were exacerbated when, as Brown relates in his notes, a key part of the outfit that linked the mules' harnesses to the wagon broke:

> While pulling through deep sand the whipple trees broke and we had to send back to Farmington for a new pair. This occurred near an Indian encampment and a little boy came over with a small pail of milk which was most welcome after using condensed milk all summer... in exchange [we] filled his bucket with corn meal. That evening Granger and I went over to watch the milking process;... a small boy or girl would [catch] a goat by the hind leg.... The mother... grabbed the goat legs, spreading them apart and spat on her fingers as she grasped a teat. She alternated spitting onto her fingers and dipping them into the milk until the goat was finished.... Sometimes there were goat droppings in the milk which she skimmed out with her fingers before sending the milk to us. Having seen the process, we lost our appetite for fresh milk but continued to give the Indians meal in exchange for the milk.[29]

It was then and still remains critical that expedition crews make every effort to coexist with the indigenous inhabitants of their field area. Finding fossils and maintaining a camp is difficult enough without antagonizing the locals; besides, the local inhabitants possess valuable knowledge about the region's roads and resources. Their assistance and advice can be a key to success, and exchanges of gifts and supplies are still common. But being polite and politic can also carry health risks related to eating food provided by the locals that may not have been preserved or prepared properly.

"After a month's hard work, under the most trying circumstances," Wortman reports, "we found ourselves with practically no results, and what was still more discouraging, with but a few scattered fragments of the animal whose remains we were so anxious to secure." Abandoning the Early Eocene (Wasatchian) deposits where they had hoped to find *Coryphodon* skeletons, the team explored a sequence of older, Early Paleocene (Puercan) sediments and finally found one productive bed of red clay that yielded fossils of several extinct mammals, including *Psittacotherium, Protogonia, Pachyama, Pantolambda,* and *Mioclamus*. This allowed Wortman to report that "the finds in the [Puercan] had amply repaid the usual outlay of labor, to say nothing of the cost of money expended." However, the venture could not yet be deemed a success. "In one of its main objects, viz: the securing of the missing parts of... *Coryphodon*," Wortman lamented, "the expedition was unsuccessful.... It was thus determined to try once more in the Wasatch sediments of the Big Horn Basin of Wyoming, to obtain the much desired *Coryphodon* remains."[30]

Brown was happy to be back to more familiar haunts, presumably buoyed further by the addition of his KU classmate Elmer Riggs. By July 10, Wortman's crew began winding their way with a newly purchased wagon from Casper toward the Bighorn Basin and "reached the collecting grounds on the Grey Bull River on the 18th."[31] The crew's optimism about finding good specimens of their quarry was justified, for in 1890 an AMNH party had found a partial *Coryphodon* skeleton, including the pelvis, a forelimb, and a hind limb.

Wortman and the crew wandered these Early Eocene outcrops for six weeks, at the end of which the leader could boast that the results were "most satisfactory." Brown had found a fine *Coryphodon* skeleton complete with the skull and lacking only the hind limbs, while Wortman had discovered two other specimens with "more or less perfect skulls." Wortman had high hopes of being able to mount a complete skeleton.[32]

Clearly, Barnum had more than passed the audition with his discovery of the expedition's crown jewel, and the crew, after also finding "the remains of the extinct horses, monkeys, creodonts, [and] carnivores," returned to Casper as winter approached in September and disbanded for the trip back to their respective homes.[33]

In his notes for his autobiography, Brown makes scant reference to the 1896 trip, except to summarize: "We secured large collections, many specimens being new to science. The winter was spent in the laboratory preparing

FIGURE 4. Riggs and Brown collecting a specimen of *Coryphodon* in the Wasatch badlands of the Bighorn Basin in Wyoming during the AMNH expedition of 1896 (17753, American Museum of Natural History Library)

the specimens for display, description, and cataloging."[34] For the moment, however, Brown had an influential ally at the museum in Wortman, having impressed him with his collecting and general field skills. He also provided his boss with a critical source of entertainment during the expedition, supplied by Brown's relatives in Carbondale: "His greatest pleasure was to read my weekly paper 'The Astonisher and Paralyzer' sent to me by my parents with other mail."[35]

A letter from Brown to Wortman reveals that Brown, supported by funds sent by Wortman and presumably approved by Osborn, lingered near Laramie long after the crew disbanded.[36] Outside of some recreational deer hunting, most of Brown's time throughout late September and October seems to have been spent evaluating the possibilities for collecting mammal and dinosaur material the following year from the region's Jurassic Morrison Formation. These 150-million-year-old sandstones and mudstones had eroded off a precursor of today's Rocky Mountains to be deposited by rivers and

FIGURE 5. Wagon at the base of the Wasatch badlands along the Greybull River in the Bighorn Basin of Wyoming during the AMNH expedition of 1896. Specimens of *Coryphodon, Hyracotherium,* and *Phenacodus* were collected from the basal exposures. (AMNH Vertebrate Paleontology Archive 7:2 B1 S36)

streams on the adjacent floodplain. Marsh's crews had first exploited the rich trove of dinosaur fossils entombed in the ancient sediments at Como Bluff near the end of the 1870s. Especially noteworthy was the suite of spectacular sauropods and stegosaurs that Marsh's collectors gleaned from these remote outcrops. Even a few minute Late Jurassic mammals, whose jaws and teeth must be viewed under magnification to discern their anatomical details, had been discovered in one of the quarries, which certainly must have piqued Osborn's interest in returning to see what else might be discovered.

Osborn had begun to lay the groundwork for collecting dinosaurs in 1895, in the wake of opening the museum's first Hall of Fossil Mammals. He wrote the trustees to advocate that the mission of his department be enlarged to encompass all the disciplines within vertebrate paleontology, including dinosaurs, which would be displayed in a new Hall of Fossil Reptiles. The trustees concurred. Once again, Osborn's basic goals were threefold. The public's fascination with dinosaurs offered another avenue to educate the masses about evolution through public exhibitions and popular articles, a service that Marsh had not exploited. From the perspective of collections and research, Osborn sought to surpass his rival at Yale and "break down Marsh's work as far as possible" by making AMNH the "world's leading center for fossil reptiles." Also, the publicity associated with dinosaur discoveries and exhibition could be extremely useful in his fundraising efforts.[37]

At the end of the field season in 1896, Brown was aided in his assessments of the Morrison Formation by one of Marsh's former collectors, William Reed, then associated with the University of Wyoming in Laramie. In his initial sales pitch, Brown informed Wortman that "there are excellent opportunities to collect. Plenty of fossils, good roads to haul over, which would be a big thing in handling these bones and good camps; plenty of wood and Mt. water. The deposits I examined are about thirty miles west of Laramie in clay and limestone. Reed and Knight have worked this bed considerable though. . . . Reed assures me however that he can take me to a much richer deposit, especially in mammal jaws about fifty miles north of Laramie." To bolster his claim and appeal to Osborn's interest in fossil mammals, Brown noted that while poking around the facility at the university, he had seen a wealth of new material. "There is a far greater field to be worked up in Jurassic mammals than I had any idea of," he concluded.[38]

Brown returned to Lawrence halfway through the term in mid-November to continue his coursework, once again living in the Williston house. But back in New York, plans were already taking shape for next summer's field season in Wyoming. In early spring 1897, Osborn made his first direct contact with Brown, in a letter containing a proposal that must have seemed like a dream come true to the aspiring paleontologist:

> Dear Mr. Brown:
> I have had considerable conversation with Dr. Wortman regarding your work during the coming summer and winter. If you would like . . . , I think we can arrange for you, at the close of your summer's collecting, to take a course in [Columbia] University towards the University degree, also to do enough work in Museum [AMNH] to help pay your expenses together with what you will earn during the present summer. It is possible by prompt application and strong recommendation from Professor Williston that you can secure a University scholarship, which means free tuition, but

I do not consider it very probable, there are such enormous demands for these places.

As Dr. Wortman wrote you, I would like a report from you regarding the Jurassic mammal beds, and although I would like you to talk over your plans with Professor Williston, for several reasons I prefer that you should not speak to anyone else about them. I . . . would like to have you enter the field, determined to secure a complete representative collection, if it is possible. Collecting as you know is extremely difficult and delicate work, but the scientific results are proportionately of very great importance. The only collections at present known are those in the British Museum and at Yale.

For next winter my idea would be to have you take the first year graduate courses in the college, which would occupy half your time, or three working days a week. The other three days I would have you give to the Museum preparing the collections which you secure in the field, and learning the museum paleontological technique.

. . . I write to ask if it would be possible for you to leave Lawrence by April 15, if this were found necessary.[39]

Within a week, Brown had written the requested "report," and Williston had sent his recommendations for the summer's agenda, presumably along with his endorsement for the scholarship. This exchange of correspondence marks the pivotal moment when the torch lighting the path for Brown's career passed from Williston to Osborn. Brown's report makes little attempt to conceal either his eagerness to accommodate Osborn's wishes or his gratitude for Osborn's offer: "I can leave here by April 15th or whenever necessary as I have my work nearly completed. . . . I am deeply grateful for your kind offer. The University and Museum work is exactly what I desire."[40]

In their correspondence, Brown and Williston laid out a plan to reopen Marsh's mammal quarry, the famous Quarry 9, south of Aurora, Wyoming, along Como Bluff. It would involve excavating the dirt with which Marsh's crew had filled it at the end of their earlier work and scraping off the overburden along the back wall to expose more of the fossiliferous bed. Brown was clearly anxious to prospect for new localities in the region, especially ones that might yield good dinosaur material: "As to the reptiles," he wrote Osborn, "I think we can obtain any amount of material. . . . I worked with Mr. Reed a few days in a quarry west of Laramie where the bones were literally packed one on top of the other, nearly all in a good state of preservation. . . . The Jurassic appears in frequent escarpments along the western and northern border of the Laramie Plains, only a few of which I have examined, all containing more or less prospects."[41]

Williston had another idea, however. "Until Brown has learned where and how to look for the bones," he observed,

> it would be, in my opinion, the wisest thing for him to reopen the [mammal] quarry, which is by no means exhausted. I have directed him to the spot so that he can not fail to find it.... The teeth occur in deposits similar to those of the Laramie; that is among waterworn fragments, small bones, etc. in sandy... matrix. Brown found a number of cretaceous teeth in Wyoming two years ago, and will know what to look for.... Teeth of mammals are never or rarely ever found in connection with the large bones of the reptiles [dinosaurs], and it will be idle to search for them among such, unless the reptile bones are wanted more than the mammals.[42]

In fact, Osborn wanted both, as Brown recalled in his notes for his unpublished autobiography: "Previous to this time, the trend of scientific thought was to consider that only mammals were to be relied on in the study of evolution and relationships. But during this summer, Professor Osborn directed me to continue collecting mammal remains; and also see what we could do with dinosaurs."[43] In his response to this directive, Brown confidently stated in a letter: "The plan suits me very well. I have talked to Dr. Williston...; he thinks I can get good men there at reasonable prices. If however you want me to do work in the reptile beds afterwards it will be advisable to have someone who knows about the subject. I have found it difficult to get men to handle 'sick' [poorly preserved] bones carefully, in fact few collectors use the care I like to see."[44] Brown went on to suggest that he engage a classmate from KU, Harold W. Menke, to help him collect the dinosaur bones. Presumably chuckling under his breath at Brown's characterization of fractured dinosaur material, yet still wishing to be supportive of his young apprentice, Osborn reassured Brown that he probably wouldn't need a trained assistant in the mammal quarry: "At the time of my visit, we will talk over the question of engaging Mr. Menke, in case you need him to handle sick bones for you."[45]

By early May Brown was in Laramie, champing at the bit to assault Marsh's mammal quarry. Osborn, in effect, had appointed Brown to spearhead the expedition, and the burden of responsibility began to weigh heavily on the apprentice, particularly once he learned from Reed that the information Williston had conveyed about the quarry was outdated and incorrect. Brown knew that his success or failure in this enterprise would indelibly

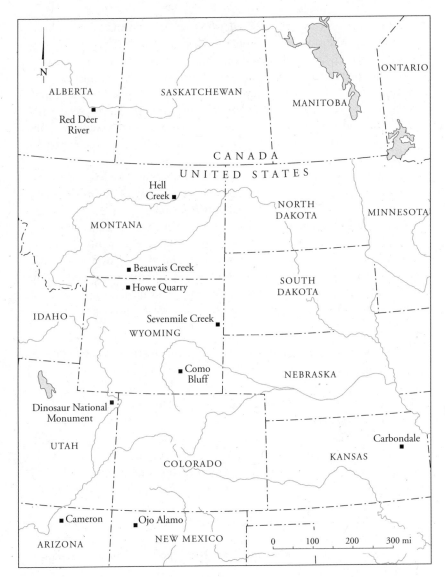

MAP 1. Western North America, showing Brown's fossil sites

tint Osborn's first impressions of his ability to manage and conduct field operations, and a life in the field was without doubt his most highly treasured aspiration. In a state of considerable consternation, Brown fired off a long letter to Osborn:

I came here in order to consult Reed, who you know did all Marsh's collecting with Dr. Williston. I learn from him that things are in a very discouraging condition up there, different from what Dr. Williston and I have written you. I feel pretty badly mixed up in the situation, that I have misinformed you somewhat. Mr. Reed... says that Marsh had another man continue work there a year after he left and he went back into the bluff several yards until they had a twenty-foot bank to face, now all this has caved in... and of course the amount of dirt you have to move rapidly increases as you go back....

I can hire a man and team for $3. per day or single man for $25. per month and board [equivalent to about $70 and $575 today].... Can buy team, wagon and harness for about $85. [approximately $1,950] at the outside which wouldn't pay many days work in excavating. You see I am completely handicapped. I don't know how much expense I dare take on myself or what other things might present themselves to your mind. Wire me immediately what to do. I want to do everything to the best advantage of the Museum. If I shall get an outfit will need more money. Meantime I shall gain all information possible and lay over on my own expense if necessary.[46]

Two days later, although still waiting for Osborn's response, Brown wrote again, but he now sounded settled on what he thought best:

It is provoking to loose all this valuable time and find things so much different from what I expected.
 ... [Dr. Williston] has evidently confused this [mammal] quarry with one of the numerous reptile quarries on the same line of escarpements....
 You advised me to use my best judgment this summer in regard to all things and I hope I have done right in delaying here until you know exactly how things are....
 Now I feel confident that it is policy to buy an outfit here which I shall do if your reply to my other letter is favorable....
 There are certainly other mammal beds there and plenty of reptile material. Shall I not collect everything?[47]

No correspondence has been located to document Osborn's response, but his report, written after the field season, indicates that he generally approved of Brown's plan:

The expedition was temporarily placed in charge of Mr. Barnum Brown, a graduate of the Kansas University, who with his assistant Mr. H. W. Menke also of Kansas, left Laramie, Wyoming, early in May for Aurora, a small station on the U.P.R.R. [Union Pacific Railroad], near Lake Como, after purchasing a suitable outfit and equipment.

The early part of the season was spent almost wholly in the opening up of the Mammal quarry which had been accurately located through the kind assistance of Prof. Wilbur C. Knight and Mr. William Reid [sic] of the Wyoming University. Later in the season, during the visit of Prof. Osborn to the camp, a skeleton of a large reptile [eventually identified as *Diplodocus*] was discovered by himself and Mr. Brown within a short distance of the Mammal quarry.[48]

To our knowledge, this was the first face-to-face encounter between the supervisor and his apprentice. A photo reminiscent of big game hunters posing with their prize trophy shows the two men positioned high on the bluff savoring success in the embryonic quarry. Osborn, seated behind the limb bone and sporting the faintest hint of a smile, imperiously stares down the camera with his deep-set eyes, while a kneeling Brown, serious but almost pregnant with pride, wields his rock hammer near the end of the partially exposed bone.

Having briefly blessed and evaluated the operation, Osborn left the enterprise, for the moment, in Brown's hands. But in his progress report of June 14, Brown once again sounded a bit overwhelmed: "Cope quarry [containing the *Diplodocus* specimen] is a veritable gold mine and I have been in bones up to my eyes. I think I shall take out a good part of a skeleton [of] beautiful bones. I have bared about thirty feet of vertebrae . . . east of the femur we located. . . . Haven't worked but two days in the Mammal quarry since locating the bones in Cope quarry."[49]

Given the prospective success and great amount of work that would be required during the rest of the season, Osborn called in reinforcements. As during the year before, Brown would team up with Wortman, who was called in from the Huerfano Basin in late June, whereupon the excavations were vigorously accelerated over a period of several weeks, during which another skeleton was encountered by Wortman and Dr. Knight not far from the first.[50] This second skeleton turned out to be an *Apatosaurus,* or brontosaur. Faced with a still burgeoning workload and a rapidly diminishing field season, Osborn requisitioned still more troops.[51] Brown was now reunited with Granger, his crewmate from the previous summer, and for the first time worked with Albert Thomson, with whom he would team up again several decades later. The pair had been directed to close up their operations in the

FIGURE 6. Barnum Brown (left) and Henry Fairfield Osborn at Como Bluff during the AMNH expedition of 1897. The quarry, shown in its initial stage of excavation, produced the first dinosaur specimen in the AMNH dinosaur collection, AMNH 223, belonging to the genus *Diplodocus,* a limb bone of which can be seen in the image. (17808, American Museum of Natural History Library)

mammal quarry at Hay Springs, Nebraska, to join the effort. Also enlisted was William Diller Matthew, who followed from Kansas close on their heels. Matthew, brilliant but soft-spoken behind his wire-rimmed spectacles, was an expert on fossil mammals, who had studied under Osborn at Columbia before finishing his graduate work and joining the AMNH staff in 1895, first as Osborn's assistant, then as an assistant curator. Meticulous in both fieldwork and research, Matthew would go on to become one of the most influential paleontologists of his time, both for his work on mammalian

FIGURE 7. A slightly later image of the *Diplodocus* quarry at Como Bluff excavated during the 1897 AMNH expedition. The woman appears to be Marion Brown, Barnum Brown's first wife, whom he married in 1904; however, no other record of her presence during this expedition is known to exist. (AMNH Vertebrate Paleontology Archive; 7:2 B1. S66)

evolutionary relationships, such as horses, and for his theoretical work on biogeography. Eventually, Matthew would become Brown's closest colleague along with Osborn at the museum.

In mid-August, Brown reported that

> Dr. Wortman, Mr. Menke and Mr. Granger are working the big Brontosaur. They have taken out the femora, one pubis, one scapula and coracoid which are beautiful bones, as perfect as any I have ever seen from the Jurassic. The femora measure five feet eleven inches in length.... At present they are boxing up the fossils and going back into the bank.
>
> I am working the [*Diplodocus*] quarry alone. My method is to expose some of the bones on top and sides of the vertebral column, cover these with paper, then run cement and plaster over them, strengthening the whole with boards, then digging underneath ..., being careful to brace the section well.... I wrap with rawhide drawn tightly and nailed to the boards, which shrinking holds all together as firm as rock.

I have turned over the account as you directed though I still manage affairs.⁵²

In his summary report, Osborn related that the enlarged crew made rapid progress, finishing the excavation of one specimen by the end of August and the other by late September. These were truly Herculean tasks, given the immense size and weight of most sauropod bones. With the completion of the larger, first specimen, some of the crew was freed up to focus on Marsh's old mammal quarry; to find fossils, however, the men would have to make "a cut of some twenty-five feet or more down through the solid clay and sandstone in order to reach the thin layer in which the Mammal remains are found.... The sandstone layer, between six and seven feet in thickness, requiring to be blasted before its removal could be accomplished.... When this was [finished] the entire layer [about 200 square feet] was taken up and boxed, one half being shipped to the Museum laboratory for examination. The remaining half was left to be examined by Mr. Menke in Aurora." Osborn went on to describe in glowing terms the procedures that Brown had used to jacket and collect the huge bones: "Many new methods were introduced and practiced which so far as I now know to the contrary have never been employed before.... So far as the skeletons themselves are concerned ... they are, perhaps, by far, the most complete and perfect of their kind that have ever been collected and will make magnificent material for purposes of exhibition."⁵³

More than sixty years later, Brown concisely described the summer's work in the notes for his autobiography. The sense of pride and accomplishment still radiated through his words:

> In the north end of the uplift at Como Bluff, I opened up the mammal quarry discovered by Bill Reed, one of O. C. Marsh's men of Yale University, and found a few mammal teeth, but I was more fortunate in discovering a partial skeleton of a dinosaur that turned out to be Diplodocus. This was the first dinosaur excavated by any Amer. Museum expedition, and here I introduced the use of Plaster of Paris in excavating fossils. Previously flour paste was used....
>
> Prof. Osborn came out in the middle of the summer, and was with me when I found the Diplodocus skeleton. Part of the skeleton paralleled the face of the cliff, then turned directly under the cliff, and had to be mined out, as coal is excavated by propping up the roof and sides.
>
> 40 years later, I went back to the site and found the hole as I had left it, with the timber in place, and little changed.⁵⁴

Brown, like Osborn, honestly thought that the young apprentice had invented the technique of using plaster and burlap to jacket vertebrate fossils. However, as James Farlow, in his notes for R. T. Bird's *Bones for Barnum Brown*, points out: "By 1880 . . . all of Marsh's collectors were covering fossils with jackets made of cloth or burlap dipped in plaster."[55] Nevertheless, at the age of just twenty-four, Brown had launched the development of what would quickly become the world's best and most famous collection of dinosaurs. And beyond that, by covering the fragile, 150-million-year-old hindquarters of his *Diplodocus* in a sturdy plaster jacket reinforced with wooden struts, he helped establish a critical collecting technique that is still used today by paleontologists around the world.

Despite having all these achievements under his belt, this child of the frontier returned to New York to face an even more daunting challenge: the intimidating environment of a major eastern university. Apparently after pulling some powerful strings, Osborn had secured the scholarship for Brown at Columbia. In a matter-of-fact conclusion to his letter of June 14, Brown confirms to Osborn that he "received notification of scholarship being granted for which I am very grateful."

Indeed, Columbia proved to be tough sledding for this farm boy from eastern Kansas. As it was put to his dean, presumably by Osborn, a year later in a letter regarding renewal of his scholarship,

> [Brown] is a member of our field expeditions to the Rocky Mountains, and is a very capable field explorer, having discovered one of the magnificent reptiles, which is now being prepared for exhibit in our Museum. Mr. Brown's opportunities in the West very naturally not being as good as those of his fellow students in the East, may account for the fact that his work during the present year has not reached a high standard. I believe, however, that he is a man who will gradually work up into a very useful position, and will do credit to the support which we may give him.[56]

Osborn had caught Brown's essence, for it would take another decade for Barnum to complete his undergraduate degree at Kansas. Fieldwork would continue to divert his fancy and take precedence over academic work in both his thinking and planning. In fact, in his final letter to Osborn from Aurora, Brown, despite having yet to enroll at Columbia, was already setting his sights further afield: "Dr. Wortman informs me that you are thinking strongly of sending someone to South Africa and I told him to write you that I wanted very much to go. I should also like to be a candidate for the South

American trip. I feel that I can do the work successfully and satisfactoryly and this winter would be a preferable time for me to make the trip even if I throw up my scholarship at Columbia."[57]

Cheeky, to say the least, especially when Osborn, one of the nation's most powerful academics, had just gone to great pains to secure the highly coveted scholarship for him. But Brown was determined to be a player. How would the imperious Osborn react?

FOUR

To Land's End: Patagonia
(1898–1900)

WHATEVER OSBORN AND BROWN may have discussed regarding the young man's desire to participate in an international expedition, it is clear that Osborn didn't grant the request immediately. Instead, Brown moved to New York City and apparently began taking an extensive roster of graduate courses at Columbia University. During the summer field season of 1898, Osborn assigned Brown to work under the guidance of another of the museum's rising stars, William Diller Matthew, who, like Osborn, was interested in the evolution of mammals.

Matthew joined the museum in 1895 as Osborn's assistant and quickly became an intellectual lynchpin in the Vertebrate Paleontology Department; he would eventually become Brown's long-serving supervisor as well.[1] Born in 1871 into a Canadian family in Saint John, New Brunswick, Matthew acquired his interest in geology from his father, who, though employed as a clerk and surveyor for the city's customs house, spent extensive time studying and publishing on the geology and natural history of the region.[2] Described as "a slightly built young man of fair complexion, with rosy cheeks, sandy hair, and brown eyes behind steel-rimmed spectacles," Matthew completed his liberal arts undergraduate degree at the University of New Brunswick in two years, where he was described as "the genius of '89."[3] His superior intellectual gifts did not taint his steady personality. He came to view himself as "half Canadian, half American," which led him to adopt "a broad, pragmatic, and tolerant view of life in an interdependent world."[4] Matthew entered Columbia University in 1889, where he came to the attention of Osborn when he joined the faculty in 1891. Matthew intended to focus on mining and hard-rock geology, but in 1893 he enrolled in Osborn's course on mammalian osteology; as he modestly wrote his family, "Prof. Osborn seems to think I am getting along pretty well, and as he is the most influential man on the Pure Science Faculty,

I want to keep in with him as much as possible with a view to reelection as Fellow next year. I don't get up an intense enthusiasm over bones, however, tho' it is a fairly interesting subject."[5] Osborn clearly had his eye on Matthew as a potential recruit for the museum, and in 1894, with Matthew's fellowship renewed at Osborn's behest, his mentor made his play to hook Matthew irrevocably on paleontology by sending him to a coal mine in North Carolina to search for Triassic mammals.[6] Although Matthew found no mammals, he did find bones of the crocodile-like phytosaur, *Rutiodon*.

The stage was set, and in 1895 Osborn hired Matthew to take on several tasks, including reevaluating the descriptions and classification of the mammalian specimens in the Cope collection, which Osborn had purchased that year, and writing labels for specimens displayed in the museum's fossil exhibition.[7]

Of course, fieldwork was also in the offing. Beginning in 1895 and 1896, Osborn had requested $10,000 from the museum's trustees to initiate a project dealing with collecting and researching the evolution of fossil horses. One outcome was an expedition in 1898 focused on rock units in northwest Kansas, southwest Nebraska, and eastern Colorado, the object of which, Matthew stated in his final report, "was the obtaining of fossil mammals, especially ancestral horses and camels, from the Miocene and Oligocene bad lands."[8] Why horses? A primary reason was that a synoptic sample of fossil horses, especially those that Cope and Marsh had already assembled, provided a compelling record to support the theory of evolution. At this time, Marsh had the best collection, one that had impressed even Huxley when he visited Yale over twenty years before, in 1876. For Osborn, that would just not do, and the AMNH horse project "would enable him to surpass Marsh." As Osborn's biographer Rainger notes:

> Osborn hoped to obtain a wealth of scientific information and to distribute it to the public on a grand scale ... [thus] fulfilling the interests of the trustees. The history of the horse family provided a means for readily conveying nature's great law of evolution. The subject illustrated the importance of [the] American fauna. Horses were also a subject of personal interest to many of the trustees. Osborn and his family owned and rode horses, and the trustees ... William C. Whitney, and Cornelius Vanderbilt were avid horse owners and fanciers.

Osborn's first appeal for support went unheeded, "but in 1897 Whitney gave $15,000 to develop a project on horses that entailed collection, research, and display.... Matthew, Walter Granger, and Albert Thomson were also directed to search for fossil horses throughout Colorado, South Dakota, and Texas."[9]

Osborn's goal was to eventually publish a monumental monograph on horses with the help of Matthew and another young paleontologist out of Princeton, James W. Gidley. However, the evolutionary view that Osborn developed differed substantially from that of Darwin, and Matthew supported Darwin's views. Biographer Regal argues that, although Osborn's father served as Henry's "role model for his outward, personal behavior and business-like approach to life," it was his mother, Virginia, who most influenced his religious views and, indirectly, his evolutionary views. She believed that "good works, personal effort and exercise of free will could all contribute to the achievement of salvation" and that without struggle, "no advance could be achieved."[10] Her son co-opted these beliefs and turned them into a modified version of Lamarck's evolutionary paradigm based on the inheritance of acquired characteristics. In essence, Regal argues, "As a scientist, Henry Fairfield Osborn subconsciously used his mother's salvation credo as part of the foundation of a theory of evolution. Working with [his professors at Princeton], and particularly Edward Drinker Cope..., Osborn began to meld religious conviction and scientific theory as they did.... He argued, in a nuanced combination of science and metaphysics, that evolutionary advance was only possible if organisms struggled against the environment to go beyond their inborn biological characteristics."[11]

Matthew rejected the neo-Lamarckian leanings of Cope and Osborn. As his biographer, E. H. Colbert states, "Darwinian evolution was Matthew's guiding star throughout his professional life."[12] Thus, even though Matthew was in charge of most of the fieldwork, as well as the subsequent stratigraphic and comparative anatomical analyses of both the AMNH and Cope horse collections, his theoretical disagreements with Osborn over the nature of evolution eventually led him to abandon the joint study and actually criticize Osborn's analysis.[13]

These conflicts did not mar the expedition of 1898, however, which was focused strictly on discovering new specimens of fossil horses and other Tertiary mammals, ones that would eventually play roles in Matthew's own publications about fossil horses and other mammalian lineages. No notebook detailing Brown's activities on this expedition survives, but the final reports make clear that the crew "met with little success [in Kansas and Nebraska], the region having been worked out by previous explorers." Consequently, at the end of June, Matthew, with the rest of his crew in tow, made a beeline for the Oligocene and Miocene badlands in northeast Colorado, "[near] the heads of Lewis, Cedar, Horsetail, and Pawnee Creeks."[14] After a month of more successful collecting, Matthew returned to the museum, leaving Brown in charge of the three-person team. A couple of letters that Brown wrote to Matthew detail continued success: "Work is progressing nicely and the collection is increasing rapidly, principally of the Oreodon family. We have found a number of Merychyns [*Merychyus*] and Merycochverns [*Merycochoerus*] skulls.... Have several good rhinoceros jaws and two or three skulls from upper beds.... Have a good stock of provisions and about $25. in cash."[15] A month later, near the end of the season, Brown again wrote Matthew: "We have taken out some fine camel material ... which will work up quickly and make a fine exhibit ... [as well as] a sabre tooth cat skull and ... a perfect carnivore marked Hyenadon [*Hyaenodon*] skull."[16]

With winter on the horizon, Brown set about packing the specimens for shipment back to the museum, relating to Matthew that the summer's collection comprised "27 boxes weighing 3451 pounds."[17] Osborn was pleased with the crew's efforts, as his summary report makes clear: "The work was extremely difficult and the results far surpassed our expectations; the principal discoveries being a Camel having the proportions of a Giraffe and a complete skeleton of an ancestral Racoon."[18]

As the result of this cooperative effort, Brown and Matthew, who was only two years Brown's senior, forged a fairly close and respectful professional relationship, which Brown would lean on in later years as Matthew rose through the ranks of the museum's administration. For the present, however, it was time for the young explorer to return to New York and resume the rigors of his coursework at Columbia.

That plan was, as we saw in the prologue, called into question on December 7, 1898, when Osborn enlisted Brown to join Princeton's Patagonian expedition, which had been initiated by Osborn's old friend and colleague, William Berryman Scott. Osborn was especially intent upon extending the scope of the AMNH mammal collections beyond the bounds of North America.[19] Fossils from Patagonia had become well known as a result of Charles Darwin's collections from the region in the 1830s during the voyage of HMS *Beagle*. Subsequently, in the early 1890s, Patagonia again became a focus of mammalian evolutionary research when the Argentine paleontologist Florentino Ameghino provocatively proposed that Argentina had served as the cradle for mammalian origins and the focal point for vertebrate evolution and distribution. Not surprisingly, this proclamation raised the hackles of many North American and European paleontologists. Both Scott and Osborn sought to sort through these issues, and Osborn was determined that AMNH would play a significant role in this trendy contemporary debate.[20]

In assigning Brown to be the sole AMNH representative on the expedition, Osborn infuriated Brown's original field mentor at the museum, Jacob Wortman, who had also lobbied Osborn for the right to participate.[21] Apparently Wortman had come to dislike Brown, primarily because he felt that Osborn was giving Brown too many important field assignments. To Wortman, this reflected a lack of respect for seniority and merit, and it soon led him to leave AMNH for the Carnegie Museum.

Osborn undoubtedly took solace in the fact that his young apprentice would be shepherded by two highly experienced field men. Princeton's John Bell Hatcher, who Osborn had unsuccessfully tried to hire away from Marsh in 1890, was the expedition leader, and his assistant was Olaf A. Peterson, a former collector at AMNH, both of whom had already worked in Patagonia.[22] Brown's path had crossed those of Hatcher and Peterson before. Brown had met Hatcher in 1894 when Williston's KU expedition bumped into Hatcher's Princeton crew in the badlands of South Dakota; and, fatefully for Brown, it had been Hatcher who lured Peterson away from Wortman and the AMNH, thus opening up the position that Brown now occupied.

It is fascinating to note the similarities in the career trajectories of Hatcher and Brown. Both were raised in the Midwest, Hatcher having been born in Illinois in 1861 and moving with his family to Iowa as a young boy. Like Brown, Hatcher became intrigued with fossils associated with Paleozoic coals, even going on to work as a coal miner to fund his college education, which began at Grinnell College. In 1881, Hatcher transferred to Yale's Sheffield Scientific, where his small collection of fossils came to the attention of Marsh, who hired him as an assistant. Hatcher eventually graduated in 1884, which made him a contemporary of Williston, and like Williston, Hatcher became disillusioned with Marsh's policy of not letting his assistants publish their own research. In 1890, when Osborn was making plans to found the Department of Vertebrate Paleontology at AMNH, an effort was mounted to have Hatcher join the staff, but he and Osborn were unable to reach an agreement. Hatcher left Yale for Princeton in 1893 to become the curator of vertebrate paleontology. Three years later, he had developed plans and raised funds for an expedition to Patagonia, the first of three he would

MAP 2. Southern Patagonia, showing fossil sites where Brown and Hatcher collected in 1899

spearhead from 1896 to 1899. On the first and third of these forays, Hatcher enlisted the aid of his brother-in-law, O. A. Peterson, and as we have seen, Osborn negotiated for Brown to join them on their last excursion.[23]

Logistically, the voyage to Patagonia was uneventful, except for a short delay caused by entanglement in the rock weed of the Sargasso Sea. Land was sighted only once en route, off the coast of Brazil near Perambuco. After a voyage of thirty days, the crew arrived in Punta Arenas on January 5.[24]

Despite the unfamiliar maritime atmosphere of the port, several aspects of this home base must have seemed comforting to Brown, who noted that its Main Street featured "a ... Bank, impressive Opera House, night club, hotels, and many large mercantile establishments." The town was the major trade and distribution center for the sheep-raising industry that dominated the region's economy, and it served as the diplomatic nerve-center for southern South America, with headquarters for many European as well as North and South American governments. This cosmopolitan mix included "English, German, and Latin merchants."[25]

The crew found the stores well stocked with "barrels of macaroni in all shapes..., rice, beans; bags of sugar, coffee and Mate tea, slabs of bacon, dried smoked beef, all cuts of fresh mutton; tallow, candles; stalls of camp equipment, harnesses, bridles, leather goods; stoves, shovels, picks and crowbars; bins walled off containing whiskey,... bottles of Italian and Spanish wines." One challenge was finding fit horses because the pampas grass was not terribly nurturing, so most horses were "weak and lack[ing] stamina." The care given by some of their owners also left something to be desired: "One horse trader... was beating [his horse] over the head with the loaded end of his quirt. 'Beat that horse once more, and you'll be a dead man,' said Hatcher, drawing his revolver; I firmly believe he would have shot the trader."[26]

Healthy horses would, as always in those days, be key to the expedition's success, especially because Hatcher and his crew would attempt a true feat in this desolate landscape. Their primary vehicle was "a Studebaker wagon, with water barrel fastened to one side, pulled by two teams." It was spacious enough to hold all the camping gear and collecting equipment, as well as "provisions to last for several months." What made the whole enterprise unique was that "ours was the first four-wheeled vehicle to cross the great Patagonian Pampas—a vast, treeless, windswept plain, apparently level, but which rises gradually from the coast to the foot of the Andes."[27]

In searching Patagonia for fossils, Hatcher's crew was following in illustrious scientific footsteps. During the voyage of the *Beagle* in 1832, Charles Darwin discovered a suite of Ice Age fossils when the ship stopped at Punta Alta near Bahía Blanca, well north of where Hatcher and his crew were working.[28] Specimens of armored mammals, called glyptodonts, giant ground sloths, and unique South American ungulates fascinated the young naturalist and helped kindle thoughts that eventually led to his theory of evolution by means of natural selection.

No highways traversed the country along the Bahía Grande, where

FIGURE 8. Miocene exposures at Lake Pueyrredon, Patagonia, 1898, as photographed by John Bell Hatcher (AMNH Vertebrate Paleontology Archive). From J. B. Hatcher, *Reports of the Princeton University Expedition to Patagonia*, vol. 1: *Narrative of the Expeditions. Geography of Southern Patagonia* (Princeton: Princeton University Press, 1903), fig. 21.

Hatcher's crew had to pass. Yet to some degree, the ground was naturally paved "with flat, water-worn stones left by the receding ocean as the country was elevated" and now punctuated by tufts of bunchgrass and low brush, such as calafate.[29] Nonetheless, hazards abounded in the form of steep-walled river canyons and volcanic cinder cones with their treacherously jagged flows of hardened lava. Brown's long-tempered skills as a teamster would therefore once again prove invaluable to the expedition's success.

Hatcher, anxious to get to a prime fossil locality that he had discovered at the end of his previous expedition, rode ahead to check in with sheep ranchers—including the Feltons, with whom Brown would later stay—along the four-hundred-mile route, leaving Peterson and Brown to follow with the wagon and other saddle horses. The crew headed north from Punta Arenas, across the Argentinean border, through Gallegos, and up to Santa Cruz, where they followed the drainage of the Río Chico northwest toward the Andes.[30]

"One day, when I was driving," Brown recounts, "I saw Peterson far ahead across an apparently solid flat piece of ground. I headed toward him, and when the four horses were on it, the entire surface, nearly an acre in extent, began to undulate. I turned the team as sharp as I could without breaking the wagon-tongue, but even then, the front wheels cut through the surface, and soft matter oozed up. Had the team and wagon gone into this mass, we would all have gone down without anyone knowing where we had disappeared. This is known in our western states as a 'Soap hole.'"[31] In terms of perils for travelers, soap holes are akin to quicksand, bogs composed of alkaline mud rather than sand.

Normally rolling along at about ten miles per day, Brown marveled at the formerly incendiary landscape as they approached the Chilean border, where thirty-foot-thick lava flows covered a plain densely pocked with now extinct volcanic craters. "I counted eight within a radius of five miles," he mused. "A wonderful spectacle this must have been in prehistoric times, when these cones gushed forth a sea of moulten lava and ashes, covering the pampa for long distances; blotting out animal and plant life."[32]

When they reached the edge of the Andes, they skirted the foothills northward toward their destination near Lago Pueyrredon, a body of water that both Hatcher and the Argentine Limit Commission had discovered independently the year before.[33] With the topography getting steeper, Brown and Peterson streamlined their operation by securing their wagon on the shoreline of a dry lakebed and continuing up through the foothills on horseback toward the Hatcher fossil locality. They met their leader about March 3 above timberline, where, quickly engulfed by a blizzard, they were forced to release the horses to find temporary shelter, while the paleontologists huddled together in a single bed for two days trying to fend off the freezing temperatures. At last the wind and snow relented, and the crew retrieved their horses from their refuge lower down the mountainside. But conditions remained challenging.[34] The only fuel they could find to build a fire was grass roots, and one man had to hold the frying pan over the fire to boil water for coffee and fry their bacon.

As soon as possible, the crew set to work prospecting for fossils at the small hill on which Hatcher had discovered numerous fossils the previous May. The fossils came from 18-million-year-old Miocene beds within a 250-meter-thick rock unit called the Santa Cruz Formation. Unfortunately, they again soon found themselves "being driven out by a snowstorm before more of the beds could be examined."[35] (In the southern Andes, such sudden storms with violent winds can be almost daily occurrences at any time of the year.) To add insult to injury, fossils turned out to be in much shorter supply than Hatcher had

FIGURE 9. Hatcher's image of the team and wagon that Brown drove for him through the snow during their 1899 Princeton/AMNH expedition (AMNH Vertebrate Paleontology Archive). From J. B. Hatcher, *Reports of the Princeton University Expedition to Patagonia*, vol. 1: *Narrative of the Expeditions. Geography of Southern Patagonia* (Princeton: Princeton University Press, 1903), fig. 31.

anticipated. Despite a "diligent search" that involved following the trace of the bed across the countryside for two weeks, the crew failed to locate another rich locality. Brown found only four fragmentary skeletons of typotheres before yet another blizzard blasted them off the mountain, forcing them to retreat.

Although the crew, before they set out into the foothills, had secured their wagon with a heavy tarp to keep "four-footed intruders" at bay, this precaution was not as effective against bipedal foragers: "On reaching the wagon, we found that the Argentine-Chilean Limit Commission had been there and taken all of our sugar, leaving a voucher for payment in Buenos Aires. Black coffee without sugar was at first a bitter dose, but by the time we again reached the coast, we no longer wanted sugar in our coffee." The loss, at least to some degree, was compensated for by Brown's marksmanship: "On St. Patrick's Day, . . . which I well remember, . . . I shot a fine big buck deer that supplied us with fresh meat during our journey. By the time it was finished the meat was so tender you could cut it with a fork. Even in the summer the wind blows cold at this latitude.... When the carcass was hung on a pole in the night winds, and wrapped in a tarpaulin during the next day... the meat was preserved without refridgeration until completely consumed."[36] We still use this tactic, even in the scorching climate of Mongolia's Gobi Desert, where we wrap the leftovers of a butchered sheep carcass in a burlap bag to keep the flies from laying eggs on it. By hanging the bag off the truck at night, out of reach of ground-dwelling poachers, it can last for several days and still produce palatable meals, as long as it's been thoroughly cooked.

The crew arrived at the coast near Santa Cruz on April 7, at which point Hatcher and Peterson began to check shipping schedules and plan the trip home. In the meantime, Brown, no doubt concerned about the lack of success in the foothills of the Andes, explored some nearby marine beds and "found a very fine skull, jaws and vertebrae of... *Argyrocetus*," a Miocene toothed whale.[37] Apparently soon after, Brown parted ways with Hatcher and Peterson south of Santa Cruz. Since Hatcher's locality had turned out to be a bust, there was no reason for them to stay longer, having already prospected other areas within the region during previous expeditions and having collected a representative sample of the fossil faunas.

It would be the last time that Brown and Hatcher joined forces in the field. Shortly after returning to the United States, Hatcher left Princeton for the Carnegie Museum in Pittsburgh, where he served as curator of paleontology and osteology for four years. Sickly by nature as a child, he contracted typhoid fever and died in the summer of 1904 at the age of forty-two.[38] In summing up his impressions, Brown wrote that Hatcher

> was a truly remarkable man, with few vices and more virtues than are found in most men. Strictly honest, he was a paradox in some respects. He probably had had a disturbed childhood, for he said he would not trust his own father unless an agreement was signed. During our association, I found him generous to a high degree, offering any part, or all of his outfit when we separated....
>
> As a worker he was indefatigable. He would ride off alone in an uncharted area, with only his blankets, revolver, and a pocket full of salt, living off the game of the land as he traveled. His geological observations were, to my knowledge, accurate; and as a collector, no one ever surpassed him.[39]

It is noteworthy that Marsh's biographers often ascribe the title "King of Collectors" to Hatcher, and Brown certainly seemed to find much in Hatcher's professional abilities to emulate, both immediately and throughout the rest of his career.⁴⁰ Although paleontological lore sometimes suggests that Hatcher abandoned his young colleague in Patagonia, Brown expressed no ill feelings about being left to fend for himself. Given Hatcher's occasionally prickly personality and Brown's preference for solitude in the field, the separation probably suited them both. In any event, Brown was now far away from home and free to follow his own instincts. Because correspondence was difficult, few letters have survived. But we do know of one letter from Osborn dated April 24, 1899, about the time that Hatcher left for home and Brown set out alone to collect along the Patagonian coast. The letter, which we have not found in the archives, didn't make it into Brown's hands for almost four months, and when it did, Brown was not pleased, for it contained a revised contract. In his response to Osborn, Brown stated: "Before signing this paper I wish to have a better understanding of its meaning. Also to discuss my relation to the museum."⁴¹

When Brown left on the expedition, Osborn and his counterpart at Princeton, William Berryman Scott, had already negotiated most of the details regarding Brown's involvement in the enterprise and how the specimens that were collected would be divided.⁴² Brown would receive a salary of at least $50 a month (about $1,700 in today's dollars), a sum that would be doubled if Osborn succeeded in raising more funding from his friends and museum trustees. Brown also received $500 in cash for expenses, along with the understanding that another $500 could be drawn from the AMNH account during the first year, as needed. Specimens collected by Hatcher and Peterson would become the property of Princeton, while those discovered by Brown would belong to AMNH, but with a couple of provisions. As originally drafted, the agreement, according to Brown, stipulated that "material new to the Princeton Collection is to be worked up and described by myself." In other words, Brown would describe and name any new species or genera, and once they were published, Scott and Hatcher would have the right to include that material in their own summary being developed, *Monograph of the South American Fauna*. However, the revised agreement that Osborn sent in April set slightly different terms, stating that Brown could publish only new "genera of vertebrate fossils that are not contained in the Princeton collection." Brown protested directly to Osborn about that modification: "I may," he said, "have any number of new species of a described genera [*sic*] and not be able to describe them under this paper [revised agreement]." Another change involved the publication of new geologic information, which Brown likewise complained about:

> Article IV reads: "That J. B. Hatcher shall have the right to publish any observations upon the Geology of Patagonia."
>
> This is also vastly different from our agreement which reads "his observations" and means a great deal to me as I look over my notebook. You can readily see the position my notes are placed in at present to say nothing of future work. I want none of Mr. Hatcher's observations but I do want my own.[43]

Brown, thoroughly isolated from discussions that involved him and apparently feeling more than a bit insecure about his prospects, decided, like the good poker player he was, that it was time to show his hand. Thus, this exchange appears to represent a seminal moment in the developing relationship between Osborn and Brown. Having studied under Williston and having just spent almost half a year traveling with Hatcher, Brown must have been aware of the stories of Marsh's mistreatment of his assistants. He must also surely have been aware of Wortman's complaints regarding Osborn's decision to let Brown lead the initial AMNH effort at Como Bluff and participate in this Patagonian expedition. Like Marsh, Osborn had pursued postgraduate studies in Europe, where the major professor was master and assistants rarely participated in key decisions or publications. Nonetheless, although Osborn could be an "overbearing taskmaster," he was also more flexible and supportive of his assistants than Marsh, often lobbying the museum's administration in their interests and citing their accomplishments in reports on departmental activities.[44] Although Osborn's name often appeared alone on papers and monographs to which his assistants had made considerable contributions, many of his assistants, including Matthew and eventually Brown, published extensively in their own names, either with or without Osborn as a co-author.

It seems likely that all of these issues were fomenting in Brown's brain as he pondered his play in the current and potentially explosive situation. In a single page, Brown diplomatically yet forcefully confronted Osborn about his performance and prospects at AMNH. In some significant respects Brown's response, including a serious confession about his academic standing, set the coordinates for the course of his career and established a foundation of mutual trust and respect that would characterize the professional relationship, and even friendship, of Brown and Osborn for the next thirty-five years:

> I have been with the Museum now three years at $50 per month and assets have just about covered liabilities. Through you, I have received many kindnesses not reckoned in dollars for which I am deeply grateful. Although I

failed at Columbia University (it's a bitter pill to swallow) my time there was not lost. I tried to cover too much and got swamped.

For me to remain with the Museum I am sure I can best serve its interests in the field where physical energy and resource are most called for. After a thorough collection has been made from the different horizons in South America, ... there is South Africa, Australia and Siberia which must eventually be represented in the American Museum. But this takes time and means, if I am the man to do the work, that I must give up other projects and interests and rely wholly on my salary from the Museum.

I do not feel justified in doing this for less than $100 per month. I know you can get any number of men for $50. For the last two years my work has been at its best and if it is worth only $50 you will do me the greatest service yet in saying so at once. I value my time higher than that. I like the Museum, its workers and the advantages it offers a combination which I might not find elsewhere. The description of new material would count for a great deal but I have no assurance that I am even a worker in the Museum except in the field. Believe me sincerely this letter is not dictated by a spirit of greed but an awakening that I must know where I am at.[45]

No record of Osborn's direct response has been found, but given that Brown continued in the service of the museum until he retired some forty years later, and continued his association afterward, it is clear that a deal was reached. And in the end, Brown's bibliography does include one solely authored paper resulting from this expedition, titled "A new species of fossil Edentate from the Santa Cruz Formation of Patagonia."[46]

Despite this distraction, Brown remained focused on his task, anxious to find more fossils to justify his trip. The fossiliferous units that interested him most, again in the Santa Cruz Formation, were exposed along sea cliffs and shorelines, as well as the adjacent arroyos that emptied into the Atlantic. From April to the following February, during the austral spring and summer, Brown focused his prospecting efforts in this region between Río Gallegos in the south and Santa Cruz in the north, with other key locations being Monte León, Cañon de las Vacos (Brown rendered it *los* Vacos), and Río Coy. From April to mid-October, Brown concentrated on the territory from the mouth of the Río Coy north to Santa Cruz.[47] He discovered that between Monte León and Cañon de las Vacos, fossils included only isolated armor scraps from the tanklike glyptodonts, but the "great number" of them suggested that the animals had been abundant.[48] Near the top of one vertical cliff at Cañon de las Vacos, he reported,

> I spotted a skeleton projecting from a rock, which I had to chisel off and lower in a sling. This turned out to be the rare *Propaleohoplophorus*—an armadillo-like creature [glyptodont] now in the American Museum.
>
> I was using flour paste and burlap to preserve the specimen, and was busily absorbed in my work [when I] suddenly looked up and saw a huge black and white condor circling overhead, swooping so close I could see his pinions directed at angles to determine his course. Fearful that he might be dangerous, I grabbed my rifle and shot him.
>
> When this great vulture dropped I spread out his wings, and they measured 14 feet from tip to tip.[49]

Brown completed his collecting north of the Río Coy about October 20, shipping "13 boxes from Santa Cruz and one from Coy to Punta Arenas for storage." In all, Brown concluded, "This collection includes a more varied fauna than that obtained by Mr. Hatcher from Gallegos, few specimens are crushed.... This will make a great exhibit."[50]

By mid-November, Brown had returned to the region around Río Gallegos and through the rest of the year focused his collecting efforts in that area.[51] Hatcher's introduction of Brown to the Feltons paved the way for Brown to use their enormous estancia as his base of operations. Felton and Brown became close friends, and Felton often accompanied the young paleontologist as he roamed the ranch land to prospect. Felton's was one of the largest estancias in this part of Patagonia, and like others was "divided into paddocks, usually about 10 miles square, part of which border the coast, or rivers, where the sea breeze melts off the snow in winter, and the sheep can find grass uncovered. There is a shepherd's house in each paddock, and his chief duty is to ride the range seeing that no sheep fall on their backs and die, and that no mountain lions are making kills."[52] On previous expeditions, Hatcher had found abundant fossils near the Felton Ranch, weathering out of the canyon walls along the Río Gallegos, but he had collected everything he could find. Fortunately, Brown discovered that the vertical walls had eroded in the interim, producing "an extensive collection of fossils, some of which were new to science."[53]

Collecting along the shoreline presented significant dangers because the fossiliferous beds extended out into the Atlantic, becoming exposed only at low tide. One day, Brown

> uncovered the skull and jaws of a rare Astropotherium [*Astrapotherium*]—the only known specimen of this genus found on the Patagonian expeditions.

The skull and jaws were in the hard seafloor, and I had to work rapidly to get them out before the tide came in. I dug around the skull, but it was so heavy I could not lift it onto my pack animal, and had to leave it in the excavation. Reluctantly, I got on my horse, and rode along the shore back to the ranch.

When I returned the next day with the shepherd to help, I expected this rare fossil would be lost, but it was just as perfect as when I first found it.[54]

This animal looks very much like a hippopotamus, though it is not closely related.

While working the tidal exposures, Brown discovered that deep holes in the seafloor supported abundant marine life, which could supplement his typical diet of mutton: "As I rode close to one, the pounding of the horses hoofs caused three large fish to leap out. I jumped off my horse to grab two, putting them in my saddle bag. Thereafter, when I wanted fresh fish . . . , I'd ride around some of these holes . . . and was always rewarded with a plentiful catch." His horse, however, never became so sanguine:

> One day I was collecting sealife in the water holes left in the seafloor . . . , when I found an octopus of pickling-jar size. . . . I was busy trying to get him into the bottle, but as fast as I got two of his eight arms in others would come out. Completely absorbed by this effort, I was unaware of the incoming tide.
>
> Suddenly I heard a whinney, and looking up saw that the sea was all around us, and up to the horse's belly. Throwing the bottle and octopus away I jumped on my horse, for the water holes were now covered and obscured, and if I rode into one of these pits I would probably lose his life—and possibly my own. This was a close call.[55]

The fickle tides were not the only danger lurking on the estancia. While prospecting one day, Brown came across the scene of a mass murder involving sixty lambs, each with its neck broken. Tracks of a large mountain lion with its two cubs disappeared over the edge of a sea cliff. Brown quickly returned to the ranch to alert the shepherd, and the pair returned to the scene with dogs, firearms, and a candle. The track

> disappeared in a waterway down the cliff which was 400 feet high. The dogs would not follow me, so we decided that he with the dogs and rifle should stay above, while I took the revolver and candle, and [descended] to the beach [to] drive the lions upward.
>
> Soon I came to the mouth of the waterway where there was much lion spoor, carcasses of dead sheep, and offal. The entrance was fairly high. . . .

I lighted the candle, drew my revolver and started in. I had proceeded a short distance when all at once there came a roar—apparently back of me. Holding the candle up, I wheeled, and saw the reflection of two eyes. I fired between them.... I heard a kicking; then all was still.

The roar of the shot resounded up the waterway, and I was joined by the shepherd.... We covered the spot with our two guns, and threw stones where the lion had been, but there was no response. The lion was dead.

(I think this terrifying experience was when I lost my hair—figuratively scared off.)[56]

Completing his survey of the exposures along the shoreline and canyons of the Río Gallegos in early January 1900, Brown worked his way back up the coast to the Río Coy, "where I got very few fossils owing to the fact that Mr. Peterson had worked there only a few months before."[57] Thus, Brown's enterprise along the coast of Bahía Grande was finished.

When Hatcher and Peterson had left for home, Hatcher sold his wagon and other equipment to Felton. Without direct approval from Osborn or a suitable wagon, Brown was loath to brave another foray to the Andes. Brown therefore reluctantly bade farewell to the Feltons and rode back to Punta Arenas, where he sold his "horses, saddle and gear, retaining only my rifle, blankets, tarpaulin and personal effects."[58] But his adventures on this odyssey were not yet over.

As he packed his substantial collections and made arrangements to ship them to the museum, he discovered that no ship was scheduled to sail for New York for another four months, "so I decided to do some exploring." In keeping with the poker theme of his expedition, Brown met a gambler at his hotel named Saltpeare, although the man freely confided to Brown that that was not his real name: "He told me he had been a former newspaper man. He always wore gloves to keep his fingers sensitive, because certain playing cards are marked on the edges. He would never play poker with me, for he said I would not have a chance with a professional gambler."[59] Saltpeare was a Dickens aficionado who could recite the whole plot of one of his novels if Brown simply mentioned a title. He also owned a six-ton ship, converted to sails, which had formerly served as a New York cutter. In lieu of passing the time playing poker, Brown enticed the gambler to take him on a cruise "around the Horn," but first Saltpeare would need to retrieve his ship, which had been stolen one night while he was playing cards. When Saltpeare inquired into the whereabouts of his cutter from one of his seafaring colleagues, "The Yankee Captain," a salvager of wrecks along the infamous coast, he learned, according to Brown, that "it was anchored right in the lee of this vessel he [the Yankee Captain] was salvaging. And he said... there was a one-armed man there, that he had some squaws on board, and that [the one-armed man] had bought rights [of salvage] and provisions from this Yankee skipper."[60]

The Yankee Captain agreed to ferry Brown and Saltpeare down to the wreck site along the southern shore of Tierra del Fuego. Their odyssey began on February 28.[61] Yet a somewhat awkward confrontation arose when they arrived at the site near Spaniard Harbor.

> Sure enough, when we reached the wreck, there was the little six-ton cutter and the one-armed skipper... and Saltpeare said, "Well, you thought you were going to get away with it, didn't you?" He [the one-armed skipper] said, "Yes, but I found I couldn't navigate any farther."
>
> Saltpeare said, "This man represents the American Museum of Natural History, and he would like to circumnavigate [Tierra del Fuego] and he would like to have you stay on board and help us operate." The one-armed man said, "I don't think I'll go with you." He was afraid he was going to be marooned down there, no doubt, for trying to steal the boat.[62]

With Saltpeare's ship back in his possession, he, Brown, and a crew member sailed south, but soon they were caught in a current that led them into the "turbulent" Straits of Lemaire:[63]

> It was icy cold and after several hours hauling on the sails I was exhausted, and went down to my bunk. Presently, Saltpeare aroused me saying they had tacked repeatedly but were being driven out to sea by the high winds, and they wanted my voice in what should be done.
> ... We decided to take a reef out of the sail, and again tack—our one slim chance of making land at all. The boat came over with two feet of the gunnel under water, but we gained on each tack.
> What a sight was presented in the pitch black night—schools of porpoises cutting back and forth across the bow, leaving a streak of silver in the phosphorescent medusa-filled sea....
> We finally reached Spaniard Harbor, anchored, and everyone fell into the bunks, completely exhausted.[64]

When the weary crew awoke the next morning, Brown noticed a black streak in the outcrops along the shore, which he thought might be coal, and took the cutter's dinghy to row in to investigate. As he contentedly collected samples, he looked up and saw that the cutter's flag was flying upside down, the agreed-upon emergency signal that would tell Brown to return to the ship if "Indians" appeared or other danger arose. Dropping his coal samples, Brown scurried to the skiff and rowed out to the cutter. Unbeknownst to Brown, Saltpeare had been watching the barometer drop, "2 points as tho' struck by a hammer.... Before I reached the boat, another storm had struck. We upped the anchor and made two turns. We tacked back and forth [trying to exit the narrow bay]. But the third time she wouldn't come around, and we went up on the crest of a wave ... we hit a rock and it took six feet of the side off. I couldn't swim—you know how cold the water is down there—and I went in [to shore] on a barrel, the other men went in on hatches."[65] As he reached land, he looked back to see only the tip of the cutter's mast sticking out of the water. Marooned like many before them, they trudged along the shore until they reached "an Austrian gold-mining settlement we had passed the day before, and stayed there for 16 days until a coastal trading steamer picked us up, and ... took us southward around Cape Horn—Land's end."[66]

While the steamer wound its way back toward Punta Arenas, Brown marveled at the landscape and its tenacious inhabitants, including the Yahgan or "Canoe" Indians and the Onas, as well as the missionaries. The Yahgans both fascinated Brown for their habits and repulsed him for their appearance and lack of cleanliness. As he put it, they "practically lived in their canoes, only coming ashore at night to sleep; hence their torsos were well developed because of constant paddling, but their legs were stunted by lack of exercise. They inhabited the land of Terra del Fuego ... where, by actual records ... , it rained 300 out of 365 days of the year. Because of the water-soaked wood it is difficult to start fresh fires, so they carry smouldering embers from camp to camp [on a bedding of dirt in their canoes]."[67] In contrast, he noted, the Onas were "related to the Tehuelches of the mainland," whom Brown had encountered earlier. In fact, it was the large size of the Tehuelches that had led Magellan to name the region Patagonia, meaning "big foot," when he encountered their footprints along the shoreline of what would eventually become Argentina.

Without further misfortune, Brown and Saltpeare arrived safely back in Punta Arenas, where Brown shipped four and a half tons of fossil specimens back to New York. True to his word, Brown's collection provided not only an important assemblage of specimens for scientific research, but also exceptional skeletons for exhibition. Even before the extinction of large dinosaurs at the end of the Cretaceous Period, 65 million years ago, South America had been an island continent, and it remained separated from all other land masses until about 3 million years ago, when the Isthmus of Panama formed. As a result, a unique fauna evolved there in isolation.

Three groups of fossil mammals comprised most of Brown's specimens. Edentates, which include modern sloths, armadillos, and anteaters, were represented by primitive ground sloths and the tanklike glyptodonts. Even today, the skeleton of Brown's early glyptodont, *Propalaeohoplophorus,* as well as one of the primitive ground sloth, *Hapalops,* draw visitors' attention in the Hall of Mammals and Their Extinct Relatives. Beyond, in the Hall of Advanced Mammals, Brown's specimens belonging to litopterns (*Diadiaphorus, Thoatherium*) and notoungulates (*Nesodon*), two groups of unusual herbivorous ungulates—of which modern relatives include horses, rhinos, deer, camels, and cattle—are on display.

By early spring of 1900, Brown had finally had enough:

> I was ready to leave Patagonia. For many months I had been out of touch with civilization. There were no cables, and mail often reached me via Liverpool. The Spanish [American] War had been fought and won, but I was happy following the life work I had chosen.
>
> I obtained money... for passage to Europe, as there was a boat due in three days, and none for New York for several weeks. When the vessel arrived there were no vacant berths, but the Captain said he would take me if I slept in the saloon.[68]

What better place for Brown? By the time the ship left Buenos Aires for Lisbon, however, he had been installed in a more respectable cabin. In Europe, traveling by train to Paris and then on to London, Brown took the opportunity to visit museums and study their extensive collections. Always adaptable, Brown spruced himself up for his triumphant return to New York, as his daughter, Frances, later recounted: "Since he had lost all his clothing in the shipwreck near Spaniard Harbor, he bought an entire new outfit in Paris, including a tall, collapsible, silk hat. Sporting a moustache and pointed Vandyke beard, he then considered himself the epitome of a cultured European."[69]

Well, at least he was now, truly, a "Man of the World" and would shortly provide his fellow inhabitants with spectacular, skeletal evidence of the most imposing predator ever to stalk Earth's far-flung continents.

FIVE

To the Depths of Hell Creek
(1900–1903)

BROWN HAD SCANT CHANCE TO flaunt his new suit and goatee on the gaudy streets of New York or regale his friends and colleagues with adventurous antipodal anecdotes. Having arrived back home from Patagonia on June 10, 1900,[1] Brown found himself, once again, ensconced in the field by early July.[2] This time Osborn sent him to the northern Great Plains along the border between South Dakota and Wyoming. On the way out, the newly minted world traveler had the opportunity to visit his family in Carbondale, where he found his father "very feeble" and so delayed his arrival in South Dakota.[3] But once he got there, the quest for Cretaceous dinosaurs that he launched would occupy his field efforts for the rest of the decade.

There were several related reasons for Brown's almost reflexive return to the field. First, the region around the Wyoming–South Dakota border was already known for its Cretaceous dinosaurs, such as *Triceratops,* which Hatcher had collected for Marsh. The exposures were also generally familiar to Brown from his 1895 expedition with Williston. In addition, other exposures south of the Black Hills in South Dakota had already produced good specimens of Cretaceous marine reptiles, such as the sea serpent–like plesiosaurs and gigantic marine lizards, the mosasaurs. Although AMNH now had representative collections of Jurassic dinosaurs from Wyoming, it did not have a good sample of either Cretaceous dinosaurs or marine reptiles. This trip represented a first step in plugging that gap.

Beyond that, while Brown had lingered in Patagonia, another round of musical chairs had shuffled the roster of major museums that had decided to join in the dinosaur sweepstakes at the dawn of the twentieth century. Foremost among them was the Carnegie Museum in Pittsburgh. Andrew Carnegie, who made his fortune in steel, was intent on educating the public about many topics, including natural history, as evidenced not only by the museum but also by his widespread funding for local libraries in the United States.[4] He was particularly mesmerized by the tales of adventurous quests by Osborn and others for immense dinosaurs. Carnegie instructed the director of his new museum, W. J. Holland, to initiate a search for a dinosaur skeleton "as big as a barn."[5] Holland immediately started lining up a competent cohort of paleontological ringers. First, he lured Jacob Wortman away from AMNH, giving him the first curatorship of vertebrate paleontology. John Bell Hatcher soon followed. By 1899, Wortman was combing the exposures of the Jurassic Morrison Formation along Sheep Creek in Wyoming, and that same year he discovered what would become the type specimen of the lean and long-necked sauropod, *Diplodocus*.[6]

Given the previously mentioned animosity that Wortman then held for Osborn, and probably to some degree for Brown, as well as the fact that Hatcher had spurned Osborn's offer to join AMNH when the vertebrate paleontology department was first established, it's not surprising that an air of competition arose. However, in the wake of the 1880s feud between Cope and Marsh, no one seemed interested in igniting another scandalous, public donnybrook over dinosaurs. All the museums simply pursued their own programs, while keeping a wary eye trained on their competitors.

Carnegie was not Osborn's only adversary in this new episode of the dinosaur sweepstakes. In 1898, the Chicago mercantile magnate Marshall Field hired Brown's former classmate and best friend, Elmer Riggs, to head up a similar dinosaur-collecting enterprise at his recently established Field Museum in Chicago (then called the Columbian Museum of Chicago). The next year, Riggs got wind of dinosaurs weathering out of the Morrison Formation in Mesa County, Colorado, through a letter from a local dentist. By 1900, he had collected the front limb of the sauropod *Camarasaurus*, as well as what would become the type specimen of the tallest sauropod then known, *Brachiosaurus*. The game was definitely on.[7]

Feeling the heat, Osborn sent Brown west. While work continued at Bone Cabin Quarry in the Jurassic Morrison Formation under the guidance of other AMNH collectors, Osborn and Brown sought to press their advantage in Jurassic material by opening a new front in the Cretaceous. The most specific and immediate goal was to find a *Triceratops* skull for purposes of both research and exhibition. Based on Osborn's directive, Brown's 1900 season would last from July 1 to October 1, with a budget not to exceed $800—a little over $18,000 in today's dollars. Brown's salary in the budget for that three-month period is listed as $300, indicating that he had secured his desired raise.[8]

By mid-July, Brown had set up operations in Edgemont, South Dakota, renting "a complete outfit with saddle horse... for forty dollars per month..." and hiring a man to cook, look after camp, and help with the heavy lifting.[9] Since Hatcher and the 1895 Kansas expedition had focused their collecting efforts on the region of southeast Wyoming south of the Cheyenne River, Brown opted to focus on the adjacent region north of that river in Weston County, about forty miles west of Edgemont.[10] Today as then, this region is a sparsely populated, broken prairie, punctuated with shallow stream courses flowing south toward the Cheyenne that have etched down into the Late Cretaceous Lance Formation. The outcrops consist of drab brown and gray sandstones and siltstones deposited across a floodplain that bordered a shallow continental seaway about 66 million years ago. These exposures, which still attract paleontologists for their fossils of dinosaurs and other organisms, are picturesque but not spectacular, owing to the relatively low relief of the topography. Among the small buttes, hearty ranchers run cattle and other livestock, which share the landscape with pronghorn, coyotes, and soaring raptors.

Although Brown discovered that fragments of *Triceratops* skulls, including their distinctive horn cores, were in abundant supply, complete skulls were not. Brown's first significant discovery was made along the tributary of Alkali Creek called Seven Mile Creek. Clearly ecstatic, he fired off a letter to Osborn on September 1, crowing that "it gives me great pleasure to announce to you the discovery of a skeleton of Claosaurus."[11] The duckbill skeleton included the skull, lower jaws, vertebral column, ribs, and some limb bones; only the tip of the tail appeared to be missing. It took Brown until the start of October to extricate the specimen from the entombing sandstone.

While completing that task, Brown continued to prospect for a more suitable *Triceratops* specimen. But the fossil gods had other intentions, for on October 4, when he was supposed to be headed home, Brown penned another fateful letter to Osborn in which he related: "I have discovered another [specimen] which I hope to call a skeleton, though the bones are disassociated and scattered over considerable ground. This is a Carnivorous Dinosaur. I have worked on it two weeks and taken out seven ribs three vertebrae... one limb bone... besides parts of Pelvis, lower jaw, more ribs and several other vertebrae."[12]

Further excavations led Brown to report in mid-October that the specimen was "much better than I knew at the last writing." Much to Brown's delight, it appeared to represent "a Carnivorous form undescribed in Prof.

Marsh's work." At first blush, Brown thought it might be "an ancestral type of Ceratosaurus,"[13] though this seems anachronistic, since *Ceratosaurus* is a distinctly Jurassic form from the Morrison. In summarizing the specimen, Brown noted that "numerous plates [scutes] varying from a half inch to six inches across, always found closely associated with ribs, formed the dermal armature. On some of these plates there are crossed markings as in *Nodosaurus,* but it does not appear on all. Numbers of other bones impossible to determine in their matrix complete the specimen. Among the bones were the teeth of *Hadrosaurus, Palaeoniscus [Palaeoscincus],* the most numerous a species undetermined, scales of fish and small bones,—all evidences of the animal's last meal."[14]

Determined to collect all he could find of the specimen, Brown continued to labor through mid-November, when a blizzard interrupted his work. Osborn was elated but worried:

> I was much interested in your letter of November 20th. I have been fearing the snow storms would overtake you but I congratulate you upon the continued success of your work. The Carnivorous Dinosaur will be of very great value to us.
>
> I think you had better pack up and come in as soon as you can.[15]

Yet Brown persevered, despite the cold and the better judgment of his team of horses, who, with winter upon them, "skipped camp last week giving me a five days hunt finding them about fifty miles from camp."[16] Even after he rounded up the team, his troubles weren't over: "I started to town this morning with one of my heaviest sections of Claosaurus weighing close to two tons; had gone about two miles when one of the heavy side timbers broke and my whole load came down; fortunately breaking no bones, hardly scratching the plaster. I got out of the wreck with only a few bruises."[17]

Dead set against ever using soft cottonwood again, Brown purchased "ten inch oak bridge timbers" and returned to camp to retrieve the remainder of his jackets. All of the collection, he reported, came from within "a radius of five miles from the center of Alkali Creek."[18] He ended up shipping thirty boxes that filled more than two-thirds of Baltimore and Ohio's car No. 40816. Some blocks were so large that the side pieces had to be removed from the door to fit them inside. With the car weighed down with not only the hadrosaur but also the immense theropod with both its lower jaws but no skull, Brown requested and received a two-week vacation to visit his family before beginning the return trip to New York in the first week of February 1901.[19] Osborn had apparently arranged for the specimens to be shipped for free, which meant that the whole seven-month escapade cost only $781.60, slightly less than the original budget.[20]

With a couple of men and another field season, Brown was confident that the elusive *Triceratops* skeleton could be found and collected in this same area on the north side of the Cheyenne River. Osborn, however, had other ideas. The following field season, beginning in mid-May, saw Brown "loaned" to the Smithsonian Institution to help a paleobotanist explore Triassic exposures along the Little Colorado near Holbrook and Flagstaff, Arizona, for fossil pine cones.[21] It's not clear why Brown received this assignment; probably, now that he had footholds in the Jurassic and Cretaceous, Osborn was anxious to begin evaluating prospects in the earliest period of the Mesozoic. Cope's collectors had combed this area decades before, and Osborn, noting the Triassic fossils in the Cope collection, may well have been sending Brown back for more. Brown reported that "great quantities of fossil wood" were available, including "over forty tree trunks standing upright with roots embedded in position" near Tanner Creek, but fossil vertebrates were scarce. He did manage to find a partial skeleton of the crocodile-like phytosaur *Belodon*, which included the femur, vertebrae, and scutes, as well as several other fragmentary specimens of fossil reptiles.

By mid-July, Brown headed north back to the Miocene deposits near Sterling, Colorado, to meet with Matthew and finish the prospecting started in 1898 as part of the horse project, although he clearly hoped to be sent back for some more reconnaissance in the Cretaceous Lance exposures in southeastern Wyoming.[22] While waiting for Matthew to show up, Brown "found a fine *Protohippus* skull and part of a skeleton," a three-toed, early relative of modern horses, as well as a partial skeleton of an ancient camel.[23] By early August, he had also collected a fragmentary antelope skull and limbs, as well as two turtle shells, one with the skull, as well as skulls of some oreodonts and rodents.[24] The antelope skull, belonging to the genus *Ramoceros*, along with a skeleton of the horse, *Hypohippus*, are still on display in the museum's Hall of Advanced Mammals. Brown shipped his collections to New York by the end of August and returned via Carbondale.[25]

As evidenced by his prodding of Osborn for time to further explore the Cretaceous deposits in Wyoming, dinosaurs remained Brown's top priority. Although Osborn agreed that the collection of Cretaceous dinosaurs was an important task, he maintained competing goals of obtaining fossil mammals for his own research.

FIGURE 10. Brown collecting skeleton of the Miocene horse *Hypohippus* near Pawnee Buttes in Colorado during the AMNH 1901 expedition (18029, American Museum of Natural History Library)

Then, in the spring of 1902, Brown's desire to return to the Cretaceous deposits of the northern Great Plains received a serendipitous boost from another of Osborn's protégés. In addition to running the museum's paleontology program, Osborn was chairman of the board for the New York Zoological Society, the group that established the Bronx Zoo. The man Osborn and the board chose to direct the design and development of the zoo was a contemporary of Brown, thirty-one-year-old William T. Hornaday. Having served as chief taxidermist at the Smithsonian Institution's Museum of Natural History for years, he was widely regarded as the nation's foremost expert in his field.

While he was at the Smithsonian, word of the diminishing bison herds began to reach the east coast of the United States. As Hornaday tells the story, "In March, 1886, the writer received a severe shock, as if by a blow on the head from a well-directed mallet. He awoke, dazed and stunned, to a sudden realization ... that the buffalo-hide hunters of the United States had practically finished their work. The bison millions were not only 'going,' but gone! ... However, a belief was expressed that there were, even then, somewhere in the

West, some unkilled bands of bison from which specimens might be taken before the last of them were swept away."[26] Under Hornaday's direction, an expedition was launched. In September, Hornaday, along with a scientific student and three cowboys, gathered in Miles City, Montana, to begin the hunt. The crew spent the next two months hunting north of the divide between the Yellowstone and the Missouri near Hell Creek. Their goal was to collect some twenty bison for the museum collections, but it was anything but easy. After weeks of futile searching, Hornaday finally spotted the ultimate prize:

> I fired at the bull. . . . Down went the great beast, head foremost, . . . he staggered to his feet, in spite of his broken leg and galloped off over the hill. . . . After a short run we again overhauled our prize, on the side of a hill, near the crest of which he once more halted and stood at bay. Thirty yards away from him I pulled up, and gazed upon him with genuine astonishment. . . . He was a perfect monster in size, and just as superbly handsome as he was big. . . . With the greatest reluctance I ever felt about taking the life of an animal, I shot the great beast through the lungs, and he fell down and died.[27]

In those days, exhibits of animals generally consisted of single specimens mounted in a static pose on a horizontal wooden base. Expressions or dynamics of the animal were rarely attempted. Hornaday would change that paradigm, in homage to the bison. Using six of the specimens collected near Hell Creek, Hornaday conceptualized and executed the first exhibit of a genre now referred to as a life or habitat group. His goal was not only to portray the diversity within the species, but also to show them as he had seen them in the wild. In his display all the animals, including his prize bull, were featured as if in natural interaction around a water hole. To make the scene more realistic, Hornaday had shipped barrels of soil and samples of turf from the prairie in the Hell Creek region, and he used the feet of the specimens to create tracks in the mud and dirt near the pool. The result was a quantum leap in exhibition technique, and it caused a sensation within both the media and the public. "It was even said that a visiting Sioux Indian commented that he believed the bison moved around in their enclosure at night, because their tracks were plainly visible."[28] From Hornaday's life group, it was a small step to add a background mural and create the exquisite dioramas that have become the cornerstone of modern natural history museums.

Today, Hornaday's bull is the focal point on the Great Seal of the U.S. Department of the Interior. It was also depicted on three different postage stamps, and it may have been used as a model for the buffalo-head nickel.

Hornaday became the designer and founding director of the New York Zoological Park, now know as the Bronx Zoo, in 1896. His primary policy goal was to promote wildlife conservation and protect endangered species, including the bison, through breeding programs at the zoo. He was also instrumental in establishing the National Bison Ranges, still operating today, which have helped bring the species back from the brink of extinction.

In 1901, Hornaday returned to the badlands of Hell Creek with famed western photographer L. A. Huffman to hunt and document the lifestyle of blacktail deer. But blacktails were not the only species to catch Hornaday's eye. As Hornaday later related, the settler with whom they stayed, Max Sieber, "found three chunks of fossil bone which when fitted together formed a hornlike mass nearly a foot long.... Then [he] took me to a spot near by where he had found the badly weathered remains of what once had been a fossil skull, as large as the skull of a half-grown elephant.... The skull was so badly weathered that nothing could be made of it, but near it lay several fragments of ribs in a fair state of preservation."[29]

Well aware of the interest in dinosaurs at AMNH, Hornaday showed Huffman's photographs to Osborn and Brown upon his return to New York, and they identified some of the remains as belonging to the long-sought prize for the museum's new exhibitions, *Triceratops*. It came as no great surprise that Hornaday had found Cretaceous dinosaur fossils in Montana, although this particular area along the Missouri Breaks had not yet been extensively prospected. Cope had prospected in Late Cretaceous exposures, now referred to as the Judith River Formation, in 1876 with the famous collector Charles H. Sternberg (whom we will meet later).[30] In the immediate aftermath of the Sioux massacre of Custer's regiment at Little Bighorn, Cope's crew kept a wary eye out for the retreating bands of warriors. Although none were sighted during the actual prospecting, which produced the first specimens of the horned dinosaur *Monoclonius,* Cope's entourage did have to dodge remnants of the dispersing tribes as the collectors returned from the outcrops to load their fossils on a steamer along the Missouri River.[31] Cope returned to the region in the 1890s to collect a specimen of a duck-billed dinosaur. But in 1902, the area of badlands just south of the Missouri and north of the Big Dry was basically as wild as ever.

In letters that followed their meeting in May 1902, Hornaday did what he could to pinpoint the locations of the fossils he had seen for Brown. One of those letters described the locality for bones found near the small town of Forsyth: "I am just in receipt of a letter from ... Mr. L. A. Huffman ...

giving the location of the dinosaur found by Mr. Harrison. I send you Mr. Harrison's map of the location, from which I am sure you will have no difficulty in locating it on one of the land-office maps of Montana. It will not, however, be quite so easy to find it on the ground! But you will manage that."[32]

Brown set out the next month by train to Miles City to investigate. Montana was, in many ways, still a vestige of the Wild West. It had achieved statehood in 1889, less than thirteen years before, and in 1900 boasted a population of only 243,000—a dramatic contrast to the burgeoning population of sheep, which totaled six million and made Montana the nation's leading wool producer. A dozen eggs cost fourteen cents, while sugar was four cents a pound, and although the population on the plains of Montana was increasing disproportionately in relation to other areas of the state, Brown was heading toward a particularly remote and untamed terrain.[33]

Brown's destination was the Hell Creek region: not the prairies then littered with bison skeletons, but the rugged ravines etched into the Great Plains by the mighty Missouri River and its veinlike maze of tributaries. First explored for the United States by Lewis and Clark and often called the Missouri Breaks, these pastel-colored badlands extend in a seemingly endless panorama for hundreds of square miles just south of the Missouri. Acutely sculpted ridges topping steeply incised ravines snake off toward the horizon, while stalwart buttes, banded in rings of yellow, green, brown, gray, and jet black, rise resolutely above the chaotic incisions. Although grasses and shrubs cling desperately to the sparse flat spots, few trees can withstand the harsh landscape. Like the landscapes of the Lance to the south, the buttes and ridges of the Hell Creek Formation were formed about 66 million years ago by long-lost streams, which flowed out of mountain ranges to the west and drained into a shallow continental seaway to the east. Today, spreads encompassing 10,000 to 20,000 acres of prairie and badlands are home to families of rugged, weather-tanned cattle and sheep ranchers, and pronghorn, deer, coyotes, foxes, and eagles still roam the Breaks, just as when Brown wandered the ridges and ravines.

Upon his arrival in Miles City, Brown quickly learned from the locals that dinosaur skeletons abounded between there and the small town of Jordan to the north. On June 17 he wrote to Osborn: "Yesterday, I heard of another specimen having been found within five miles of Forsyth. I rode out to the place finding the location and the debris, the skeleton having been destroyed by souvenir hunters. It was a Claosaurus.... The vertebrae had been broken off and packed away while as near as I can learn the skull [and neck] vertebrae were broken off... and sold to the Smithsonian Institute about four years ago. It seems a Baptist Sunday School teacher had stolen the skull and sold it."[34]

Clearly, Brown already faced competition in this fossil-rich frontier. But none seemed to be around at the moment, and by June 19 he had located Harrison's "dinosaur."[35] Unfortunately, this and another of Hornaday's specimens in this area turned out to be badly deteriorated mosasaurs from the brown marine mudstone called the Bearpaw Shale that underlay the dinosaur-bearing beds. But that was a small setback, for this was not the area that most interested Brown. He immediately set about equipping himself for a summer-long assault on Hell Creek, musing in a letter to Osborn over the best logistical approach: "Whether to buy or rent an outfit. The best figure I have been able to obtain is six dollars per day for rent of an outfit while it is equally expensive to buy, but I rather think I shall buy three horses and rest of equipment.... I shall try to rent an outfit with understanding that rent goes on payment of purchase if I wish to buy after two weeks."[36]

As Brown left Miles City, the myriad tasks involved in managing the expedition were taxing his organizational skills, as he apologetically confessed to Osborn: he had written his boss from Miles City, "but discovered the letter still in my pocket yesterday so I will try a new one."[37]

By July 7, Brown and his crew had reached Jordan, bustling with a new post office and store, as well as a saloon, restaurant, and hotel. His primary colleague on the trip was one of Osborn's doctoral students, Richard Swann Lull. Tall and giftedly athletic, Lull was six years Brown's senior. He had worked as part of Osborn's field crew in 1899 at the Bone Cabin Quarry in Wyoming while Brown was in Patagonia. Despite Lull's seniority in age and academic qualifications, Brown had more field experience and clearly headed up the Montana crew. Consequently, some tension existed between the two. Brown also considered Lull to be rather pompous. Nonetheless, the two saw it through, and as the result of this trip, Lull would become especially intrigued with horned dinosaurs, eventually publishing several scientific papers and finishing Hatcher's long monograph about the group.

Arthur Jordan, the pioneer founder of the newly minted frontier town, met Brown and Lull when they came through. As Jordan recalls in his autobiography,

During the first days of May, 1902, Barnum Brown and Lull... arrived with their outfit and went to work in the breaks searching the different strata. I visited them a number of times during the summer and watched them at their work. I had seen many petrified bones in my ranging through the breaks, but these scientists knew where to dig to obtain perfect specimens of the huge prehistoric monsters. It was very interesting for me to view the specimens.

Those times in Jordan were wild, and every shady character that could not stand the spotlight of civilization drifted in and around the new town, always ready to have fun or start trouble, and a few of them desired to pick on the scientists and their men. I did everything I could to avert trouble and succeeded fairly well at that. Only once did they annoy Professor Lull when he came in town to get his mail. But one day, when one of their helpers drove in for the mail and a few supplies, I saw that trouble was brewing. I cautioned the man to slip out of town as quietly as he knew how, and he was well up the hill... before they discovered he was gone. They ran for their rifles and began to throw lead all about him. The fellow made his team do their best in going over the hill and away from those drunken, demented morons. Thereafter, the fellow gave the town of Jordan a wide berth.[38]

By July 12, Brown and Lull had safely reached a ranch north of Jordan, formerly owned by Hornaday's old friend Max Sieber. The road to the ranch had dropped down off the parched plains into the sizzling summer maze of the Breaks (now part of the Charles M. Russell National Wildlife Refuge), and success in the crew's quest quickly followed:

> We are now camped seventeen miles from the Missouri River on the head of Hell creek. Pulled in here to the old Sieber ranch Wed night. Dr. Lull and I started out Thursday morning and I located two good prospects one in clay which we have not worked. The other is a Triceratops or Torosaurus....
> ... there may be plenty of fossils here for we have prospected only the one day....
> The country greatly resembles Lance Creek in Wyoming along the head of the breaks but the main canyons are certainly *bad* lands, almost impossible lands I might say....
> The great drawback to this region is the distance to freight fossils. It is over a hundred and thirty miles to Miles, the only available point.[39]

Meanwhile, Osborn was supervising the preparation of a dinosaur specimen at AMNH that Brown had collected previously, and he was not pleased.

In a diplomatically stated yet nonetheless direct admonishment, Osborn wrote Brown on July 25:

> I am very glad to learn that you are at last finding some fair prospects. I trust that you will uncover some of these bones thoroughly, and ascertain positively whether they are worth collecting.
>
> You will be very much disappointed to learn that the Dinosaur which you collected with so much care and labor has proved almost valueless. . . . It will perhaps yield two or three bones of value. The skull proves to be entirely crushed and unrecognizable.
>
> This seems to warn us that we should certainly examine material a little more carefully in the field before taking it up in a block and going to the heavy expense of freight shipment to the east. I know you sent the specimen to us after the best possible methods; but it should have received a more careful examination.[40]

Brown, however, did not yet know that he was in the doghouse. What he did know was that he had a solid chance of finding good specimens of *Triceratops*, since it was among these same rugged ridges and ravines that Hornaday had located his fossils of that animal. And happily, success soon arrived. On July 19 he reported to Matthew and Osborn: "I have been rather fortunate in locating a *Sterrholophus* [= *Triceratops*] skull in fairly good condition, basal portion in fine condition, not crushed. Front very punky . . . top of horn cores gone . . . but will be able to make a fine exhibition specimen of it."[41] This news immediately restored Brown to Osborn's good graces. Osborn indicated in his next letter that he was pleased to learn about the *Triceratops* skull, "which I trust is the first of the series, since your letter leads me to believe that you have struck what Marsh used to call the 'Holy Ground.'"[42]

Brown received Osborn's admonishment in mid-August and humbly conceded: "I greatly appreciate your criticism and every pound of matrix that we can possibly remove from these specimens will come off though it takes a great deal of time from prospecting." He then went on to describe the crew's progress in collecting both the Hornaday specimen and the new *Triceratops* skull and lower jaw. Impressive though these were, however, he had an even more dazzling discovery to report. In the next paragraph of the same letter, Brown played an ace for the ages to counter Osborn's rebuke: "Quarry No. 1 contains the femur, pubes, [partial] humerus, three vertebrae and two undetermined bones of a large Carnivorous Dinosaur not described by Marsh. . . . I have never seen anything like it from the Cretaceous. These bones are imbedded in flint-like blue sandstone concretions and require a great deal of labor to extract."[43]

With this brief, matter-of-fact mention, a somewhat chastened Brown announced the discovery that would catapult both him and the animal he had found into the cult of celebrity. The bones from Quarry no. 1 belonged to the most intimidating carnivore yet found to have walked the Earth, *Tyrannosaurus rex* (see Appendix 2 for Brown's account).

Brown immediately recognized the significance of the find. But he also realized that such a gargantuan skeleton would be difficult to excavate and transport 130 miles over the rut-riddled roads to the railhead at Miles City, and he was anxious to keep costs manageable. Accordingly, when Brown informed Osborn that he had spent $275 (about $6,240 in today's dollars) on three hearty horses, along with a new wagon and essential camp equipment, he quickly added that he could resell them all for "nearly full value." To underscore his efforts at parsimony, Brown noted that lumber and plaster were extremely expensive, especially plaster from Miles City at five dollars a barrel, so "I go as economical as possible . . . [and] use flower paste wherever practicable."[44]

Jordan's small store was ill equipped to handle all the demands of its new paleontological customers. Stout lumber and plaster were staples of fossil collecting, essential for constructing sturdy casts around the bones to protect them during transport. It's a time-honored technique that even we have used in extracting more modest fossils from Hell Creek. But in Brown's day, there were no jackhammers and backhoes to help excavate large specimens. The bulk of the overburden entombing the skeleton had to be blasted away with dynamite, as Brown recounted on September 3:

> We are still at work on quarry no. 1 [on the] Carnivorous Dinosaur. . . .
> . . . The bones are separated by two or three feet of soft sand usually and each bone is surrounded by the hardest blue sandstone I ever tried to work in the form of concretions.
> There is no question but what this is the find of the season so far for scientific importance. . . . It is necessary to shoot the bank and as it is not accessible to horses work goes slow.[45]

(Archival photos taken during the excavation show Quarry no. 1 perched near the top of a steep sandstone cliff that formed the eastern bank of Hell Creek. Although the walls of the quarry are now eroded away, along with the track that Brown used to haul the blocks out in his wagon, the cliff still

FIGURE 11. Brown's crew using horse team and scraper to remove overburden from Quarry no. 1, which produced the type specimen of *Tyrannosaurus rex,* at Hell Creek, Montana, during the 1902 and 1905 AMNH expeditions (18172, American Museum of Natural History Library)

stands. One can even walk around the area and identify the positions from which various photos were taken. As one sits atop the bluff on the opposite side of Hell Creek, it isn't hard to imagine Brown and his crew pounding endlessly on the concretions, scrambling for cover during the blasting, and spurring the team as it strained to pull the laden wagon out of the quarry. Nearby, the stones forming the foundation of Max Sieber's old dugout cabin lie just across Hell Creek on the western bank, and the ruins of two other cabins lie immediately to the north and south. Built after Brown's expedition, they housed the Engdahl and MacDonald families, whose sons still owned the surrounding ranchland in the late 1970s and early 1980s.)

During the excavation, other logistical difficulties included injuries and illnesses, in the face of which there was little to do but tough it out. In the 1902 season, for example, Brown's physical efforts were painfully complicated by a bad case of gout, "such that he could barely ride his horse and needed help each night to get into his bed. It was extremely painful for him to work in the quarry, but did not last long."[46]

FIGURE 12. Brown collecting specimen of *Triceratops* at Quarry no. 2, Hell Creek, Montana, during the 1902 AMNH expedition (AMNH Vertebrate Paleontology Archive, 7:2 B3 S4)

With the snows of fall and winter approaching, Lull returned to New York, and Brown began to have members of the crew transport some of the fossils back to Miles City, where Osborn had made arrangements for an empty box car. Meanwhile, Brown continued prospecting for new fossils around Hell Creek, discovering four "crocodile" skeletons in the beds above the dinosaur-bearing rocks. One of these specimens, belonging to the genus *Champsosaurus* (which looks like but is not closely related to crocodiles), is on display in the museum's Hall of Vertebrate Origins; two other skeletons from this expedition, representing *Triceratops* and the duckbill *Anatotitan*, occupy the central island in the Hall of Ornithischian Dinosaurs.

By the first of October, winter was in the air. Although Brown found the change in weather to be "cool and exhilarating," it was nonetheless time to head home with his hefty treasure. On October 13 he wrote to Osborn: "Just arrived in Miles with rest of fossils. There are about nineteen boxes in all, haven't finished packing, aggregating between ten and fifteen thousands of pounds weight. Rate from Miles to New York is $3.13 per hundred which at lowest tonnage means over three hundred dollars freight."[47]

Not even Osborn would quibble at that amount; it was a piddling price to pay for what would become the most famous dinosaur of all time. Excluding shipping costs, total expenditures for the crew's travel, equipment, supplies, horses, and labor came to $1,345, or a bit over $30,000 in contemporary U.S. currency. It would take two to three years for the specimens to be prepared and analyzed before two scientific publications authored by Osborn would announce the discovery of *Tyrannosaurus* to the world.

Back in New York, plans were afoot for the next year's season, and in early May 1903 Osborn appointed Brown to head up "the northern expedition, especially for Reptiles in Wyoming, Dakota, and Montana," adding: "I am especially desirous of finding a new region for Jurassic Dinosaurs."[48] By May 18, Brown had arrived in his old haunt of Edgemont, South Dakota, having rented an outfit for $50 per month and hired a cook, with the intent to explore the Late Cretaceous marine shales of the Pierre or Niobrara Formation, about twenty miles south of the city.

The region proved to be a mother lode for marine reptiles, especially mosasaurs.[49] Over the two and a half months Brown spent, mainly in a thirty-square-mile area south of the Cheyenne River and north of Pine Ridge on Alkali Creek, he garnered no less than ten specimens, including a complete fifteen-foot-long skeleton and at least three other skulls. In jest, Brown bragged to his boss that "at present rate, we will need a Mosasaur Hall."[50]

In addition, Brown discovered four partial plesiosaur skeletons, although much to the disappointment of both him and Osborn, none included a skull, which Osborn indicated in correspondence was "one of the greatest desiderata," for both research and exhibition purposes.[51] Early on, Brown also proclaimed "another ten strike in [finding] some fragmentary Pterydactyl bones. They are merely fragments but unmistakable, crushed together."[52] Eventually, Brown excavated the four distal bones of one finger in the wing.[53]

Near the end of his stay south of Edgemont, discussions were under way regarding Brown's potential participation in another international expedition. In a letter of July 3, Brown made a direct play for being the museum's solitary representative: "Regarding the South African expedition I should say that from my experience in South America as well as in the States I find that I can do more effective work at ... much less expense to the Museum alone than with a party for the simple reason that I am not encumbered with a lot of paraphernalia and can go at will. By looking over the record I think you can verify the statement. The work is harder for me because I put in more hours and work harder... but I don't mind that."[54]

AMNH's interest in South Africa was catalyzed by the recent discoveries of Robert Broom, a Scottish-born paleontologist who had trained as a medical doctor before moving to South Africa in 1896 to start a practice and collect fossils in the rich Permian-Triassic sediments of the Karroo Desert. By 1903, Broom was amassing a spectacular collection of ancient synapsids, early relatives of mammals, as well as a few dinosaurs.[55] These discoveries greatly interested Osborn and his colleagues, including Brown.

By mid-July, Brown was on the move westward toward Buffalo, Wyoming, to follow up on a lead about a magnificent Jurassic dinosaur specimen supplied by a Mr. Sparhawk. Unfortunately, it turned out to be a bust, consisting of only a partial sauropod limb: "When sifted down to direct facts this man says he saw only two bones, ... the rest of the skeleton was purely imaginary. Sparhawk himself is a bar-room inhabitant and general wild horse scheme promoter with views of large sums of money in the future."[56]

In the wake of this disappointment, Brown set out to survey the Jurassic exposures along the flanks of the Bighorn Mountains from south of Buffalo northwest toward Pryor, Montana. In his wanderings, Brown had particular success at Cashen's Ranch along Beauvais Creek, where he found "the remains of a plated Dinosaur resembling *Stegosaurus,* in which, however, the plates differ from those figured by Marsh," as well as a partial skeleton of a small *Champtosaurus* [presumably *Camptosaurus*]" and "four limbs [and seven vertebrae] of a large Morosaur [*Camarasaurus*] in good condition."[57] At the time, this part of his excursion fulfilled Brown's goal to find new exposures with Jurassic fossils because he thought they belonged to the Morrison Formation. However, the age of these striking, pastel-colored strata that form the low ridges along the creek would eventually be reassigned to the Early Cretaceous Cloverly Formation. Brown and his crews would return to this site numerous times in the next several decades, and one day make a startling discovery that would eventually clarify the evolutionary origin of birds.

Later in July, Brown again pressed Osborn regarding the South African trip, explaining that he needed clarification, but this time for a very different reason: "My fiancée is pressing me for some knowledge of my movements during the winter for she requires some little time for preparation for our marriage.... Can you let me know soon whether there is a probability of my going to Africa this winter. I hope the trip will come to pass for I am anxious to take it."⁵⁸ Although understanding of Brown's concerns, Osborn did not provide a definite answer.

By mid-September, Brown was taking some well-deserved down time with his family in Carbondale before returning to New York. On his way back east he made an unscheduled stop in Wilcockson, Arkansas, triggered by a letter from Osborn describing a fossiliferous Pleistocene site.⁵⁹ The fossils were preserved in a crevasse deposit associated with a shaft for a zinc mine, and Osborn was keen to sample the fauna of the apparently forest-dwelling mammals. Finding the shaft too unstable to excavate without removing the overburden, Brown observed the in situ exposures but limited his collecting to the adjacent dump piles, recovering "jaws and skulls [as well as postcranial remains] of 315 individuals representing 36 species" including rodents, cats, sabertooths, dogs, foxes, raccoons, bears, deer, birds, lizards, and snakes.⁶⁰

By October, Brown was back in New York, where his life was about to change dramatically. Not only was he preparing for his upcoming nuptials, but in early November he also received the sad news of his mother's death at age sixty-three.

SIX

Love
(1903–1906)

IT'S NOT CLEAR EXACTLY WHEN Barnum Brown first met his match, except that it occurred when he was a part-time student at Columbia. Yet the site of this mutual conquest was duly recorded in notes for his aborted autobiography. Brown lived in a house owned by Dr. Herman E. Meeker, who ran a medical practice on East 67th Street. Meeker's home doubled as a boarding house, where he rented rooms to a select set of denizens, which included "myself and a Miss Marion Brown, whom I was later to marry."[1]

Fortunately, Barnum and Marion's daughter, Frances, is more accommodating in her description of her mother. Marion Raymond Brown was the daughter of a distinguished lawyer and educator in Oxford, New York, and by graduating from Wells College in 1898 with a degree in biology had proved to be no educational slouch herself. She immediately went into teaching biology near New York City while working on her master's degree at Columbia in her spare time, especially during the summer. Upon finishing that coursework, she was offered another teaching position at Erasmus Hall High School in Brooklyn, the city's most prestigious public school for science students. Frances recounts:

> At some point... in all this Marion and Barnum met and fell in love. The courtship had its complications, though, because Barnum lived in a rooming house near the museum. Marion roomed in a house near Erasmus Hall.... Subway service was not the problem, however, when Barnum would escort Marion home from an evening party or outing. The difficulty lay in the fact that Barnum would promptly go to sleep as soon as he got on the train home. Slumbering peacefully, he would ride far past his Manhattan 79th Street stop, never waking at the end of the line and the beginning of the return trip to Brooklyn. How many times he woke up in the gray dawn right back in Brooklyn, Barnum was not willing to specify.[2]

FIGURE 13. Brown's first wife, Marion, washing matrix for Pleistocene fossils near Conrad Fissure in Arkansas during the 1904 AMNH expedition (AMNH Vertebrate Paleontology Archive, 7:2 B3 S36)

Eventually, Barnum's conundrum was solved when the couple was married at St. Paul's Episcopal Church in Oxford on the chilly, sun-splashed day of February 13, 1904. Due to the chill, Marion's wedding apparel included a pair of bright red socks, which she pulled over her white satin slippers to keep them clean. But a minor problem ensued:

> The wedding procession had started, and the bride was just ready to step down the aisle on her father's arm . . . [when] she remembered the socks. With a muffled giggle and two vigorous kicks, she got rid of them just in time. Not that Barnum would have cared if she had appeared at the altar in red socks. . . .
>
> Many friends of both the bride and groom often spoke later of Marion's blond, almost ethereal beauty as a shaft of sunlight through a stained glass window struck her happy face.[3]

Probably in part as a consequence of her professional interest in biology, Marion would accompany Barnum on his next field excursion in 1904. One advantage of her participation is Marion's remarkable unpublished chronicle of the trip, entitled *Log Book of the Bug Hunters—1904*. In effect, the trip seemed to serve up equal portions of scientific expedition and extended honeymoon.

Departing on May 29 by train, the newlyweds traveled to Marion, Indiana, to inspect a nearly complete mammoth skeleton found and collected by locals, who were opportunistically charging an admission price of ten cents to see it laid out in a shed.[4] The evolutionary history of elephants and their kin, broadly termed the Proboscidea, was a favorite field of research for Osborn, as well as a main thrust of his collecting and exhibition agenda. Osborn spent more than three decades supervising the preparation of a two-volume monograph on the group, with much of the scientific legwork being done by Matthew and other subordinate assistants on his staff.[5] The first volume was published shortly after his death in 1936, and the second in 1942. On his expeditions throughout the United States, Mexico, India, and Burma, Brown would play a significant role in collecting a number of the specimens that Osborn included in the study. In 1904, however, the project was still in its infancy, and the museum was greatly interested in purchasing the specimen in Indiana for Osborn's research and possible exhibition.[6]

Once in the field, the couple spent three weeks collecting in the shales of the Niobrara or Fort Pierre Formation near the head of Moss Agate Creek, southwest of Edgemont, South Dakota.[7] On June 11, Barnum finally found what he had unsuccessfully sought the previous year. Marion fairly beamed: "Barnum and I tramped over the Bad Lands ... and located a Plesiosaur skeleton. After dinner under one of the two trees in this vicinity, we continued work and found there was a skull. Only 3 Plesiosaur skulls have been found up to the present time and Dr. Williston has all those so we are much pleased."[8]

Although conditions were tough, with days hovering near a hundred and nights laced with lightning and torrential storms, Marion seemed quite satisfied in her new role. She especially enjoyed the flora and fauna:

> June 15, this morning when we were at breakfast a yellow-headed blackbird flew down near us. He was a gorgeous fellow, with yellow head and neck, white coverts and the rest of his body glossy black. Twice today he came back to the same spot and busied himself running after bugs....
>
> The little white wing blackbirds are smaller ... , and their tertiary wing coverts are white. The female is brown, with little lines of white and the same white tertiary wing coverts.... The males have a sweet song which they sing only when on the wing.[9]

Like Barnum, Marion also had a keen sense of humor, and she relished a good yarn, such as this one from their field assistant and cook, Leroy Parkin: "Roy told us a story this morning worthy of being recorded in the palaeontological records. He told . . . a rancher about the Mososaurs which he had been helping Barnum to find last year. After he had explained as well as he could what Mososaurs were, the man said 'Yes, I see one of them darned things swimming down in Hat Creek the first year I was here.'"[10]

In all, they gleaned four plesiosaur skeletons, two with skulls, and three more mosasaur skeletons from the rolling exposures of shale. On June 24, they headed off to the Black Hills and Wind Cave National Park near Hot Springs, South Dakota, for some "R and R," though by July 1 they were again on the move, having arrived in Billings. There, they procured a wagon and supplies, then headed south toward the Cashen Ranch and the Crow Reservation. They were powered by two temperamental mules, whose propensity for kicking kept Marion on her toes.[11]

The couple's festivities on the 4th revolved around a dance celebration that the Crows mounted each year at Pryor Gap. A meadow about 600 feet in diameter surrounded by sheer walls of rock housed dozens of tepees and other tents, including a large central one adorned with an American flag "flying . . . wrong side up." Nearby stood a smaller Medicine Tent. The morning's festivities featured slow-paced dances by women decked out in radiant costumes complete with beaded belts. Later, men and women, each forming half of an immense circle, danced together before adjourning to the Medicine Tent for consultations with the shaman while "several women cooked weird looking mixtures over the fire." The afternoon finale, heralded by men galloping through camp on horseback, starred brilliantly painted warriors, "one . . . navy blue with yellow spots," and several with "war bonnets of eagle feathers." The formal festivities ended with gift-giving involving "blankets, calico, [and] many bottles of soda."[12]

On their first day in the badlands, about twelve miles southwest of the Cashens' place, Barnum found the prize of this leg of the expedition, a complete skull and lower jaws of *Teleorhinus,* a Cretaceous crocodile. The specimen is still on display in the museum's Hall of Vertebrate Origins. Meanwhile, Marion noted the proliferation of insects, including "small and large green flies, deer flies, horse flies, bot flies, . . . mosquitoes too numerous to mention. . . . Cicadas, crickets, and grasshoppers make the air ring with their music."[13] Barnum recounts how his wife utilized one of these campmates to supplement their limited diet: "[There] is a stream of pure, cold,

FIGURE 14. Crow Indians at Pryor Gap, Montana, during the Crow July Fourth celebration in 1904 (AMNH Vertebrate Paleontology Archive, 8:1 Brown Album)

melted snow water, . . . and this year it was filled with speckled trout. Marion improvised a hook from a safety pin, baited with grass hoppers, and caught several messes of good-sized beauties. Cooked in the frying pan on our sheet-iron stove, they were a welcome change from canned food."[14] Her good luck no doubt reminded Barnum of his own similar good fortune at nearby Yellowstone Lake on his foray with his father. After two weeks of roaming the exposures around the Cashen Ranch, during which they discovered a "complete turtle and fragmentary material of a new genus of crocodile," the couple headed back to Billings to reunite with Leroy Parkin, resupply, and head out toward Hell Creek.[15]

Their whirlwind tour north of the Yellowstone River began on July 17 with their mules Jack and Jane and "two saddle ponies" leading the way through clouds of mosquitoes past Pompey's Pillar to Forsyth.[16] Although abundant exposures of the Fort Union Formation rimmed the river, Barnum found few fossils, as temperatures soared to 104. By the 26th, the party headed northwest up Big Porcupine Creek over the divide separating the drainages of the Yellowstone and Missouri. Once over the pass at Hole in the Wall, the crew was assaulted by Mother Nature. As Marion described it: "July 29, a terrific storm came up about 4 A.M. with hail. Leroy came into our tent and he and Barnum tried to hold it up but it blew over and threw them down into the water, then about 4 in. deep under our cots."[17] To which Barnum adds: "Roy in his underclothes was soaked, shivvering so that his teeth chattered. I told him to get into my cot and under the blankets. When the rain stopped and it was quiet again, I said to Marion, 'Now old Girl, if you ever get obstreperous I'll tell of the time I saw a man in bed with you.'"[18]

By August 1 they were camped north of Jordan near Snow Creek, where they "found a triceratops skull, hardly worth taking, . . . and part of a carnivorous skeleton." Despite the dearth of fossil finds, Marion was entranced by the "high pinnacles, cathedral spires, and fortresses of fantastic shapes" that the erosive power of the great Missouri River and its countless tributaries had hewn into the landscape of the Breaks.[19]

Pressing on, the troupe headed west to prospect in the exposures of the Judith River Formation and Bearpaw Shale along the Musselshell River. There, Marion was dazzled by the beauty of fossil ammonite shells, distant relatives of the living nautilus, in which "the colors of the mother of pearl were as beautiful as opals."[20] Barnum's luck finally improved on August 9 with the discovery of an immense duckbill skeleton in the Judith River, including "pelvis, vertebral column, and limb bones."[21] Meanwhile, Marion, tapping into her abilities for anatomical observation, recorded a milestone in the life of a small pet she had acquired: "Old Mrs. Toad had 13 little ones born this noon. . . . They are wrapped in a very thin membrane which dries in a couple of minutes, and then they are up and running around. They are funny little things about 1/20th the size of the old toad. The pineal eye is very distinct. The cord comes from the abdomen . . . and spreads out in the amnion. . . . The old toad never paid the slightest attention to them after they were born."[22]

Returning to Billings, the couple shipped their specimens to New York and hopped the train to Denver for a leisurely excursion to Pike's Peak, where they rode the narrow-gauge cog railway to the top, powered by its "queer little engine with its hind wheels lower than its front to make it level on a grade."[23] Marion was struck by the scenic views of the Rockies with their pristine mountain lakes and lushly forested slopes, as well as by the panorama that unfolded below, which included the Garden of the Gods. As the engine labored past timberline, the honeymooners donned their overcoats, which Barnum had apparently brought in order to avoid the fifty-cent rental fee. Upon reaching the peak amid a snow squall, Marion marveled at the alpine flora with its tenacious lichens clinging to the granitic boulders and the delicate flowers huddled tenuously in between.

From Colorado, the newlyweds set out southward by rail toward Gallup, New Mexico, intent on finding more fossils. Outfitting was difficult: "Finally got [a team] for $2.50 [per day], and a wagon from Mr. Kenny for 50 cents." On August 19, they rumbled northeast out of Carmen toward Pueblo Bonito on a somewhat suspect road, which from Marion's perspective "proved to be very minute for the first 2 or 3 miles and not much good thereafter." But soon she was diverted by the radiant buttes of tilted red sandstone east of Gallup, which slanted down toward the bed of the arroyo and were covered in fragrant cedars. That night under the stars, however, she encountered a common desert annoyance: "I got into an anthill after dark and got covered with big black ants so that I could not get them out [of my clothes]. They gave a fierce bite."[24]

Wandering among the Navajo and their cedar-log hogans, Marion took notes on how the women wove their strikingly patterned woolen blankets. On August 23 they came in view of the "great yellow wall" of Chaco Canyon and by midday arrived at Pueblo Bonito to set up camp. Barnum had previously befriended the owner of the trading post near the site, John Wetherill; although he was off escorting a group of students around the region, Barnum was able to secure a load of hay for the horses. Marion delighted in exploring the semicircular ruins of the Anasazi, which included an impressive seven-story structure nestled at the base of the sheer sandstone wall and circular estufas that intertwined with the rectangular rooms. She rendered two diagrams to document the structures and wrote two full pages of notes.

Barnum spent the next week diligently assaulting the outcrops in the region, a sequence of rock formations that include the Late Cretaceous Fruitland Formation, Kirtland Shale, and Ojo Alamo Sandstone. The Ojo Alamo exposures are essentially the same age as the Lance and Hell Creek strata that Brown had been working in southeastern Wyoming and eastern-central Montana; presumably Osborn and Brown wanted to compare the fauna of dinosaurs at these different latitudes. On August 27, Brown discovered what would become "the type skull and jaws of the [duck-bill], *Kritosaurus navajovius*" in the Late Cretaceous outcrops of the Ojo Alamo Formation.[25] That specimen is now on display in the museum's Hall of Ornithischian Dinosaurs. In the same paper that describes and names *Kritosaurus*, Brown also describes the rock layers in the Ojo Alamo for the first time.[26]

On the way home, Barnum assessed some Pleistocene fossil deposits associated with mines in Arizona, and in mid-September Barnum introduced

Marion to his family in Carbondale. They then continued on to the Ozarks, where Barnum hired three men to help him dig out the Conrad Fissure, the fossiliferous deposit of rubble in a cave near Wilcockson that he had evaluated the previous year. With the aid of dynamite, the men excavated a shaft "twelve feet long, seven feet wide, and twenty-five feet deep." The Pleistocene carbonates yielded "ten complete and many fragmentary skulls of rodents and carnivores, about 1000 jaws, thousands of limb bones and vertebrae, representing nearly forty species.... These remains show a large fauna of extinct and many living species..., including bears, sabre-tooth tigers, lions, deer, musk ox, hogs, foxes, wolves, beaver, rabbits squirrels, mice, rats, weasels, shrews, bats, birds, snakes, lizards, toads, and frogs."[27]

Perils abounded in Arkansas. "The Conrad Fissure," Barnum noted, "was in a forest of tall trees with poison ivy over most of them. Our men who were immune to the ivy... left short stems... which I did not see as I put my hands on the rocks to jump down.... I was soon covered with ivy juice—fortunately not in my eyes.... That night... Marion had to wrap each finger separately to keep them from sticking together. This kept me at home for several days."[28] Shortly thereafter, Marion faced an even more dire threat. Lounging outside the tent one afternoon, watching for Barnum's return, Marion was unaware that something was in turn intently watching *her*. As Barnum hiked over the hill toward camp, their daughter later reported, he spied, "on the grass beside her, with its head poised to strike, a large copperhead. Whipping out his revolver, Barnum called softly to Marion to sit perfectly still. Unquestioning, she did, and Barnum killed the copperhead with one shot.... Decades later, Barnum remarked that Marion's instinctive reaction to his unexplained command was an example of the perfect rapport between them."[29]

Having avoided disaster, the newlyweds wound up their nearly five-month honeymoon by returning west to Arizona's Petrified Forest and Grand Canyon. While at the Petrified Forest, they happened to meet the famed orator and presidential candidate William Jennings Bryan. (Bryan would later gain fame as the anti-evolutionist lawyer in the Scopes trial, and Brown would eventually assist Osborn in trying to counter Bryan's views by looking for fossils that could help document our human origins in relation to other primates.) When the Browns arrived at the forest's guesthouse for lunch, they found it already occupied by Bryan's party. Not wishing to intrude, the young couple laid out their picnic under a nearby tree, but when Bryan realized the Browns were outside, he magnanimously insisted they come in out of the heat, which they no doubt gratefully did.

After wandering among the petrified trees, where Marion collected a horned toad for her biology class, both parties headed toward the Grand Canyon. There, Barnum and Marion secured a front-row seat for one of Nature's greatest shows by setting their camp right "on the edge... where it was 1000 ft. down to the first landing. In the morning we joined the Bryan party and rode mule-back down the trail to the river. Mr. Bryan on his mule was a perfect replica of the Democratic emblem."[30]

The Browns and Bryans extended their association on the train homeward to New York, and when Bryan's presence became known to other passengers, their car was mobbed with folks pleading for the legendary orator to speak. Bryan obliged by giving what Barnum called a vibrant account of the land they were traveling through, including its history from the conquistadors to the present. Barnum considered it a great "privilege to hear this man speak."[31]

Apparently, Barnum and Marion's endorsement of evolution came up during their conversations, and Bryan, not surprisingly, expressed a different view. As the train neared New York, Marion invited the Bryans to come to their apartment for a visit; Bryan gently but decidedly deferred by teasing, "Mrs. Brown, I will be delighted to visit you when you are able to tell me the connection between horned toads and politics."[32]

When Brown resumed work at the museum, he and Osborn faced a troubling paleontological puzzle. In the years since 1902, the tyrannosaur specimen had been meticulously prepared in the museum's laboratory by Richard Swann Lull, Peter Kaisen, and Paul Miller, who with hammers and chisels judiciously chipped away at the tenacious concretionary shroud in which the bones were entombed.[33] As the preparators neared the bone itself, they used awls and other small tools to gently pry off the remaining matrix. Once the fossils were exposed, shellac was applied to harden the mineralized bone and to glue the naturally fractured fragments together.

What emerged, however, proved to be incomplete. Despite Brown's diligent work in Quarry 1, only parts of the skeleton, including "both jaws and portions of the skull, vertebrae, ribs, scapula, humerus, ilium, pubis, ischium, [and] metapodials," had been recovered.[34] One surprising aspect of the hind limb bones was that, despite their large size, they were hollow, just like bird bones.

As the 1905 field season approached, Osborn began to prepare a scientific description of the bones.[35] In the published paper—in which he acknowledged that Brown had discovered the specimen and collected it with Lull, but did not list either man as a co-author—he proposed the new genus and species *Tyrannosaurus rex* to describe the specimen Brown had found along Hell Creek in 1902 (AMNH 973). He also created a separate genus and species, *Dynamosaurus imperiosus,* for the skeleton of the large carnivorous dinosaur that Brown had found in Weston County, Wyoming, in 1900 (AMNH 5866). One basis for their separation was the numerous scutes of dermal armor found near the ribs of 5866, which Brown believed belonged to that animal.

Osborn and Brown longed for more evidence, however. They now wondered whether all the bones preserved in the 1902 tyrannosaur quarry had actually been exposed and collected, and determined that reexamination of the site was imperative. In May 1905, therefore, Brown once again set out for Hell Creek.

Brown and Leroy Parkin rendezvoused in Billings, but with rain making travel to the northeast impossible, they instead set out for another short spell on the Crow Reservation to the south. Brown almost didn't survive the trip. On June 5, he wrote Osborn of an encounter with an unexpected adversary on the sparsely traveled tracks: "We were out only three days when the team [was] frightened by a motor cycle, pulled the picket pins and ran into a barbed wire fence. One was badly cut on the shoulder and cannot be used as a wagon horse for two or three weeks. Have been forced to put in one of the saddle horses to pull the load."[36]

With only two crocodile jaws to show for their efforts,[37] Brown returned to Billings to resupply before rolling north to the Musselshell then east along the Big Dry to Hell Creek, finally reaching the same ranch that had served as his base of operations three years earlier on about June 17.[38] There he learned that a competing crew under William Utterbach from Carnegie had visited the region earlier that year, but apparently had not found any fossils to their liking. He also encountered competition from some of the local ranchers, who, having witnessed Brown's determined search for skeletons in 1902, kept an eye out for bones as they roamed their lands. The Sensiba family, for example, had written the museum to offer it a specimen they had found for the "wild" sum of $8,000—almost $175,000 in today's dollars. Brown recommended that contact with them be limited, lest the museum appear "anxious."

But all those complications were quickly forgotten, for work in the quarry was producing unexpectedly rich results:

> Yesterday we struck a concretion containing a large skull bone and today found another.... The large concretion weighs six or eight hundred pounds.... These concretions are at least six feet back in the bank from the ones taken

out in 1902 and I had very little hopes of finding any more but now I will make a fifteen foot cut in the bank which will take at least three weeks with powder and horse power for it is a solid sand bank....

Please... send me 1/2 doz. short heavy chisels of best steel tempered for these hardest concretions also doz. crooked awls without handles.[39]

Osborn, pleased, urged Brown to make a deep cut to search "for more of that animal, every portion of which will be valuable."[40]

At the end of June, on a trip to Jordan, Brown did a little prospecting and found a "shattered" skull of *Triceratops*. Despite the missing horns, he suggested, it could still be prepared at the museum and make a "fine specimen."[41]

But the crew's main focus continued to be the tyrannosaur quarry, where two concretions were uncovered that contained "the other femur... [and] a fragment of femur about the size of the small humerus that we suppose belongs to Deinodon [the informal name they were using at the time for *Tyrannosaurus*]."[42] One block weighed "4,150 pounds when shipped. It took six horses to haul this block from the quarry and four to haul it to [the railhead at] Miles City."[43] Brown went on to advise that the publication describing the new carnivore be delayed, feeling that "the small humerus may belong to another animal."[44]

By July 15, despite the torrid summer heat, the pit had expanded to enormous proportions through the use of dynamite and a horse-drawn scraper. Brown reported:

> Am now at work on the second large cut in the [tyrannosaur] quarry which is 100 ft long, 20 ft deep and 15 ft wide; hard sandstone which has to be blasted before it can be plowed. This is a heavy piece of work but [tyrannosaur] bones are so rare that it is worth the work.
> I have made the important discovery that the large limb bone thought to be a femur is a humerus....
> This will change our idea entirely of the structure of Deinodon.[45]

Osborn responded with skepticism: "Your letter is interesting. I have just described the big dinosaur under the name *Tyrannosaurus rex*. I hardly believe it possible that the humerus you have found belongs to this animal... considering the many points of close resemblance between this animal and *Allosaurus*. I hardly think it possible that it has a long fore arm."[46]

A full three weeks later, Brown and his crew were still struggling in the quarry, and Brown was becoming concerned that time was running short.

FIGURE 15. Hauling a 4,100-pound block containing the pelvis of type *Tyrannosaurus rex* out of Quarry no. 1 at Hell Creek, Montana, during the 1905 AMNH expedition (AMNH Vertebrate Paleontology Archive, 7:2 B3 S45)

The operation had been quite successful, but he was laboring under Osborn's quite reasonable direction to glean every bone possible from the concretionary outcrop. Thinking ahead, Brown began to plan for contingencies. "The work on Tyrannosaurus is being pushed rapidly," he wrote Osborn.

> I have a force of three men besides myself and it is imperative that I supervise this work till completed, that is till this 20 ft cut is taken out and then if the bones continue into the bank as I think they will I shall let a contract to have the upper 20 ft of sand removed this winter for that will be cheaper than keeping an expensive outfit on the ground. There are about 600 cubic yards of sand stone above this cut which will cost 80 [cents] per cu. yd. and this figures out cheaper than removing it myself for labor is high here.
>
> I trust this meets your approval and I should like your opinion immediately. It means a great deal of expense but this is the rarest animal in the [Hell Creek] and I have never seen another fragment of it anywhere.
>
> Up to date I have taken out femur, humerus, scapula ? metapodial, several ribs, skull bones, both lower jaws, illium, and some bones not identified.
>
> There is no question regarding the association of the humerus for the only other animal in this quarry is the little carnivore. . . .
>
> The specimen is entirely distinct from the carnivore secured in Wyoming [in 1900] and will prove a different genus as I remember the shape of the teeth

and femur in that specimen. Moreover I have not found any plates... and the other is a plated dinosaur. I am confident that it is a mistake to combine the remains of the two.[47]

Two weeks later, on August 22, Brown's excavation was nearing completion, since no more bones had been found leading into the cut bank. The team had toiled under the scorching sun for two months. At the time, Osborn was contemplating a trip out to Hell Creek in order to celebrate Brown's success and see the quarry for himself. Although he was happy to accommodate his boss, this raised an unanticipated issue for Brown, and he felt obliged to come clean about the identity of one of his crew members:

> Mrs. Brown did accompany me from my home to camp and has done the cooking for the outfit this summer reducing our living expenses about half.
> I did not discuss the matter with you for it seemed a purely personal matter with me as long as I performed my duty without any added expense on her account and the Museum has certainly been the gainer.[48]

In the end, Osborn was unable to visit the site. Regarding the debate over the tyrannosaur's forelimbs, he reported simply that the "paper has been delayed, so that I have been able to refer to the parts you have recently recovered."[49] All told, the quarry, finally exhausted, had yielded the major bones of the limbs, a few vertebrae and ribs, as well as several cranial elements, including a premaxilla and both lower jaws.

Toward the end of the season, Brown focused on the specimen for sale by the local ranchers. His strategy of feigning disinterest was paying off. He had intentionally not taken time to personally inspect the specimen, citing his "pressing" responsibilities at the tyrannosaur quarry. As a result, Brown reported to Osborn on August 22 that the asking price had precipitously dropped from $8,000 to a "low" $2,000. He was determined, however, to drive an even harder bargain for his boss.

In September, he inspected the specimen firsthand as it lay in the ground and confirmed that it was a hadrosaur, or duckbill. The complete vertebral column was clearly visible, along with one femur and one side of the lower jaw. Brown was confident that the rest of the skull and limbs were also present below the surface, "but I thought it wise not to show up any more bones until I had made the purchase."[50] His final round of negotiations reduced the price to $200 (about $4,350 in modern currency), with an additional $50 due at the start of excavation, which he planned for his next trip. In the meantime, the specimen would be buried under sediment to protect it from discovery and weather-related damage. What turned out to be a nearly complete duckbill skeleton was therefore secured for about 3 percent of the original asking price. Brown was not only Osborn's best dinosaur collector but also his best purchasing agent.

In addition to the tyrannosaur and duckbill skeletons, prospecting had identified several poorly preserved skulls and one nice forelimb of *Triceratops,* a lower jaw and hind limb of a hadrosaur, and, north of the Missouri, another hind leg of *Tyrannosaurus*. In all, the collection comprised twenty-one large boxes of fossils. Total trip expenses totaled $1,557, just short of $34,000 in today's currency.

When not occupied with excavation work and prospecting, Brown had been carefully noting the sequence of rock layers, or stratigraphy, of the exposures throughout the region. In his last letter of the season to Osborn, he revealed the presence of an important stratigraphic pattern: "I am fully convinced after several years of work that the Lignite [coal-bearing] beds are separate and distinct from the Ceratops [or dinosaur-bearing beds]. I have yet to find a dinosaur bone in the lignite beds."[51] Not only did this observation help Barnum and his crews find dinosaur fossils, but the boundary between these layers would also prove to mark a momentous milestone in the evolution of life: the Cretaceous-Tertiary extinction event marking the end of the Age of Dinosaurs 65 million years ago.

By the end of 1905, Osborn had published his paper naming and describing *Tyrannosaurus rex,* with Brown's AMNH 973 designated as the type specimen. Thanks to Brown's return to Quarry 1, the skeleton now included "the jaws, portions of the skull, vertebrae, shoulder girdle [including the tiny humerus], abdominal ribs, pelvis, and hind limbs."[52] With the formal publication complete, the existence of the "Tyrant King of the Dinosaurs" could be announced to the public—and so it was, in grand fashion. On Sunday, December 3, 1905, a full-page article appeared in the *New York Times*'s first magazine section. Both Osborn and Brown were prominently quoted. For Osborn, the coverage simply added to his already considerable notoriety, but for Brown and *Tyrannosaurus,* it was a debut destined to engender fame in both scientific and public circles. No doubt, Brown's banner year served as a salve for the wounds he had suffered during his aborted academic career at Columbia. It also placed the young collector firmly back in Osborn's good graces. For as Osborn pointed out to his young new star, "The results of our season elsewhere have been very disappointing."[53]

SEVEN

Loss

(1906–1910)

BROWN RETURNED TO HELL CREEK in the summer of 1906 to collect the hadrosaur the museum had purchased the previous year. It was clearly Brown's desire to bring Marion again to serve as cook, along with his assistant Peter Kaisen. Kaisen did come, but there was a glitch with Marion's participation. Brown had raised the matter with Osborn shortly before his departure, but apparently someone on the AMNH staff complained about wives accompanying crew members into the field. Before granting permission, therefore, Osborn wanted to check with the museum's director, Hermon C. Bumpus.[1] In the meantime, Brown and Kaisen set out.

Documents reveal that the skeleton, sometimes referred to as "Sensiba's mule," had a complex history of ownership. It had been discovered by two locals, Oscar Hunter and Gus Colan, in perfect condition, with the dorsal spines on the vertebrae exposed, but Hunter and Colan "investigated the find by kicking off everything in sight."[2] Eventually, the Sensiba brothers traded a "six shooter" for the specimen and recovered most of the missing fragments. It took Brown's crew three weeks to excavate and collect the skeleton, which included most of the backbone, one femur, and most of the skull and lower jaws. Today the majestic skeleton resides, along with its "mate," in the renovated Hall of Ornithischian Dinosaurs.

Regarding Marion, Osborn diplomatically wrote to Brown on July 17 that he "needed to take all our parties into consideration and be prepared to say to each of our field explorers that they may or may not be accompanied . . . by their wives." With Bumpus and Brown's senior colleague W. D. Matthew concurring, Osborn concluded: "It does not seem to be in the best interests of the American Museum . . . that the explorers if married should be accompanied by their wives"—though he added that "I will leave that decision to Mrs. Brown and yourself."[3] Brown responded with equal tact:

I am gratified... to have you leave the matter of Mrs. Brown's presence in camp to our discretion. The whole matter resolves itself into my fitness to conduct this work, and that must be determined by the results of my efforts. It is not a question of expense to the Museum, for it is receiving her services free which if done by a man would cost forty-five to fifty dollars per month. Regarding the work done, the Museum receives eleven to fourteen hours each day whether Mrs. Brown is here or not and my time at least is undivided. So far as I can see, it is a matter for us to decide. I should not like to leave the Museum but would certainly rather than [be] separated from my wife from two to five months every year.[4]

There is no correspondence to document whether she joined the crew that season, but the work clearly continued. Further prospecting yielded two other hadrosaur specimens, including one with skin impressions on the tail "which will give us an accurate knowledge of this creature's external appearance."[5] Also discovered were a soft-shelled turtle, ornithomimid limb bones, a "new" ceratopsian, a crocodile skull, mammal teeth and bones, and a collection of Tertiary leaves.

That summer, Osborn published a scientific article in which he revised some of his conclusions about *Tyrannosaurus*. Again serving as sole author, he nevertheless acknowledged that Brown "cooperated with me in all the details of description and measurement." Brown's further efforts at Quarry 1, moreover, meant that the type specimen (AMNH 973) now comprised "the jaws, portions of the skull, vertebrae, shoulder girdle, abdominal ribs, pelvis and hind limbs."[6] One troubling aspect of the tyrannosaur skeleton, he noted, was the proportionally miniscule arm bones: "The humerus is so small that grave doubts were entertained as to its association with this animal. These were finally set aside for three reasons: (1) the humerus is hollow, proving that it belonged to one of the Theropoda; (2) the head of the humerus fits into the glenoid cavity [shoulder socket] of the scapula; (3) while absurdly reduced as compared with the femur... it nevertheless is provided with very stout muscular attachments... which proves that it served some function, possibly that of a grasping organ in copulation."[7] Today, paleontologists still argue about how the arms of *Tyrannosaurus* functioned, although Osborn's scenario regarding a role in copulation is no longer seriously invoked.

Osborn further argued that the *Tyrannosaurus* skeleton, AMNH 973, and the *Dynamosaurus* skeleton, AMNH 5866, belonged to the same genus,[8] which by rules governing biological names would be *Tyrannosaurus*, since he had described *Tyrannosaurus* first. Thus the name *Dynamosaurus* became invalid. Another continuing problem involved the bony scutes of armor that Brown had found with 5866. Because the sandy, channel-fill deposit that contained AMNH 5866 also contained a fragment of *Triceratops* frill and a partial hadrosaur jaw, the association of the armor plates with 5866 could not be established with certainty. This kind of problem often arises when specimens are discovered in ancient stream channels, where the bones were jumbled by the water currents before becoming buried and fossilized. Given this complication, Osborn still held out the possibility that the two specimens might represent different species, though he was skeptical that the armor plates found with 5866 actually belonged to that animal, concluding: "It is difficult to imagine why this carnivorous Dinosaur should be protected by any form of dermal armature, unless possibly against attacks by members of its own family."[9] To be sure, many parts of *Tyrannosaurus* remained unknown, including a complete picture of what the skull looked like.

In October, at the end of the 1906 season, Brown told Osborn: "I have material for a geological paper on the Montana Laramie [rock unit] which I want to publish [when] I return. This will give the results of my three years of work up there with a sketch map of the region which is badly needed for all existing maps are so faulty that towns and prominent Buttes are thirty and forty miles out of position."[10]

To cap off the 1906 holiday season and ring in 1907, AMNH gave the people of New York and the world a New Year's present. On December 30, trumpeted by a nearly full-page article in the *New York Times*, *Tyrannosaurus rex* made its public debut in the museum's Hall of Dinosaurs. However, only a part of the animal's skeleton was revealed: a photograph in the article shows a man standing dwarfed between the skeleton's huge hind legs beneath its massive hips. For those of us familiar with the present-day mount, which includes all of the animal's bones, this image of the headless and tailless icon seems quite odd. Nonetheless, it represented the first step in bringing what would become the world's most famous dinosaur out of the shadows of deep geologic time and into the glare of our modern global stage. Sadly, this momentous event was undoubtedly tempered for Brown when he learned of the death of his seventy-four-year-old father in early January.

Although Brown had done a reconnaissance in 1906 farther east from Hell Creek along the Big Dry and south of Miles City to evaluate possibilities for the following year, there would be no field season in 1907. For the first time in a decade, Brown stayed in New York all year. The first seven months he spent working primarily on the two *Anatosaurus* mounts for exhibition,[11] and he caught up on research, writing articles on the Conrad Fissure collection and a review of the ankylosaur family. It is also probable that Brown took a break from fieldwork because he wanted to spend more time with Marion, who was pregnant with their first and only child. Their daughter, Frances, was born on January 2, 1908.

Brown also followed through on his commitment to publish a paper on the stratigraphy of his field site in the Missouri Breaks, noting in his year-end report that "the last two weeks in April were given to preparation of [the] 'Hell Creek Beds' bulletin, 22 pages, published in October."[12] In this article he not only coined a new name for the latest Cretaceous, dinosaur-bearing sediments, which underlie the earliest Tertiary, coal-bearing sediments; he also, more importantly, set the foundation for understanding the transition between the end of the Age of Dinosaurs, or Mesozoic Era, and the beginning of the Age of Mammals, or Cenozoic Era. During the rest of the twentieth century, this sequence of rock layers produced the most complete terrestrially based paleontological record of the faunal and floral changes that marked the extinction of nonavian dinosaurs and the eventual rise of our own mammalian dynasty. The Hell Creek and Tullock Formations, as they are now called, also produced key evidence related to the hypothesis, first proposed by the Alvarez team in 1980, that the impact of a large asteroid or comet played a role in that episode of extinction.

The Alvarez hypothesis was based originally on fossil records preserved in marine rocks, especially in Italy. The key piece of evidence was an unusually high concentration of iridium preserved in the thin clay layer that separates underlying Cretaceous limestones from the overlying Tertiary limestones. Iridium is rare in most rocks of the Earth's crust, because it tended to filter down toward the mantle and core as the planet cooled and consolidated. Asteroids and comets, however, which more closely reflect the original bulk chemical composition of the solar system, have relatively high concentrations of iridium. The Alvarez team therefore interpreted the high concentration of iridium in the marine boundary clay as representing fallout from the impact of an asteroid.

But the marine rocks that triggered the impact hypothesis did not contain dinosaurs. Thus, Brown's rock layers of the Hell Creek and Tullock Formations became a scientific touchstone for investigating dinosaur extinction. For the last three decades, scientists including paleontologists and geologists have vigorously debated whether the Alvarez hypothesis represents the best or only explanation for the extinction of nonavian dino-

FIGURE 16. Portrait of Frances Brown, only child of Barnum and Marion Brown, from the 1929 yearbook of Wells College (Wells College Archives, Long Library, Aurora, New York)

saurs 65 million years ago.[13] Field crews from the University of California Berkeley were among the first to document the presence of high levels of iridium at the Cretaceous-Tertiary boundary between the Hell Creek and Tullock Formations. Further evidence of the impact has since been found at Hell Creek, including fractured mineral grains that were blasted out of the Chicxulub crater when the asteroid or comet hit along the coast of what is now Yucatán.

Although nonpaleontologists tend to favor the impact scenario for explaining the mass extinction at the end of the Cretaceous, most vertebrate paleontologists that have worked on the problem maintain that other momentous and contemporaneous geologic events also played a role. The massive eruptions of flood basalts that formed the Deccan Traps in India, for example, began a few million years before the impact and reached their peak

FIGURE 17. Blasting overburden covering the bone layer in the quarry at Big Dry Creek, Montana, which in 1908 yielded the specimen of *Tyrannosaurus rex* mounted at the American Museum of Natural History (AMNH 5027) (AMNH Vertebrate Paleontology Archive, 7:3 B4 S39)

at the end of the Cretaceous, just before the impact occurred. These eruptions could have produced many of the same "killing mechanisms": acid rain and short-term cooling, followed by long-term greenhouse warming, all due to the release of particulates and toxic gases. The retreat of shallow seas off the continents and back into the major ocean basins may also have altered the climate enough to kill off the dinosaurs. The debate goes on, and every year paleontologists return to Brown's outcrops at Hell Creek and other places around the world to seek new evidence regarding the extinction of nonavian dinosaurs.[14]

Back in 1908, both Osborn and Brown were determined to fill the gaps in their knowledge of *Tyrannosaurus*. Another expedition was planned, therefore, in which Brown would move his Hell Creek operation about thirty miles to the east along the Big Dry, where he had found a couple of prospects during his 1906 reconnaissance.

Brown and Kaisen left New York in June, and although the first few reports back to Osborn were not encouraging, the two prospectors pursued their quarry with determination. It was not all work and no play, however, as Brown's field notebook makes clear; the entry for June 21, for example, describes a get-together with local ranch families: "All were invited to dinner by Mrs. Twitchell. Bessie Willis . . . rode in for dinner and we had a lively time. Bessie is belle of the Dry and a lively lass. We planned for a celebration here on the fourth of July."[15]

On July 1, after evaluating a hadrosaur specimen he had targeted in 1906 and deciding not to collect it, Brown discovered fifteen tail vertebrae and other bones from an animal he could not immediately identify. The vertebrae, he noted, were "somewhat crushed laterally but are too long for either [a hadrosaur] or Triceratops."[16] Brown decided to move camp near the new specimen, and on July 3 he matter-of-factly stated in his notebook: "Found Tyrannosaurus lower jaw and back of skull near one of the buttes. Will take it."[17] At last he knew he was on the right track, and the timing was perfect, with Brown's favorite field holiday just around the corner.

It was time to celebrate Independence Day with the rest of the nation. As the instigator of this particular extravaganza and the only unmarried female in the region, "Bess Willis rode all over the country collecting [phonograph] records."[18] Barnum, as one might expect, was determined to hold up his own end of the bargain at Twitchell's party:

> This was the day of the great celebration on the Dry. About 11 o'clock people began to come in buckboards and springwagons decorated with flags. There were about 30 all told and ten babies. . . . Everything was placed on the table and people took plates and ate in various parts of the house like a regular social. Had chicken, all kinds of cake, salad and ice cream and lemonade. About three o'clock some one suggested that we . . . start the dance so we cleared the dining room and started the phonograph. I took pictures. . . . In the evening we put up a swing and had an early supper after which dancing

FIGURE 18. Celebrants at the Fourth of July party held at the Twitchell ranch in 1908. Bess Willis is third from the right in the back row, in front of the American flag. (5:3 B2, American Museum of Natural History Library)

began in earnest. The children dropped off to sleep one by one and were stored on the beds and floor of the third room. There were not enough beds for all the women so I brought most of mine up from the camp while I took the horse blankets and extra tarp. I danced till twelve finding some very good partners and one bad one. Some of the people danced all night till seven o'clock Sunday morning then had breakfast and drove to their homes.

Bess Willis went all the way through, for it was her party.[19]

Parties such as this, along with socializing in the local bars, often serve an important paleontological purpose. Quite commonly, conversations provide leads to dinosaur bones that ranchers have seen weathering out of the ground, and a round or two of drinks can easily prompt an invitation to come out and have a look.

On July 8, Brown wrote Osborn that the vertebrae found on July 1 looked promising: "At last I have some good news to report. I made a ten strike last week finding fifteen [tail vertebrae] connected; running into soft sand. We have made a six foot cut with pick and shovel and found bones continued in sandstone concretions so have moved camp to the specimen and borrowed scraper and plow for a big cutting."[20]

Fortunately, this specimen was not located in as precarious a setting as the original tyrannosaur quarry, and work progressed quickly. Thanks to Kaisen's assistance, Brown also had more time to prospect for other specimens. Yet despite the good news, all was not going smoothly in camp. Finding a trustworthy cook, for example, had proved a challenge. In his letter of July 8, Brown reported that he had hired a good man named Johnson to be the cook and teamster, replacing a hired hand from Miles City who had "found the work too heavy when it came to the pick and shovel so he left us last week."[21]

Although the bones in the new quarry were partially embedded in sandstone, the rock was easily chipped away, and within a week enough diagnostic parts were exposed for Brown to make a positive identification. On July 15, he announced to Osborn: "Our new animal turns out to be a Tyrannosaurus.... We have the complete vertebral column except the tip of tail, skull and lower jaws, complete pelvic girdle in position and the [neck] ribs.... So far no limbs have appeared.... The bones are in a good state of preservation."[22]

At long last, Osborn and Brown could revel in having almost the whole animal, for this specimen, unlike the 1902 skeleton, included the animal's four-foot-long skull, its jaws studded with six-inch-long, serrated teeth. In his July 30 response, Osborn was ecstatic: "Your letter of July 15, makes me feel like a prophet and the son of a prophet, as I felt that you would surely find a *Tyrannosaurus* this season.... I congratulate you with all my heart on this splendid discovery.... I am keeping very quiet about this discovery because I do not want to see a rush into the country where you are working."[23] Celebration was in order, and on the 21st Bessie threw another party for Barnum, again featuring chicken and ice cream.[24]

Meanwhile, although the expedition's scientific agenda was moving rapidly toward a successful conclusion, the domestic arrangements within the camp remained far from ideal. Retaining a reliable cook continued to bedevil Brown, as he lamented to Osborn on August 1: "I had everything running well in camp but our cook drove the team in yesterday and today he was so drunk that I let him go. Was sorry to lose him for he was a good cook and a good worker but I won't have a man in camp that I cannot trust to [drive to] town with the bones. Hope we may have someone who can cook without burning the water when you come out."[25] Osborn, who had mentioned that

FIGURE 19. The skull (center right) and vertebral column (extending off to the lower left) of the 1908 *Tyrannosaurus rex* skeleton (AMNH 5027) as it was found in the quarry along Big Dry Creek, Montana (AMNH Vertebrate Paleontology Archive, 7:3 B4 S34, and American Museum of Natural History Library)

he was contemplating a visit to the site to see the marvelous specimen for himself, playfully responded that he hoped Brown could find a good cook because he knew that he would "enjoy the badlands much more if the water was not burnt."[26]

On August 10, in anticipation of Osborn's arrival, Brown confidently stated: "I am sure you will be more pleased with our new Tyrannosaurus when you see what a magnificent specimen it is. The skull alone is worth the summer's work for it is perfect. We finished boxing it and moving it out of the quarry Saturday. This block weighed about 3000 pounds while the lower jaws weigh about 1000 lbs."[27]

Osborn arrived on August 26 and spent about a week examining the quarry and prospecting with Brown. Their biggest problem involved getting the heavy casts containing the tyrannosaur bones out of the quarry and across to the north bank of the Missouri. There, the nearest rail link was located in Glasgow, northwest of Fort Peck where General Miles and Sitting Bull had obtained their supplies—forty-five miles from Brown's camp. The transport would require several extra men and horse-drawn wagons, because Brown wanted to ship the whole collection at one time and load it on a train going straight through to New York; that way, the boxes would only have to be handled once. Most of September and the first week of October were spent building boxes for the bone-filled casts and packing the specimens in crates. The weather was quickly deteriorating, however, and by the last week of September storms of rain and snow began to complicate Brown's efforts. Getting the boxes from the quarry to the Willis ranch, which Brown was using as his base of operations, proved to be the weak link because there were no established trails through that section of badlands. Twice the wagon got so stuck that Brown had to abandon it temporarily in order to round up extra horses and pull it out. But finally, on October 8, Brown was able to depart for Glasgow with a convoy carrying the five loads of fossils pulled by sixteen horses, covering the forty-five miles without incident by the end of the next day. On October 10, the boxes were loaded on the train, which began the journey to New York on the 12th, where they safely arrived.[28] The costs of this expedition to secure the first well-preserved skull of *Tyrannosaurus* totaled $1,187, or almost $25,000 in modern currency.[29] In addition to the tyrannosaur skeleton with the skull, Brown and Kaisen found the back of the skull for another tyrannosaur, as well as the skull, jaws, and shoulder of *Triceratops*, foot bones of an ornithomimid, and a predentate (ornithischian) dinosaur.

After examining the three most complete specimens of *Tyrannosaurus* (AMNH 973, 5866, and 5027), along with some other, more fragmentary specimens, Osborn once again revised his thinking concerning the genus. He published his conclusions in 1912, stating that all three specimens represented *Tyrannosaurus rex*, which remains the accepted interpretation today.[30] Although Brown did not appear as an author on any of Osborn's publications, as we've seen, Osborn typically went to great lengths to be sensitive and evenhanded with his assistant. Within the context of the tyrannosaur papers, no documents exist to suggest that Brown complained about not being included as an author. Indeed, Osborn clearly acknowledged the efforts of Brown, Matthew, Lull, and even the preparators in all of his papers on *Tyrannosaurus*. Apparently, that approach represented the agreement between Osborn and his subordinates. As Rainger notes, "The department contributed indirectly to much of Osborn's work in vertebrate paleontology. While Osborn examined specimens and constructed classifications and phylogenies for many families of fossil vertebrates, staff members did the 'dirty work' of the science. Wortman, Peterson, Brown, and in later years

FIGURE 20. Brown's crew using block and tackle to crate the 4,000-pound cast containing the pelvis of the 1908 *Tyrannosaurus rex* specimen (AMNH 5027), Big Dry Creek, Montana (18341, American Museum of Natural History Library)

Albert Thomson and Peter Kaisen collected, cleaned, and prepared specimens that were the basis for Osborn's descriptive and taxonomic studies. Other members of the staff were frequently called on to aid in the production of Osborn's publications. Matthew was generally the middleman in those arrangements."[31] In the end, although Brown was not a coauthor on the *Tyrannosaurus* articles, we do know that he went on to publish not only his geological research, but also other important research about nonavian dinosaurs and other vertebrates.

This whole sequence of discoveries and studies makes the questions "Who discovered *Tyrannosaurus rex*?" and "When was *Tyrannosaurus rex* first discovered?" somewhat complicated to answer. Clearly the holotype, or original bearer of the name *Tyrannosaurus rex,* is the 1902 specimen, AMNH 973. However, two other specimens were discovered earlier, and only later assigned to that genus and species. One, as we've already noted, is AMNH 5866, which Brown found in Wyoming in 1900 and Osborn originally named *Dynamosaurus.* This specimen was transferred to London's British Museum of Natural History in 1960. The other was part of a vertebra found by Cope's collectors and described by Cope in 1892 under the name *Manospondylus gigas*. It now resides at the American Museum of Natural History (AMNH 3982). Osborn reassigned that specimen to *Tyrannosaurus rex* in his 1916 study.[32] One other possible tyrannosaur specimen that was collected earlier than any mentioned above recently came to light in 2003. It is an isolated tooth (YPM-VP 4192) discovered in 1874 by Arthur Lakes in the Denver Formation in Colorado, which he sent to O. C. Marsh, who placed it in Yale's collection. Marsh did not publish on the specimen, perhaps because the teeth of large theropods are not terribly distinctive, so he did not recognize it as a new animal. However, based on its size and the beds in which it was found, it may well be the first specimen of *Tyrannosaurus* ever collected.

Back in New York, Osborn and Brown contemplated how best to mount the two most complete specimens of *Tyrannosaurus*, AMNH 973 and 5027, for exhibition.[33] Osborn instructed a departmental artist, E. S. Christman, to sculpt a scale model of every bone in the animal's skeleton connected with flexible joints, to facilitate the evaluation of various possible poses and postures. Raymond L. Ditmars, the Bronx Zoo's curator of reptiles, won the contest with his proposal for the poses. Brown set the scene thus: "It is early morning along the shore of a Cretaceous lake four million years ago." (We now know, thanks to radioisotopic dating techniques unavailable in Brown's day, that 65 million years ago is more accurate.)

> A herbivorous dinosaur *Trachodon* [a duckbill] venturing from the water for a breakfast of succulent vegetation has been caught and partly devoured by a giant flesh eating *Tyrannosaurus*. As this monster crouches over the carcass, busy dismembering it, another *Tyrannosaurus* is attracted to the scene. Approaching, it rises nearly to its full height to grapple the more fortunate hunter and dispute the prey. The crouching figure reluctantly stops eating and accepts the challenge, partly rising to spring on its adversary. The psychological moment of tense inertia before the combat was chosen to best show positions of the limbs and bodies, as well as to picture an incident in the life history of these giant reptiles.[34]

Unfortunately, the skeletons were too large to fit both in the existing exhibition hall, so in 1915 a single skeleton (AMNH 5027) was mounted in the now famous erect or "Godzilla" posture, a portrayal that would wow visitors from around the world for the next eighty years and fire the curiosity of numerous future paleontologists.

Yet the perils surrounding these *Tyrannosaurus* specimens were not over. At the outbreak of World War II, the American Museum of Natural History sold the 1902 skeleton to the Carnegie Museum in Pittsburgh for $7,000 (about $96,000 in today's dollars) after Yale failed to respond to an earlier proposition.[35] Brown noted the sale in a memoir titled "Discovery, Excavation, and Preparation of the Type Specimen *Tyrannosaurus rex* (AMNH No. 973). Discovered 1902, Completely Excavated 1905": "Sold to Carnegie Museum 1941 . . . after we had made casts of the limb bones. The transaction was accomplished because the American Museum was afraid that German airships might bomb this [the American] Museum and destroy the second *Tyrannosaurus rex* skeleton now mounted here [AMNH 5027] and that at least one specimen might be preserved."[36] (See Appendix 2 for full transcript.)

Fortunately, both skeletons survived. During the renovation of the fossil halls in the 1990s, we remounted the 1908 *T. rex* skeleton to reflect a more anatomically accurate posture. It was a daunting assignment, since each bone had to be removed from the old upright mount, conserved, and remounted in the new, more animated posture prescribed by recent research. It took two years to accomplish, a period replete with unremitting worries over the welfare of this priceless specimen. But our crew did a spectacular job, and today Brown's skeleton stands ready to pounce on prey.

Brown returned to the Hell Creek region in 1909 with Kaisen to finish prospecting in the region and to collect a *Triceratops* skull found along Rock Creek at the end of the 1908 season.[37] There is no mention of Marion accompanying them (Frances was only one and a half years old), and Kaisen is reported to have hired a cook.

Kaisen found a fairly complete postcranial skeleton of *Triceratops* along Sand Arroyo, which, together with the skull and other specimens found previously, encouraged Brown to think that a composite skeleton of the animal could at last be mounted for exhibition.[38] Ironically, *Triceratops,* the dinosaur that triggered Brown's expeditions to Wyoming and Montana and which some now call "the cattle of the Cretaceous" due to the abundance of its partial remains, proved to be much more elusive for Brown than its rarer nemesis, *Tyrannosaurus.* Kaisen and Brown also discovered another specimen of a new "Orthopodus dinosaur," thought to be the same as the one found in 1906, along with a skeleton of the crocodile-like *Champsosaurus.*

Toward the end of the season, Barnum crossed the Missouri and surveyed exposures of the Hell Creek Formation as far west as the "Carb Hills" at UL Bend, opposite the mouths of Squaw Creek and the Musselshell.

Although "many isolated bones were found," he actually had better luck in the underlying marine Bearpaw Shale, finding "some very interesting fossil Crustaceans..., securing several genera that are new."[39]

After closing the season in Montana in late September and shipping twenty-one boxes of fossils to New York, Brown set off for Alberta to evaluate prospects for future work. Earlier work by Canadian geologists had established that dinosaur fossils were present in the region, so both Osborn and Brown were interested in evaluating the area's potential for producing new Cretaceous dinosaurs. Along the way, at Havre, Montana, he conducted a brief reconnaissance of the Judith River Formation exposures along the Milk River, then proceeded on to Didsbury, north of Calgary. Hiring a team, Barnum drove east to the Red Deer River, where he was wowed by large expanses of Late Cretaceous badlands littered with dinosaur bone exposed in the remote canyon's walls.[40] His field agenda for the immediate future set, he returned to Oxford, New York, Marion's hometown, for a well-deserved week of vacation.

It had been an incredible decade of discovery for Brown, but in the end, tragedy loomed in the shadows of success. As his daughter recounts, by 1910 Marion had returned to substitute at Erasmus Hall near Prospect Park:

> One day in the early spring..., Marion wheeled the baby over to the park as she often did on nice days. Somewhere, somehow virulent scarlet fever germs were floating around which both mother and baby picked up. The baby, Frances, developed the illness first, but Marion came down with it so soon that it was obvious that both had taken it from the same source. The baby was desperately ill but eventually recovered. Marion died at the end of five dreadful days. In that dreadful time, Professor Osborn... told Barnum to call in specialists from anywhere in the country or in the world at the museum's expense, just so Marion could be spared. But in those preantibiotic days, even specialists could only watch helplessly while this beautiful, gifted, needed young woman gasped her last breath on April 9, 1910.
>
> Marion's shocked and grief-stricken parents, both then in their middle sixties, told Barnum that they would take the baby and raise her, that they could not have anyone else have all that was left to them of Marion. Barnum, torn with grief and anger at cruel Fate, agreed that that would be the best solution for the daughter he now, perhaps subconsciously, blamed for his wife's untimely death. Long afterwards, as a mature woman, Frances could understand how her father reasoned in that desperate time: a daughter was expendable; a wife was not.[41]

There is no evidence, other than Frances's account, that Barnum actually felt that his daughter was responsible for his wife's death. What is documented is that, with characteristic stoicism and steadfastness, a deeply appreciative but gravely grieving Brown acknowledged Osborn's sympathy in a poignant letter that must have cemented their close and respectful personal bond throughout the rest of their lives:

> I want to thank you for the efforts to help keep my beloved wife alive, your kind letters of sympathy and financial aid—all expressions of truest friendship for which I cannot find words to express my appreciation.
> I shall always be most deeply grateful.
> The baby continues to improve but still shows kidney complication. Her grandmother is anxious to have her at home so I have planned to go with them to Oxford on Thursday May 12th.
> The funeral service will be Tuesday May 17th at 4 o'clock and I shall return to my duties Wednesday.[42]

Although we have no other letters or notes in which Barnum expresses his grief regarding Marion's death, there can be little doubt that he had lost the love of his life. Even though the couple had scant opportunity to share experiences, it is clear that he relished her company, especially within the context of fieldwork, an activity that he had always and would always value above all other professional and most personal pursuits. The few photos we have of Marion in the field portray her with either a radiant smile or a gaze intensely focused on her immediate task. That she exulted in being in the field with Barnum is obvious from the entries in her logbook. That he adored her dexterity in the field is evident in his persistent advocacy to his superiors of her right to join him on his expeditions, punctuated by his bold declaration that "I should not like to leave the Museum but would certainly rather than [be] separated from my wife from two to five months every year." Overall, their natural abilities were remarkably symbiotic. She was an accomplished student and graduate from two prestigious universities with a natural interest in biological sciences. Barnum "told his daughter more than once," Frances later wrote, "that it was her mother, Marion, who had the brilliant mind, not her father. He agreed that his was a good mind but that Marion's was far better. Both were fine scientists." Frances then revealed another tragic incident that the couple had endured together: "Had the male embryo that Marion lost in a fall down a flight of stairs at their boarding house been born and lived, he might have inherited [those] scientific genes."[43] But they were different, too. Marion had grown up on the East Coast, far from the wonders of the fossil-rich, western frontier, and it was Barnum's great joy to introduce her to his world, a gift she valued deeply.

Barnum would not remarry for almost fifteen years. In that time, Frances rarely saw him as he wandered the world in search of fossils. Marion's death and Barnum's flight to the field to seek solace left scars on both father and daughter that would take decades to heal.

EIGHT

The Canadian Dinosaur Bone Rush
(1910–1916)

WITH FRANCES SAFELY ENSCONCED in the care of Marion's parents, Brown was, as his daughter later put it, "free to seek numbness from pain by throwing himself into the hardest work he could find."[1] He now had set his sights on a poorly known fossil field in Alberta, where a few dinosaur fossils had been discovered by Canadian geologists surveying for mineral and other natural resources; otherwise, little was known scientifically about this remote and rugged region. The territory had been of great interest to Lewis and Clark, who hoped to discover tributaries flowing south into the Missouri River that would extend the boundaries of the Louisiana Purchase, providing access to the fertile prairies and fur-rich mountains of the northern Rockies. As it turned out, such rivers did not exist, and the area remained isolated and unexplored for more than a century.

The first dinosaur bones collected in the Red Deer River Valley were found by a surveying party led by the Canadian geologist Joseph B. Tyrrell in 1884.[2] In 1888 and 1889, Canadian geologist Thomas C. Weston and a local rancher fashioned a river-running rowboat to more efficiently prospect for dinosaur bones, but, despite an abundance of bones, collections had to be relatively small due to the small size of the boat. Another successor in the Canadian Geological Survey, Lawrence Lambe, followed in Weston's path down the Red Deer in 1897 and 1898. He discovered more specimens, which he researched and illustrated in collaboration with Osborn.[3] Their studies culminated in the publication of two companion papers in 1902.[4] In his paper, Osborn designated the Belly River Series as the basal terrestrial rock unit in the local sequence. Above it lay the marine Pierre–Fox Hills Group. Next came the Edmonton Formation, another Late Cretaceous terrestrial unit. Capping the section was the terrestrial Paskapoo, which lacked dinosaurs and was considered to be Early Tertiary in age. Thus, two different Late Cretaceous rock units were identified that contained different faunas of dinosaurs, the older Belly River Series below and the younger Edmonton Formation above. In subsequent years, further subdivisions and changes in nomenclature were made.[5] Today, the Belly River Series in the Red Deer region is called the Belly River Group. It has been split into three formations: the Foremost at the base, the Oldman in the middle, and the Dinosaur Park at the top. They range in age from about 79 to 75 million years old. The overlying marine unit, which Osborn called the Pierre, is now called the Bearpaw Formation. The Edmonton Group, which overlies the Bearpaw, has been divided into four formations; from lowest and oldest to highest and youngest, they are the Horseshoe Canyon, Whitemud, Battle, and Scollard, and they range in age from about 72 to 65 million years old.

The dinosaurs from both the Belly River Series and the Edmonton Formation were recognized to be slightly older than those Brown had collected in the Hell Creek Formation in Montana and the Lance Formation in Wyoming.

It was with this rudimentary knowledge that Osborn and Brown set their sights on the Red Deer region at the end of the first decade of the twentieth century. These dinosaurs would fill a gap, both geographically and chronologically, in the museum's collection between the Late Jurassic specimens from the Morrison Formation and the Latest Cretaceous from the Hell Creek and Lance Formations. Osborn and Brown were confident that new genera and species of dinosaurs could be discovered through a series of larger and longer expeditions than Lambe and his predecessors had been able to mount. Osborn would once again seek to utilize whatever skeletons Brown found for both research and exhibition, the goal being to promote both scientific knowledge and public literacy about evolutionary history.

In 1909, the interest of Osborn and Brown was further piqued when a rancher from the area visited the AMNH and saw the impressive displays of dinosaur skeletons mounted in the museum's public exhibition halls. He met with Brown and other museum officials to tell them that similar large fossil bones littered the canyon walls on his ranch along the Red Deer River. Proffered an invitation to visit by the rancher, Brown, with Osborn's support, quickly organized a reconnaissance tour. Then, in the spring of 1910, he and Peter Kaisen set out on the first of several major expeditions to see what they could find. These endeavors would trigger what has been called the Canadian Bone Rush, involving field crews from both the United States and Canada,

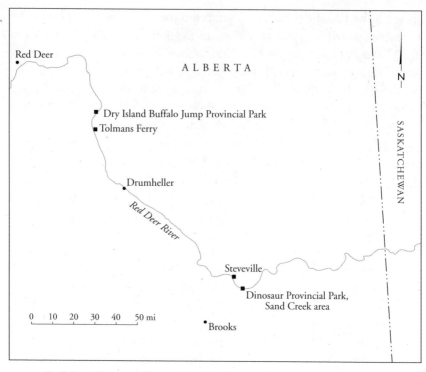

MAP 3. Red Deer River in Alberta, Canada, showing fossil sites where Brown collected between 1910 and 1915.

which continued throughout the teens of the twentieth century and involved dozens of paleontologists and collectors.

The contributions of Kaisen to Brown's enterprise on the Red Deer are difficult to overestimate. In addition to discovering numerous specimens, Kaisen played a primary role in their excavation and preparation. He often managed the operations during times when Brown was occupied elsewhere, either professionally or socially. Happily, Kaisen set out a colorful record of the crew's activities in a diary that he faithfully wrote in every day of his long field seasons between 1910 and 1914. In 2010, paleontologists led by the Royal Tyrrell Museum's Darren Tanke will retrace the route that Brown's crews followed, and the flatboat that they will run the river on will be named in honor of Peter Kaisen.

Two other field assistants merit a general mention for their efforts during the Canadian Dinosaur Bone Rush. William E. Cutler logged four summer field seasons, including one for Brown's crew. He eventually participated in expeditions for the British Museum to the famous dinosaur locality of Tendaguru in Tanzania, where he succumbed to malaria at the age of forty-two. Clayton Sumner Price parlayed his experiences along the Red Deer into a distinguished career in art, eventually settling in Oregon where he became an influential modernist painter.

The steeply sloping bluffs and badlands of Red Deer Canyon are visually arresting, with layers of beige and gray sandstones and mudstones punctuated by starkly dark beds of ironstone, black lignite, and coal. But in contrast to Montana's Missouri Breaks, the meandering Red Deer and its tributaries have yet to scour down as deeply into the ancient sediments of the Late Cretaceous floodplain and seafloor that form today's grassy plains. Standing less than a mile away from the canyon, one might not even guess that any break in the landscape exists. Yet upon reaching the precipice, magnificent panoramas materialize, tied together by the turbid ribbon of the river.

In 1910, roads in the Red Deer region were either nonexistent or impassable, and motor vehicles were rare. Brown therefore devised a new way to prospect and collect the dinosaur skeletons exposed on the river's steep bluffs. Taking inspiration from Weston and Lambe, as well as Lewis and Clark, the crew planned to construct a flat-bottomed boat four yards wide and ten yards long. A tent pitched on deck would serve as the kitchen, complete with a sheet-iron stove, its smokestack stuck through the top of the tent. Two large oars, twenty-two feet in length, were attached to the front and back of the boat for steering. The crew would float down the river, stopping to prospect whenever promising outcrops came into view. The fossils would be stored on the boat until the end of the season's trip, when they would be unloaded and shipped by train back to New York. In essence, Brown had devised a mobile field camp, which could be moved at will downriver as circumstances demanded.

The expedition began with Kaisen traveling out to the Willis Ranch east of Jordan to collect the hadrosaur found at the end of the previous season, while Brown took the train to Ottawa to survey the collections at the Victoria Memorial Museum. He was warmly received, reporting that "officials of the Canadian [Geological] Survey extended every courtesy, giving maps, publications, and much data valuable for future work."[6]

He then joined Kaisen along the Big Dry, where the collection of the nearly complete hadrosaur skeleton was accomplished in record time, beginning on July 7 and wrapping up on the 24th. The only mishap occurred in Glasgow,

FIGURE 21. Flatboat on the Red Deer River of Alberta, Canada, showing the AMNH field crew "netted for prospecting"—protecting themselves from mosquitoes—during the 1912 expedition (18547, American Museum of Natural History Library)

where they loaded the eight cases of fossils on one train for New York and the rest of their equipment on another for Alberta: "A small cyclone struck the town destroying several buildings," Brown wrote. "There was no loss of life. Peter was holding one end of a tarpaulin over the wagon and it played snap the whip with him—no damage to any of us."[7]

From Glasgow the crew set out by train to Red Deer, while the wagon and team proceeded by way of Calgary driven by their teamster A. E. Davenport.[8] Tales about the trip in the annual reports and correspondence are supplemented by Kaisen's terse diary entries. On the road, he paid particular attention to the quality of their lodgings: "Monday, July 25, we arrived at Shelby at 5 A.M. and went to Sullivan's Hotel, but that was a filthy place. At noon we bought some lunches at the grocery store and ate on the depot platform. It was better than the hotel. . . . Got to Lethbridge at 8 P.M. Went to Alexandria Hotel—fine hotel and very reasonable."[9]

On August 1, Osborn wrote Brown informing him that the museum trustees had promoted Matthew to a curatorship in the Department of

Vertebrate Paleontology.[10] Osborn himself was elected an honorary curator in the same department, presumably as a result of becoming president of the entire museum. These bureaucratic machinations boded well for Brown, as Osborn intimated that both Walter Granger, Brown's field partner in 1896 and 1897, and Brown would be promoted to positions as assistant curators within the next several years.

In Red Deer, Brown enlisted four carpenters to cut the lumber for the flatboat and accompanying small rowboat. Using such a cumbersome vessel would be a bit of a gamble. The locals who ran the river had informed Brown about some stretches of challenging rapids, especially where large blocks of the steep bluffs had tumbled down into the river, restricting its flow. But Brown also knew that there were few places in the canyon where roads would allow him to haul heavy plaster jackets out to the railroad, and he was determined to have enough space to store the fossils and equipment as they moved between the ferry sites. It took only three days for Brown's crew to assemble the two boats, a rowboat and the flatboat, which were launched on August 3. Brown was pleased with his armada, boasting that "with our present load of [3,000 pounds, the flatboat] draws only four inches of water."[11]

Brown had originally planned to build the boat in Content, sixty miles to the south of Red Deer, but a lack of lumber in that town required that the boats be built in Red Deer.[12] As a result, the crew was forced to run the perilous stretch of the river to the north called "The Canyon." To assist in the passage, Brown brought along an "experienced riverman," Charles Bremner. It was a good thing, too, because as Brown reported: "We had some little excitement up the Canyon when we hung up on a boulder in the rapids. . . . [But] by shifting cargo and levering the stern we were soon on our way again."[13]

Brown's spirits were clearly buoyed by the pristine landscapes. "No more interesting or instructive journey has ever been taken by [this] writer," he wrote. "Habitations are rarely discernable from the river, and for miles one travels through picturesque solitude unbroken save by the roar of the rapids." Such scenes no doubt served as a potent salve as he began to recover from Marion's death.

Along this northern stretch, the 250-foot-high canyon walls were composed of Early Tertiary sediments of the Paskapoo Formation. At a slide just above Erickson's Landing, prospecting revealed beds of fossil clams and other invertebrates, as well as a layer of "pebbles" that yielded several minute teeth and a few jaws of mammals.

Upon reaching more serene waters near Content, the crew began their search for dinosaur bones in the surrounding bluffs represented by the Late Cretaceous Edmonton Formation. With the river cutting down through the slightly tilted layers of rock, they were essentially heading back in geologic time as they floated downriver. "In the long midsummer days, at latitude 52 degrees," Brown wrote, "there were many hours of daylight, and constant floating would have carried us many miles per day; but frequent stops were made to prospect, and we rarely covered more than 20 miles per day.... habitations are rare along the river, and for miles we floated through picturesque solitude, the silence unbroken save by the noise of the rapids."[14]

The first find of a hadrosaur humerus and several vertebrae was quickly followed by the discovery of a locality about thirty square feet in extent that yielded "one complete hind leg,... a series of caudal [tail] vertebrae, jaws and skull bones of *Albertosaurus*," the smaller cousin of *Tyrannosaurus*. The crew also found several limbs of "Ornithomimus," which, if truly an ornithomimid, would now probably be called *Struthiomimus*. However, our Canadian colleagues Philip Currie and Darren Tanke suspect that Brown's ornithomimid may in fact be a juvenile albertosaur—a reasonable conclusion, given that this site is now known as the Dry Island *Albertosaurus* Bonebed. Brown also found a ceratopsian skull in a coal layer in the same region, though he did not collect it. Canadian paleontologists recently excavated this specimen, which now serves as the type of *Eotriceratops xerinsularis*.[15]

Drifting down the canyon to Big Valley on August 9, the crew discovered a hadrosaur forelimb and a skull-less ornithomimid skeleton.[16] Ornithomimids, or "ostrich-mimic" dinosaurs, are now well known to paleontologists, but that was not the case in 1910. Marsh's crew collected the first known ornithomimid specimen from the Latest Cretaceous Denver Formation in Colorado, and Marsh described it in 1890. Later, in his 1902 study with Osborn, Lambe described fragmentary remains, including a well-preserved hind limb, from the Belly River Series. Brown's headless skeleton confirmed that ornithomimids are relatively long-legged, long-armed, lightly built, bipedal forms that belong to the group of carnivorous dinosaurs called theropods. Its lack of a skull—due no doubt to the delicate nature of the animal's hollow-boned skeleton, which makes it more susceptible to disarticulation and breakage, thus lessening the chances for fossilization—led to a frustrating, multiyear quest to find one.

While Brown went downriver in the rowboat to find Davenport and their wagon team, Kaisen barely averted calamity: "August 14.... Wind came up and blowed the boat upstream as far as the rope allowed it. I had quite a time saving the tent with one oar in the water and when the boat started upstream the oar got caught on a stone and being pushed against the tent I just got hold of it in time."[17]

By September 5 they had reached Beaver Island, where, in addition to more hadrosaur material, they located a nearly complete new ceratopsian species, including fragments of the skull, which Brown eventually named *Leptoceratops*. They also discovered a "perfect, uncrushed skull of *Ankylosaurus* [?*Edmontonia*] with a long section of the tail, humerus, separate plates, and ribs." Wet weather had plagued the operations, however, as Brown lamented in a letter of September 12: "Seventeen days in August and seven so far in September of rain, sometimes sleet and very cold."[18] Still, Brown could be justifiably effusive about the results to Osborn:

> The fossils are so numerous that I doubt if we can possibly work farther down the river than Tolman's [Ferry] this season.
> This is without doubt the richest Cretaceous deposit in America... and [we] have not yet reached the richest field that I examined last year....
> With our numerous boxes on board containing such a variety of creatures we are living in a veritable ark.[19]

In early October, Brown rowed downriver to conduct reconnaissance for the following year's work, while the rest of the crew excavated one final hadrosaur near Tolman's Ferry. By mid-October they had beached the flatboat and hauled the twenty-six boxes of fossils, weighing 6,500 pounds, to the nearest railhead at Stettler for shipment to New York. In addition to the specimens previously mentioned, the cache included a large ankylosaur tail club.

The suite of dinosaur fossils from the "Lower Edmonton" convinced Brown that this formation was, as Osborn and Lambe had proposed "older than the Hell Creek and Converse County Beds [Lance Formation in Wyoming] of the United States.... Only two fossils had previously been collected from this formation, both skulls of *Albertosaurus*, now preserved in the Ottawa Museum."

In late October and early November, Brown returned to New York by way of Carbondale and Oxford to visit his family and young daughter.[20] While Brown was in Carbondale, W. D. Matthew wrote to inform him of his new title, with associated time off: "I have filed a formal application for leave of absence for you of two weeks from Oct. 24th; the third week to which you are entitled as Assistant Curator, I suppose you wish to hold in reserve."[21] The annual report shows that Brown now commanded the princely salary of $150 per month, or about $37,000 per year in present-day dollars.

Uncharacteristically, Brown's crews did not collect every single bone from some of the quarries. One of these sites, at what is now called Dry Island Buffalo Jump Provincial Park, contained the bones of several *Albertosaurus*, including juveniles. In 1997 Phil Currie, then of the Royal Tyrrell Museum in Alberta, used Brown's photographs of the region to plan his fieldwork. By matching up the horizon lines with one of Brown's images, he was able to relocate the quarry, and from 1998 on, this area has been excavated, resulting in the collection of many *Albertosaurus* specimens, some even corresponding to missing pieces in AMNH collections. Parts of rusted tin cans were also uncovered, presumably discarded by Brown's crew.

In early January 1911, Brown was back in the field—this time heading south. On a first leg in Texas and Mexico he primarily evaluated Pleistocene deposits, which yielded mammoths and other proboscideans for Osborn's great monograph on that group; he then swung back through Mississippi in March to collect specimens of primitive whales, reported to be *Zeuglodon* and *Dorodon*. By the end of March he had moved down through Florida, evaluating more Pleistocene localities, before sailing for Cuba in early April.[22]

There, his activities, under the guidance of Dr. Carlos de la Torre, focused on collecting fossils in caves and mineral springs that might help establish whether Cuba and the other Caribbean islands were ever connected by land bridges to either Florida or Mexico. Although his exploration of the caves produced a few Pleistocene mammal fossils, his success was far better in the hot springs at Baños de Ciego Montero.[23] He began by hiring a local resident to dive to the bottom and bring up handfuls of mud, a few of which yielded fossils of crocodiles, mammals, and turtles. A series of major pumping operations followed, first by hand and finally by a gasoline-powered machine, to empty the water out of the spring. This resulted in the recovery of a "great profusion" of "jaws and bones of [giant ground] sloths [*Megalocnus*], skulls of crocodiles and alligators, and parts of turtle shells," as well as fossils of rodent jaws and two species of birds.[24] Brown's skeleton of *Megalocnus* still greets visitors to the museum's Hall of Mammals and Their Early Relatives. Returning by ship from Havana, Brown arrived in New York on June 13.

Just two weeks later, he headed back north to Alberta via Carbondale, arriving on July 13, and met up with Kaisen, who had arrived on July 9.[25]

Over the winter, Brown had paid his teamster Davenport to remove the overburden on the ankylosaur quarry, so he spent his first three weeks collecting the remaining bones, which included "a femur, humerus, both lower jaws, ischium, fibula, three ribs, and a number of plates."[26] In addition to quarrying, the crew occasionally prospected; in Kaisen's case, the goal was more than just fossils: "August 2, we took up a foot of a [hadrosaur] . . . then we prospected the rest of the day. I picked and ate 4 different kinds of berries—strawberries, gooseberries, rasberries, and saskatoons."[27] Afterward, in the course of resealing the flatboat, the crew discovered a hadrosaur skeleton tentatively named *Podischion*, which Kaisen worked on while Brown went to meet Osborn and museum trustee Madison Grant for a run down the river in a rowboat, specially commissioned by Osborn and "loaded to the gunwales," to examine the Paskapoo and Edmonton exposures.[28]

After checking in at camp on September 2, the trio continued their journey downriver to "the mouth of Berry Creek, 250 miles below Red Deer," to conduct reconnaissance in the Judith River–equivalent Oldman Formation.[29] Brown and Osborn almost did not return. Usually, the trustee sat in the front, with Osborn in the rear and Brown rowing in the middle. Suitcases, blankets, and provisions filled the intervening spaces, so "there was not much freeboard above the water."[30] Initially, the trio made speedy progress, skimming along on the swift current. Then, seduced by the seemingly docile river, the trustee lobbied for a "go at the oars." Osborn rebuffed him, warning: "Grant, Brown knows this river better than you do, you don't need the exercise, and our boat is heavily loaded." But Grant continued to plead, and Brown finally switched places with him. All went well until a bend in the river where a work crew had constructed a temporary bridge supported by wooden pilings, which greatly increased the turbulence and accelerated the current. While Grant struggled mightily to control the boat,

> the front end struck a spile [piling], whipping the back end around against another spile, tipping the boat so that it began to fill.
> I jumped over Grant, pushed against the spile and we shot through. We landed quickly as possible to dry things before continuing our journey and Professor Osborn said, "Grant, let this be a lesson to you, but this is one experience I will not mention to Lulu [Mrs. Osborn]."

Despite their harrowing encounter with the river gods, the trio delighted in discovering "a lower jaw and scapula of *Monoclonius* and a fine skull of

FIGURE 22. Brown's crew perched on the bluffs above the Red Deer River in the Ankylosaur Quarry, about 3.5 miles below Tolman Ferry, during the 1912 AMNH expedition in Alberta, Canada (18542, American Museum of Natural History Library)

an armored dinosaur related to *Ankylosaurus*" in the Oldman exposures near Berry Creek.[31] Thus, in addition to securing more collections from the Edmonton, Brown and his crew extended their search downriver from the roughly 72-million-year-old beds of the Edmonton Formation into the 76-million-year-old rock layers of the Belly River Beds, which yielded a rich assemblage of dinosaurs. By amassing collections from the two different rock units, Brown could now build upon Lambe's work and provide two enhanced "snapshots" of how nonavian dinosaurs evolved between five and ten million years before their extinction at the end of the Cretaceous Period.

With Osborn and his mate safely on the train home, Brown returned to the camp, where Kaisen had completed excavating the apparently complete hadrosaur. Further prospecting yielded a fragmentary skull of *Albertosaurus*, as well as a hind limb and another skull of the "new" hadrosaur.[32] To celebrate their success, the crew retreated from the river for a little R and R—though according to Kaisen the trip back was a bit rough: "October 2, started to rain last night. Kept it up all day. We went to Postells [Postill's] for dinner. Got there at 7 P.M. And left at 2 A.M. Had a good time but a bad time coming back to camp. Dark and had to slide down the badlands."[33] By the end of the first week of October, Brown had shipped twenty-two boxes of fossils weighing 7,300 pounds off to New York and returned, again via Carbondale and Oxford, to the museum.

The next expedition to the Red Deer in 1912 would require extreme measures to cope with numerous hardships. Of the thousands of photos taken during the dozens of early dinosaur expeditions mounted by the American Museum of Natural History, perhaps the most curious is of Brown's field crew standing on the flatboat in the river with their heads completely covered by hats and masks of cloth. In correspondence, Brown revealed the reason for this field innovation: "Work has gone rapidly but under trying conditions. I cannot approximate the number of mosquitos but every person who moves about is forced to wear a net over the face, gloves and a coat or extra heavy shirt. I have never experienced anything like it."[34] Thus, despite the romantic image that these expeditions conjure—of lazily floating downriver plucking magnificent dinosaur skeletons from the banks—the reality is often quite different.

During the first two years that they explored the badlands along the Red Deer River, Brown's crew had the whole region to themselves, but that now changed. Up to this point, Brown's expeditions had been sanctioned by the Geological Survey of Canada; as a result of Brown's success, historian E. H. Colbert writes, "cries began to be heard in protest against the invasion of western Canada by Yankee bone hunters, who had been robbing the Dominion of its paleontological treasures."[35] The Survey therefore decided to enlist a famous family of fossil collectors, Charles H. Sternberg and his sons, to compete with Brown and collect a representative suite of dinosaur fossils for Canada.

Although born in central New York, Sternberg, like Brown, spent his late adolescence on a farm in Kansas, where he often encountered rocky outcrops rich with fossils.[36] He attended Kansas State Agricultural College, where he studied under O. C. Marsh's lieutenant, Benjamin Mudge. Unable to join Marsh's 1876 field crew, which was already fully manned, he appealed to E. D. Cope, who sent him for a season of collecting in the chalk beds of western Kansas. From there, Sternberg went on to establish himself as a premier freelance collector of dinosaurs and other fossil vertebrates. Many of his adventures and experiences are eloquently chronicled in his 1909 autobiography, *The Life of a Fossil Hunter*. Along with his three sons, George, Charles

M., and Levi, Sternberg discovered and sold specimens to many major museums throughout the United States and the world, including AMNH.

Beginning with the 1912 field season, the "Canadian Dinosaur Rush" was on. Overall, the competition between Brown and the Sternbergs was friendly. Although each crew kept a close eye on the activities of the other, they also occasionally visited each other's camps. Another account of the tension that occasionally surfaced between Brown's crew and that of Sternberg can be found in the 2001 paper by David Spalding. Osborn, for his part, was perturbed by the need to compete, especially since George Sternberg was joining the AMNH expedition. In fact, it was apparently George who alerted his colleagues at AMNH of his family's participation, writing to Matthew that "Papa and the boys will soon come up in this part of the country for the Ottawa Museum."[37]

Despite the competitive tensions simmering just under the surface, both crews found plenty of bones to keep them busy, and some of the fossils they found were truly spectacular. Kaisen, crew member George Olsen (a preparator at AMNH), and the cook, Bob Reid (who doubled as an excavator), launched the flatboat from Tolman's in early June, and Brown and George Sternberg met up with them soon after.[38] The crew soon took advantage of a gift delivered by the river:[39] On the night of June 16, the water level rose about seven feet, presumably as a result of storms upstream. Caught in the current was a small flatboat, about sixteen feet long and eight feet wide, that had apparently been snapped from its moorings. Kaisen leapt into action, scurrying along the bank with Reid for about a mile downstream until finally they were able to catch it and secure it. With the aid of Brown and Olsen, they labored doggedly to haul the boat back to camp, a task that proved to be "quite a job against the current." They then set about refurbishing its deck, which created plenty of space for their fossils and gear. Later in the season, however, the runaway flatboat almost brought disaster; as Brown relates, "At Drumheller, I ran too close to an island where a cross-current stranded the flat-boat. We levered it off after a while and went into camp forgetting that we may have strained the planks. Next morning... Olsen called that we were sinking and before we could dress our 'Titanic' was at the bottom in three feet of water. Things were in general damp but we saved everything by spending a day drying out."[40]

As the season unfolded, Brown, ever happy for an opportunity to socialize, was determined to celebrate the Fourth of July—a fact that seemed to test Kaisen's patience. While Kaisen hoisted his "little Old Glory" on the boat and remained with Olsen in camp, Brown left after dinner and "went to the dance at Hard's Farm," not returning until supper the next day. Indeed, Kaisen's notes often reflect frustration with Brown's extended absences. According to Darren Tanke, Brown may have established a romantic attachment with the wife of a local lumberman, Roy Hard, though no actual proof exists. We do know that Brown frequented the Hard home, and Kaisen frequently mentions the family in his field notes.[41]

While the rest of the crew quarried—including George Sternberg, who arrived on July 7—Brown focused on prospecting. In all, he lists thirty-six major specimens, including fifteen hadrosaurs (such as *Saurolophus*), nine ceratopsians, three ankylosaurs, four ornithomimids (presumably *Struthiomimus*), and one plesiosaur.[42] Although plesiosaur specimens are uncommon in these nonmarine sediments, they are not really unusual. Paleontologists now suspect that they swam up active river channels either in search of ballast stones or to shed parasites. In this early phase of the expedition, Brown downplayed the competition with Ottawa, reporting to Matthew: "The Ottawa party are somewhere ... [on the river] in the Edmonton formation approximately at Drumheller twenty miles below which does not disturb me as that is in the lower part of the beds [which produce] chiefly quarry [i.e., bonebed] specimens. As long as they are there I shall concentrate the whole party in this formation where the exposures are best."[43]

In another burst of innovation, Brown commissioned the construction of a motorboat, eighteen feet long with a three-horsepower engine, which proved "quite satisfactory though it would have been better with a larger engine."[44] On August 11 he boasted to Matthew, who had now become his direct supervisor:

> With the current it makes about twelve miles per hour and against the current about four.
> It rates about 100% better than the model that was on exhibition at the motor boat show. I paid ... $300 [about $6,000 in modern currency] for the boat delivered and having seen those at Edmonton know that we can sell it for that when we are through with it.[45]

Yet despite its high-tech attractions, Barnum had a love-hate relationship with his new toy—not too surprisingly, given the state of motorboat engineering at the time. While generally extolling its virtues in a September 22 letter to Osborn, Brown added that he was struggling to master the boat's mechanical eccentricities, saying that "it shows more varied moods than a woman, but I still retain a Christian spirit toward it."[46] Kaisen, in contrast, maintained more of a "hate-hate" relationship with Brown's toy from the start, as well as with Brown's navigating abilities: "August 2, we tried the motor [boat] but the first thing Brown did was to steer it on a rock and bent the shoe, so it caught the propeller. After fixing that the spark plug was broke. That was the beginning. I don't think the thing will work a damn upstream. It could not go up this morning."[47] Despite such setbacks, in early August Brown had somewhat ostentatiously motored downstream to visit the Ottawa crew, reporting to Matthew that, as expected, the Sternbergs were occupied with quarry work that would take a good deal of time to complete. This circumstance pleased Brown immensely because it meant they were not prospecting in new areas. Aside from one "very fine" specimen of *Albertosaurus,* Brown judged that the other specimens in his competitors' quarries would "serve the Ottawa Museum very well but we could not use [them]."[48]

On September 12, leaving Kaisen and the rest of the crew to quarry in the Edmonton, Brown set out with Olsen in the motorboat to prospect in the Oldman, 150 miles downstream.[49] Kaisen had a going-away gift for the duo: "We had some fun with a bad ham we tried to put in with Brown's stuff. We got it in, but George Olsen found it and threw it ashore."[50]

Brown's strategy in this move was essentially to leapfrog the Ottawa crew, in order, as he put it to Matthew, to "countermand the best material in both formations."[51] The prospects that he might succeed seemed promising, since the competing crew had just discovered an incomplete hadrosaur skeleton, which raised Brown's hopes that they would spend the rest of the season at their current field site. Matthew, clearly hoping to keep the rivalry from deteriorating into conflict, replied: "Glad to hear that [Charles] Sternberg is getting good material from the Edmonton. The better material his party secures the more friendly will the attitude of the Ottawa people be towards our collecting in the field. And I believe we can count on any material that they secure being made good use of—not buried away in a cellar but prepared, studied and exhibited."[52] (Then as today, deciding what and how much to collect in richly fossiliferous regions can be problematic. In addition to time constraints and costs, space in museum collection areas is often restricted, and the number of preparators available to clean the specimens for research and build the mounts for exhibits is often limited. So the desire to collect good specimens before they are destroyed by erosion must be balanced by other pragmatic concerns.)

Upon reaching the Oldman exposures that he and Osborn had surveyed the previous year, Brown set about excavating a skeleton they had marked for later collection. It turned out to be spectacular, "the skeleton of a crested dinosaur [reported to be *Prosaurolophus*] complete with the exception of the tip end of tail and front legs which are fragmentary. The underside of this skeleton was covered with skin impression[s] apparently continuous and I hope complete on the one side. The bones are in excellent state of preservation and may be exhibited as a panel mount showing the bones in relief on the right side with skin impression[s] on the opposite side."[53] Indeed, this magnificent specimen, now called *Corythosaurus*, still astounds visitors in the Hall of Ornithischian Dinosaurs at AMNH.

Because soft tissues such as skin, muscles, and organs decompose more quickly than bone, specimens with fossilized skin impressions, such as Brown's *Corythosaurus*, are relatively rare. The carcass must be rapidly buried after death in a generally anoxic environment to prevent predators and microbes from causing destruction or decay. Although in recent decades many new dinosaur skeletons with skin impressions have been uncovered, such as embryonic sauropod specimens from the Late Cretaceous of Patagonia, such discoveries remain uncommon.[54] Nonetheless, they do give us a good sense of skin texture and pattern, if not color.

Brown's luck did not stop there, however, for "before the . . . skeleton was prepared a complete uncrushed skull of [*Monoclonius*] was found near-by. This is the finest Ceratopsian skull in our collections."[55]

As winter set in, Brown, though excited about future collecting in these badlands, felt pressed for time, confiding to Matthew that the Belly River strata were so fossiliferous that he had to "travel with my eyes shut so as not to see more that I want [to collect] this fall."[56] By the end of October, the crew had shipped at least twenty cases of fossils; the cost of this five-month expedition, including Brown's bumped-up salary as an assistant curator of $175 per month, totaled $4,671 (roughly $42,000 in today's currency).

As the 1913 season got under way, relations continued to be cordial between the two competing groups. In mid-June, Brown visited Lambe at the Geological Survey of Canada in Ottawa and found him to be "very courteous," opening up boxes containing the type of *Monoclonius dawsoni* that Brown wished to examine for his research. However, he also lamented to Matthew: "Mr. Lambe tells me that his party is to work in the [Oldman]

FIGURE 23. Barnum Brown excavating the skeleton of "Prosaurolophus" (now *Corythosaurus*) (AMNH 5240) during the 1912 AMNH expedition to the Red Deer River, Alberta, Canada (18552, American Museum of Natural History Library)

Beds this year also. It's a pity but nothing can be done except beat them to the best specimens."[57]

The quarters in the Oldman, though spaciously spread out in three patches of badlands along fifteen miles of the river below Steveville, were rather cramped by Alberta standards—especially in the eyes of Brown—due to the competition between the two crews for prime collecting areas. AMNH, by virtue of Brown's work during the previous two seasons, had laid claim to the first patch, on the left bank, which extended three miles below Steveville.[58] The Ottawa crew occupied the exposures on the opposite bank. The third patch, also on the right bank and farther downstream, was apparently open to both crews.

By the time Brown joined Kaisen and George Sternberg in camp half a mile below Steveville, the crew was already excavating two ankylosaur skeletons and one of *Monoclonius*. Then, in a letter of July 6 to Osborn, Brown reported that Sternberg was leaving to join his father's crew, accepting a position sponsored by the Canadian Geological Survey.[59] Although he may have been peeved, Barnum, who respected George and his skills, made no objection, as long as the survey agreed to reimburse AMNH for the costs of George's trip out. It probably irked Brown more that the Ottawa crew began to copy his logistical tactics, one-upping him even in the motorboat department. As Kaisen attested, the Sternberg crew had shown up at Brown's camp on July 1, Dominion Day, with both their houseboat and a new five-horsepower motorboat.[60] Kaisen described Sternberg's well-endowed machine as "working fine and going at a good speed upstream." The Ottawa crew then taunted Brown by ferrying Kaisen that evening to Steveville, where a holiday celebration, cheerily characterized by Kaisen as a "big drown," ensued with almost two hundred revelers. (Today, that's difficult to imagine because Steveville was pretty much nonexistent from 1943 to 2002, and today it consists essentially of one ranch house and its outbuildings.)

Although visits between the two crews continued, Brown's reports to the museum, such as the one of July 6, reveal that tensions were apparent:

> Sternberg's sons were up to camp in the boat today and report finding an Albertosaurus skull in fair condition.
> There [are] plenty of exposures here for all of us for this year but I am really provoked that the Ottawa people should follow our footsteps so closely.[61]

Although Osborn sympathized, wishing to avoid a row similar to that of Cope and Marsh, he counseled compassion. In the end, he said, there was little choice but to "take it good naturedly, and remember that after all this is British territory and also that the Canadians showed us the way"—to the Oldman exposures, at any rate.[62]

By mid-July Brown had resigned himself to losing George, though he told Osborn that the young Sternberg had performed in a much superior fashion than the previous season. Nonetheless, Brown was relieved that, while prospecting and collecting, "I can now turn around without the Ottawa people knowing it."[63]

Also raising Brown's spirits was the discovery of the posterior portion of an ornithomimid specimen that continued into the outcrop, which boded well for finally finding a skull of the animal. Osborn was jubilant: "At last you have the *Ornithomymus* by the tail," he wrote on August 3. "In my long series of observations on your work as a collector I have learned to regard your tail-end discoveries as very suspicious. I surely hope it will prove to be so [i.e., an entire skeleton] in the present case, because *Ornithomimus* will be one of the wonders of creation when found."⁶⁴ Alas, it was not yet to be, as Brown reported with exasperation in his next letter, admitting that "it is the most elusive beast I have ever pursued."⁶⁵ A short while before, he had written to Matthew: "It beats all what becomes of their skulls."⁶⁶ For Brown, however, success was nonetheless manifest in the crew's discovery of several other theropods, presumably *Albertosaurus*, "one nearly as large as Tyrannosaurus with . . . a skull that will restore complete."⁶⁷

Brown now set his sights downriver, on the last of the three sets of badlands in the Oldman: "It is my object," he wrote Matthew, "now to locate all of the best material in this lower exposure before the Sternbergs move down the river. . . . They have several good specimens but I do not know what they are except the carnivore skeleton [*Albertosaurus*]. The old man [identity unknown] told me they are not to tell what they had found."⁶⁸ In mid-July Brown made his move, stopping at Sternberg's camp to negotiate a deal over collecting rights in different patches of the badlands. Kaisen dutifully "took photos of fleet there—Sternbergs had Canadian flags up and I had stars and stripes floating."⁶⁹ Brown secured rights to collect near Sand Creek, where the Oldman exposures were "more extensive than elsewhere and there we secured the bulk of our material." Meanwhile, Sternberg's crew "moved down below the Old Mexico Ranch skipping our territory."⁷⁰

In all that season, Barnum's crew collected an astounding forty-four specimens, mostly skeletons or skulls, including nine ankylosaurs, fifteen ceratopsians, ten hadrosaurs, the one ornithomimid, five albertosaurs, and another plesiosaur. This collection filled seventy-six crates, weighing in at 30,000 pounds and commanding two-thirds of a boxcar, with costs for the expedition totaling $3,562—almost $73,000 in modern U.S. dollars.⁷¹

Brown now focused on the rush to prepare and publish the material, alerting Matthew in early September that "if you can see a way clear for additional laboratory force we will need it on this collection. The Ottawa people will try to secure as many types as possible and we have several new skulls this year."⁷² Osborn and Matthew recognized the importance of the competition, with Osborn urging Brown "to assemble as many notes as you can in the field so that a preliminary Bulletin can be published as early as possible."⁷³ After swinging through Carbondale and Oxford, Brown returned to the museum about the end of October for a busy winter of work.

The 1914 expedition got off to a somewhat rocky start. On the way out to the Red Deer, Brown stopped off near Medicine Hat to conduct reconnaissance on some Oldman beds northeast of Irvine, though these turned out to be "not promising."[74] In the meantime, Kaisen and Bob Reid, along with a couple of other assistants, began excavating skeletons found and covered at the end of the previous year. With Brown not there, however, the competition closed in a bit too close for comfort, as Kaisen recorded in his journal on June 24. Floating downriver to size up the Sternberg operation, he discovered them working on an exquisite ceratopsian skull—"right in on our working [area]." The previous Sunday, the Sternbergs had assured Kaisen that they would not come near Brown's territory, and this infringement prompted Kaisen to tell them that "the word of a Sternberg was no good." Infuriated, Kaisen spied members of the Ottawa crew prospecting even closer that afternoon. Kaisen surreptitiously worked his way around the outcrops in an attempt to confront them once again, but they somehow slipped his trap.[75]

Without Brown to back him up, Kaisen, alarmed and exasperated, reported to Matthew: "I worked on all of our most important prospect[s] so Sternberg should not take them. . . . He is working on our ground . . . seems like they are going to stay right here with us. . . . It is bad but I don't see that I can do anything but hang on to what I got. . . . Those specimens I have worked on of course they will not take."[76] Here, Kaisen was invoking an unwritten rule among paleontologists, based on common courtesy, not to touch or collect specimens that are discovered and partially excavated by another colleague, at least without their explicit approval. Kaisen's undercurrent of discontent filtered down to Brown in late July, who echoed to Matthew: "Sternberg and his party are just below us [on the river] and have taken out some fossils from our territory but we have at present no serious disputes. They have no regard for ethics of bone digging."[77]

To some degree, the tense mood in camp was lightened by the collection of some animals not yet fossilized, as Brown later recalled:

> The men started to work one morning and saw a Canadian Lynx mother followed by three little kittens. . . . The men gave chase and the mother ran while each man dropped his hat over a kitten and brought them back to camp. [The cook got] chicken wire to build a cage. . . . We were bothered by mice in this camp and we had [the cook's domestic cat] and her three kittens with us to catch the mice. . . . The old house cat and her kittens would walk around the cage sniffing . . . apparently curious about these new arrivals. Finally we put the domestic kittens in the cage and they played with the Lynx babies as though they were . . . the same. . . . At night we took the domestic kittens out to be with their mother. In the morning they were ready for another frolic . . . The cage had been placed near the cook's tent and [at] night the Lynx

[kittens] growled so that the cook couldn't sleep. Next morning he was furious and said "If those damned cats growl again tonight... I'll turn them out." I told him [not to] for I wanted to send them to the Bronx Zoological Garden. But sure enough... the cats growled all night and the cook turned them out.... But riding out I found one a short distance away and enticed him back to the cage with a fish which they dearly loved. Then I built a box in which to ship it and when enclosed he became a demon. Nevertheless he went back to the Zoo.[78]

In the meantime, Brown and Kaisen had laid claim to more spectacular specimens, including skeletons of *Corythosaurus* and *Monoclonius,* a skull and jaws of *Albertosaurus,* and the long-sought skull of an ornithomimid, complete except for the top.[79] The narrow, streamlined skull proved especially unusual in that the jaws were toothless, a characteristic that seemed to belie the fact that ornithomimids are theropods, a predominantly carnivorous group. Osborn was elated and took the skull as confirmation that Brown's persistent search for such rare members of the fauna would be rewarded through discoveries providing significant new scientific insights.[80] Matthew also sent his congratulations, though he ended his note with sobering news: "As I write, the great European war has just burst, with some uncertainty still as to whether Great Britain will be drawn in. We are fortunate to be out of its direct scope, although doubtless it will affect our affairs in various indirect ways. I can only hope it will not involve the finances of the Museum in any way."[81]

Seeking reassurance, Brown in his next letter questioned Osborn as to the potential effect of the war on the museum; he also mentioned the possibility of a reconnaissance trip at the end of the season, paid for by the Calgary Oil Company, to evaluate prospects for petroleum along the Athabasca River.[82] A month later, Brown reported that their take had risen to nine dinosaur skeletons, along with two other hadrosaur skulls, a skull of the crocodile *Leidyosuchus* (though probably it was a champsosaur), and three turtle shells.[83] He planned, he said, to load the crated fossils on a boxcar and ship them on about September 21. For insurance purposes, he recommended that the collection be valued at $20,000—roughly $400,000 in today's currency. Such arrangements to protect the fossils during shipment are still made today, the value generally being based on the estimated cost of returning to the field to obtain a similar assortment of specimens.

But Brown was not yet done. A week later he reported to Matthew: "Fortune has again smiled on us with a fine Ornithomimus skull and jaws....

FIGURE 24. Brown's crew hauling their collection of fossils out of the Red Deer Canyon from their camp at the mouth of Sand Creek during the 1914 AMNH expedition (18493, American Museum of Natural History Library)

We are also uncovering Hypacrosaurus skeleton."[84] Since *Hypacrosaurus* is not found in strata this far down the Red Deer, the specimen was more likely *Corythosaurus*.

The year's fieldwork ended on September 27, after the crew completed a road out of the river canyon to transport the bounty: eighty-three cases containing twenty-five specimens, including three albertosaurs, two ankylosaurs, nine hadrosaurs, three ornithomimids, three ceratopsians, three turtles, and one champsosaur. Total expenses came to $3,363 or a bit more than $67,000 in current U.S. currency.[85]

With the season at a close, Brown took a run down the Athabasca for the oil company to evaluate the region's prospects for petroleum, thereby launching a side career in geologic consulting that would last the rest of his life. These explorations often combined corporate reconnaissance with prospecting for fossils on behalf of the museum and gathering intelligence for the U.S. government, and, as we will see, took Brown to such far-flung places as Ethiopia, India, Pakistan, Burma, Greece, Cuba, and Central America, as well as throughout the western reaches of North America.

With such a large collection from the Red Deer exposures now housed at the museum and mounting pressure to prepare that material for research and exhibition, Peter Kaisen sat out the sixth expedition to Red Deer in 1915; in his stead, Brown hired two outside assistants, A. F. Johnson from the Steveville region and William Stein from Wyoming. Johnson picked up in mid-May where the crew had left off the previous year in the Oldman exposures near Sand Creek, with Brown joining him at the end of July, and the whole season was spent along that tributary of the Red Deer. The Sternberg crew, meanwhile, split up, with "four working in the [Oldman] formation around and through our territory and four above Drumheller in the Edmonton formation."[86]

Brown's crew continued to find rare specimens, such as a horned dinosaur much smaller than *Monoclonius,* an adult just twelve feet long. But Brown was most pleased about one of the ankylosaurs: "The rarest specimen of all and most desirable to complete our collection is part of a skeleton of Palaeoscincus [now *Edmontonia*] consisting of a fine skull and jaws with dermal throat plates in position, a good pelvis, a few well preserved limb bones, ribs and vertebrae."[87]

World War I weighed on the crew and museum staff, as Matthew wrote to Brown: "The war is going to be a long and exhausting struggle, and the longer it lasts the harder we shall be hit by it."[88] Brown took these worries to heart, using the war metaphorically to describe the difficult conditions along the Red Deer: "It has rained all over this part of Canada as never before—the Red Deer River is out of banks most of the time; ten feet of water was running over our last year's camp site and came to where our fossils were parked last year. Mosquitoes are fearless of smoke, ferocious and in numbers equal to the Kaiser's army."[89]

Brown had a great deal of difficulty closing out the season in late October because specimens worth collecting kept showing up "till it seemed we would never finish though working Sundays and every day from lamp light to lamp light."[90] Fifteen major specimens were added to the overall take, including six hadrosaurs, three ceratopsians, three ornithomimids, and two ankylosaurs, and the crew filled sixty-five cases, which made for "nearly a full car[load] ... about four-fifths as much as last season."[91]

Brown summed up the take of the combined expeditions in a 1919 article for *National Geographic:* "As a result of the work ... in Canada, the American Museum Expeditions have collected 300 large cases, or three and one-half [train] carloads of fossils, two-thirds of which are exhibition specimens, including twenty skulls and fourteen skeletons of large dinosaurs, besides many partial skeletons. This material represents many genera and species new to science."[92] No less than twenty-two dinosaur skeletons that Brown and his crews collected along the Red Deer River still grace the dinosaur halls of the American Museum (see Appendix 1).

It had been just nineteen years since Brown discovered the first dinosaur specimen for AMNH at Como Bluff. By 1915, largely as a result of Brown's efforts in Wyoming, Montana, and Alberta, the museum now possessed the world's leading collection of dinosaurs.[93] Most of the more spectacular skeletons were painstakingly mounted on iron frameworks or embedded in supporting partitions of plaster and placed in the AMNH exhibition halls, where they continue to astound and educate visitors from around the world.

The rugged exposures along the Red Deer River still yield important paleontological relics from the ancient past. Paleontologists from the Royal Tyrrell Museum, situated on the banks of the Red Deer River in Drumheller, Alberta, team with scientists from universities worldwide to find new skeletons in the region every year. In 1955, a large chunk of the Red Deer exposures was set aside to form Dinosaur Provincial Park, about ninety miles southeast of Drumheller, which affords the specimens protection until they are excavated. In addition, the park plays host to the Royal Tyrrell Field Station, which not only supports the field operations of the Tyrrell staff but also houses exhibition galleries informing the public about the rocks and fossils of the park.

Brown's discoveries continue to resonate. Today, paleontologists still recognize eight species of nonavian dinosaurs based on holotypes that Brown and his crews collected from the Dinosaur Park Formation, which he called the Belly River Beds: three horned dinosaurs, *Chasmosaurus kaiseni* (AMNH 5401), *Monoclonius cutleri* (AMNH 5427), and *Styracosaurus parksi* (AMNH 5372); two duckbills, *Corythosaurus casuarius* (AMNH 5240) and *Prosaurolophus maximus* (AMNH 5386); a small carnivorous theropod, *Dromaeosaurus albertensis* (AMNH 5356); and a dome-headed pachycephalosaur, *Ornatotholus browni* (AMNH 5450), which probably represents a juvenile *Stegoceras*.[94] The collections of Brown's crews from the Edmonton Group allowed Brown to describe four other new species as well: the horned dinosaurs *Anchiceratops ornatus* and *Leptoceratops gracilis* and two duckbilled dinosaurs, *Hypacrosaurus altispinus* and *Saurolophus osborni*. Today, Brown's old quarries are being located, based on photos he took and using as evidence objects that he and his crews left behind, such as plaster, scrap lumber, and newspaper used to make the specimen casts.[95] This project is being conducted in conjunction with GPS survey work in Dinosaur Provincial Park, to preserve not only the locality data for the specimens but also their stratigraphic context, which allows for a more comprehensive understanding of when the animals lived and how the fauna evolved some 75 million years ago.

NINE

Cuba, Abyssinia, and Other Intrigues
(1916–1921)

HAVING LEARNED A GREAT DEAL about the sequence of geologic formations and their faunas north of the Canadian border along the Red Deer River, 1916 found Brown once again focusing his efforts south of the border, primarily in northern Montana. His goal was to determine how the sequence of formations there correlated with those of the Red Deer region and of the Hell Creek badlands just south of the Missouri River. His primary assistant in this quest was again Peter Kaisen.

On July 27, Brown reported to Osborn that, having taken the opportunity "to get a car cheap," he had examined the region between Fort Buford, just across the Montana–North Dakota border at the confluence of the Missouri and the Yellowstone, and west to Malta and the area near Lewiston.[1] Brown suspected that the boundary between the Hell Creek Formation and the overlying Fort Union Formation, the lowest part of which is now called the Tullock Formation, probably represented the Cretaceous-Tertiary boundary, since the Fort Union Formation lacked dinosaur fossils. But to be sure, he wanted to find supporting paleontological and geological evidence. Unfortunately, none of the "fine" Fort Union exposures produced any mammal fossils either, as Osborn had apparently hoped they would; nonetheless, Brown's reconnaissance put to rest any

> lingering doubt... that the Hell Creek Beds marked the close of the Cretaceous, on account of the apparent conformity of the Fort Union Beds in the Hell Creek Country. I can now state positively that they do mark the end of the Cretaceous. The unconformity is clearly seen along the Great Northern Railroad at Calais and for a mile east of that siding where the Hell Creek Beds finally disappear to the east under the clearly defined Fort Union.

West of the Musselshell River to Lewiston the entire Cretaceous series is exposed—Hell Creek Beds, Fox Hills, Bear Paw, Judith River, Claggett, Eagle and top of Colorado well defined in sequence. . . . they rest conformably on the Morrison which in turn overlies Paleozoic limestones.[2]

Countering Brown's poor luck finding fossils, Kaisen and his assistant, collecting in the Two Medicine Formation west of Sweet Grass, had amassed "16 boxes of bones, with the best specimen a partial skeleton of *Brachyceratops*" that, however, was missing its skull.[3] This collection was supplemented by other finds, including two duckbills, *Hypacrosaurus* and "Trachodon," a vertebral column of a carnosaur, and the pelvis of an ankylosaur similar to *Palaeoscincus*. Brown concluded that the Two Medicine Formation and its fauna, which lacked large ceratopsians and contained different hadrosaurs, was older than the Belly River Beds along the Red Deer River in Alberta and the Judith River Formation in Montana.[4] Today, though, and as the United States Geological Survey asserted at the time, these three formations are considered to be essentially the same age; the Two Medicine simply represents a different depositional environment.[5]

Back in the Hell Creek region, Brown told Osborn with some amusement that he felt like Rip Van Winkle when he asked about fossils that had been collected there because "people tell me of a 'man who years ago took out big mastodons in the breaks'"[6]—referring, no doubt, to his own discoveries of *Tyrannosaurus* and other dinosaurs in the first decade of the century. This would be the last time Brown worked in the region for over a decade, for, as Matthew wrote in his annual report for 1917, although "three field expeditions were planned at the beginning of the year, . . . the uncertainty as to how the entry of the United States into the world war might affect the affairs and staff of the Museum made it advisable to postpone fieldwork."[7]

In 1917, Brown played a wartime role consulting for the U.S. Treasury Department.[8] The fact that few details of his activities were documented is a harbinger of Brown's more shadowy side, which would emerge in the 1920s and later decades. In dossiers and year-by-year summaries of his activities that he prepared for his aborted autobiography, Brown describes his "War Assignment" for Treasury as "establishing depreciation and depletion of oil properties for taxation purposes."[9] Apparently, he accepted the role after turning down a foreign service post with the navy. The war effort not only established professional ties between Brown and the U.S. government, but it also advanced his career as a geological consultant for the mining and oil industries. He had already followed up his 1914 stint as an oil field consultant by working for Osage Oil Company in 1915, and he would soon extend his geological expertise into other international arenas.

To wit, in 1918, back on the AMNH staff, Brown mounted another expedition to Cuba, not only in search of Pleistocene mammals but also to evaluate potential mining sites for copper. The paleontological work focused on cleaning out the fossiliferous sediments in the hot spring at Baños de Ciego Montero, where Brown had worked in 1911.[10] Aided by Charles Falkenbach and a more formidable 6.5-horsepower pump, Brown sought to drain the spring and surround the main vent with a concrete basin in order to excavate the fossils. Although the basin was successfully constructed, they discovered that the weight of the water in the basin forced open new auxiliary vents nearby, resulting in a kind of "whack-a-vent" succession of futile drainage operations. Finally accepting the truth, Brown resigned himself to a three-hour-long draining exercise every morning. It was exhausting work, especially given that temperatures rose to well in excess of 100 degrees, with high levels of humidity. At one point Thomas Barbour, a herpetologist from Harvard's Museum of Comparative Zoology, visited the site; his report to Brown's supervisors at AMNH caused alarm over the crew's health. Matthew wrote Brown on April 30:

> I saw Dr. Barbour not long ago.... He thinks you have been working too hard.
> The only point is this—*your health and Charlie's must override any other consideration.* As a mere matter of policy it would be the worst sort of mistake to risk damage to either one, but especially to yours, as you are past the age of easy recuperation from damage.[11]

Having received no response from Brown, Matthew reiterated his concern on June 1, emphasizing that "tropical climates are not to be monkeyed with, and from what Dr. Barbour told me I judge that you have been getting rather run down."[12]

Yet Brown persevered, amassing a remarkable collection, even though articulated skeletons, a primary aim of this expedition, proved elusive. On April 20, Brown listed his results thus far, which included three fine skulls of the giant ground sloth, *Megalocnus,* and eight pairs of *Megalocnus* jaws; a possible *Mesocnus* skull; one jaw of *Microcnus* (another sloth); three perfect alligator and caiman skulls; as well as a "very comprehensive collection of small mammal, bird, frog and lizard bones, crustaceans, river shells, various fruits, leaves and trees," some of the plant fossils "look[ing] like they were of

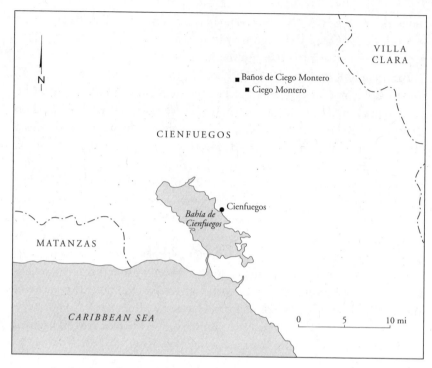

MAP 4. Southern Cienfuegos Province, Cuba, showing Brown's fossil site at Ciego Montero

last year's growth."[13] Brown reasoned that the lack of articulated specimens came about because the fossiliferous sediments were deposited during prehistoric floods, in which an arm of the river overflowed the bank and inundated the hot spring's basin, leaving behind mud, fish, turtles, and debris.[14] In June, ignoring Matthew's pleas to wrap up work and come home, Brown went on to excavate more fossiliferous deposits at a cave near Soledad.[15]

Brown was still not finished, though. He apparently stayed in Cuba until October, when he went "to Pinar del Rio to examine two copper properties. Received $50 per day this work and advised Co. they were through after spending $40,000 on one property."[16] In all, Brown received $1,027.16 for his geologic consulting report, a sizable sum approaching $14,000 in today's currency. But that didn't keep him from fretting that in Havana, "Now room and bath cost $3 per day. Food $3 per day if careful."[17] Clearly, Brown felt it necessary to augment his relatively meager salary museum salary through consulting whenever possible. A detailed report entitled "The Occurrence of

FIGURE 25. Brown and an assistant (presumably Charles Falkenbach) draining the 98°F Chapapote Spring at Baños de Ciego Montero in order to collect Pleistocene fossils during the 1918 AMNH expedition to Cuba (AMNH Vertebrate Paleontology Archive, 8:2 B7 S26)

Petroleum in Cuba" still resides in the Vertebrate Paleontology Department library.

While in Pinar del Río, Brown also collected fossil fishes encased in septerian nodules, which the local inhabitants call *quesos* because of their resemblance to heads of cheese. Preparation of one such nodule in the 1960s resulted in the discovery of an exceptionally well preserved pterosaur. This occurrence, along with associated fossils, shows that these beds were deposited in a nearshore environment and are about the same age as the famous Solnhofen limestones in Germany that produced *Archaeopteryx* (the earliest known bird), pterosaurs, and a host of other Late Jurassic marine vertebrates and invertebrates. Someday, nodules from Pinar del Río may well yield a fossil bird as well.

A letter in the archives written to Matthew from Pittsburgh makes it clear that Brown was still evaluating oil field production in December 1918: "I have been up to my ears in this work which on the whole is productive of scientific data although operators are not as a rule aware of it. The Appalachian field has been turned over to me and I have a good sized corps of assistants which allows some time to plot out results: it is very interesting to determine before hand the final production of an oil or gas well after the first year of production."[18] This project may have been for the U.S. Treasury Department, but Brown's description is not explicit. In any event, we do know that Brown rejoined the AMNH staff on May 25, 1919,[19] though he was still splitting his time between oil consulting and fossil collecting in Oklahoma and Texas. "I have arranged," he wrote Matthew from Oklahoma,

> to visit a Permian exposure in the southern part of Oklahoma where oil men report fossils that are without doubt Dimetrodon and Naosaurus [fin-backed early relatives of mammals]. Parts of skulls and skeletons are reported.
> The Empire Oil Co here has one of the finest equipped laboratories and best geological departments I have ever been privileged to see. They are making use of paleontology in a way never before attempted... in microscopic work. The complete analyses of well cuttings are as illuminating as gross surface exposures.[20]

Microscopic analyses of rock fragments recovered from the drilling process are now commonly used both to identify microfossils preserved in the rocks, which aids in stratigraphic correlation, and to assess important lithologic characteristics, such as porosity. Brown's interest in such technological and methodological innovations reflects the pioneering vision for which he has long been respected within the paleontological community. As we will see, he would eventually initiate new procedures not only to prepare and analyze fossils in the lab, but also to find new fossil localities in the field.

During this period, while Brown was still single, the first compelling evidence of Brown's propensity for womanizing arises in museum and personal correspondence regarding a legal case, the first mention of which is in a letter from Brown to Matthew on December 28, 1919. Brown, who was still conducting fieldwork in Oklahoma, writes:

> I am at a loss to know how my standing at the Museum has been affected by this blackmail case and whether I shall be wanted there again. At any rate, it seems to me best to settle the business in or out of court before I come to the museum to take up work. I know what a lot of gossips there are there.
> I wish you would take it up with Prof. Osborn.
> It is purely blackmail and I doubt if the woman would dare go to court and yet you know how innocent parties can be humiliated by being related to a party in such a case.
> I put my case in the hands of a good lawyer who is inclined to let the

woman have all the time she wants to defeat herself. Meantime I have attempted to keep too busy to think of it.[21]

Matthew responded on January 6, 1920: "I do not think there is any occasion for you to worry or trouble yourself in any degree about this blackmailing case affecting your standing here. I do not myself see why you should let it affect your plans for returning to work at the Museum; I have, as you asked, referred this point to Professor Osborn, and he will advise you of his opinion."[22] Matthew also followed up on his promise that same day, writing to Osborn: "I gather that this suit is getting on Brown's nerves. I have written to him saying that I saw no reason why it should affect his standing with us or his plans for returning to work here. It might help if you could write him a line also."[23]

Osborn penned a short note on Matthew's memo, which he then apparently returned to Matthew: "Approved. Express my sympathy. Best for him to clear up law case first before he returns." The specific nature of the suit is not known; however, Wann Langston, a retired paleontologist at the University of Texas in Austin and a personal acquaintance of Brown's, has said that he understood it to involve a breach of promise regarding a marriage proposal.

The filing of the case, along with Osborn's wish that Brown not return to the museum until it was resolved, may well have been a catalyst for a remarkable series of international expeditions to Africa, Asia, and the islands of the Aegean that Brown conducted, in the service of oil companies, the U.S. government, and the museum, between 1920 and 1924. In addition to removing Brown from the scene of the alleged offense, this sojourn freed the museum's best collector to roam the world in search of fossils to both augment the collections and advance Osborn's research into the evolution of mammals and the origin of humanity; it also supplied the U.S. government with important intelligence. It also no doubt fed Brown's wanderlust. As it was when he traveled across the West with his father, he was always keen to see something new.

An opportunity to lead a truly unprecedented international expedition landed in Brown's lap in 1920 when the Anglo-American Oil Company purchased the rights to prospect for petroleum in Abyssinia in what now comprises the arid highlands of southwestern Ethiopia. Making this enterprise doubly enticing, he arranged to take a leave of absence from his primary job as a curator, with the understanding that he would be able to look for fossils for the museum while in Abyssinia.

Despite his initial consternation, after retaining a lawyer Brown was apparently reassured that he could participate in the Abyssinian expedition and get back to New York before he would be required to appear in any legal proceedings. The Anglo-American Oil Company was an outgrowth of John D. Rockefeller's Standard Oil Trust, which, established in 1888 and based in London, represented Rockefeller's first step in developing an international network of oil companies.[24] In 1911, the U.S. Supreme Court mandated the breakup of the Standard Oil Trust into thirty-three different companies to address issues of monopoly in the oil industry. One of those spin-offs, Standard Oil of New Jersey, acquired Anglo-American. (In 1951, Anglo-American changed its name to Esso; in 1972, Standard Oil of New Jersey changed its name to Exxon.)

Since 1859, when the first viable oil well was drilled in northwestern Pennsylvania, the Standard Oil Trust and other companies had struggled to meet the rapidly growing demand for petroleum products. In 1860, global production was a modest 500,000 barrels; by the end of the next decade, that figure stood at 20 million barrels, and by 1920, production had climbed to 450 million barrels. But demand rose just as fast, leading to governmental concern that the United States could run out of oil by 1930. That fear triggered aggressive prospecting for new sources of oil both within the United States and around the world, and Brown became one of those prospectors when he went to Abyssinia.[25]

The field trip that Brown led was formally called the Dudley Expedition, presumably after one of the sponsors. As leader, Brown was supported by Captain Harry F. Moon, who kept a detailed daily journal, George H. Herring, George W. Powell, and an interpreter, Johannes Semerdjibashian. A "head man, cook, and 38 boys" acting as guides, guards, and attendants filled out the ranks.[26]

Brown's task, however, was daunting. Not only had no Westerner ever ventured into this vast African outback, but the "nation" of Ethiopia, composed of a mixture of Muslims, Coptic Christians, and animists, was ruled by a triumvirate of royal and religious figures along with local chieftains. Consequently, the politics of business associations and transactions, especially ones on the scale of this oil survey, were extremely complex and difficult to negotiate and implement.

For Brown, the basic problem seemed to be that part of the region he expected to have access to had been sold to another oil company, though Brown's employer had paid some 50,000 pounds for what was understood to be exclusive rights. The circumstances and individuals associated with this situation are difficult to sort out, but as Brown matter-of-factly stated, the minds of Abyssinians and Westerners work differently and he had never before encountered such a high level of intrigue. This observation reflected the Western attitude at the time regarding such dealings. Coming from nations based more formally in the rule of law, Brown and his contemporaries could not fathom that a foreign potentate would sell the same entity twice.

To try to resolve the dilemma, Brown and his colleagues started at the top. It took several days, but finally they managed to arrange an audience with the three rulers. As Brown wrote in an article for *Natural History*, one, Zaoditou, was the daughter of the Ethiopian emperor, Menelek II; the second was the prince of Harrar, Ras Tafari, who would eventually become Emperor Haile Selassie; and the third was the region's *abuna*, or archbishop of the Coptic Church. Brown was prepared to pay tribute to all of them:

> Following the usual mistaken custom of organizations that visit Ethiopia, the Dudley Expedition bore rich gifts to the rulers—a diamond necklace for the Queen [valued by Tiffany's at $50,000, or about $1 million in today's currency], an electric light plant for the palace, diamond watches, gold watches, a cinema.... I say "mistaken custom" because frequently such gifts arouse suspicion of the rulers. At the presentation I am sure that Europeans smiled as our party rode to the palace (Gibby)—men in full evening dress astride donkeys under a hot midday sun! Hordes of spectators lined the roadsides, as curious to our eyes as we were to theirs.[27]

Brown's daughter, Frances, paints an amusing visual picture: "Barnum was a good six feet tall, and the other ten men were far from short. The donkeys were small and low-slung, so that each rider had great difficulty not only keeping his top hat on his head but also his feet and legs from dragging right along the none-too-clean street. Barnum later vividly recalled how much their legs ached when they finally got to the palace and the way they hurriedly wiped off their shoes before entering for the audience."[28] Once at the palace, the gates were thrown open to reveal "throngs of white-draped figures" that Brown's party had to wade through as they were led to the royal chamber. Queen Zaoditou lounged among a mass of sumptuous cushions, "muffled to the eyes." The Prince Regent Tafari sat nearby, attended by a host of his ministers and officers, while an entourage of diplomats stood in rapt attention throughout the chamber. To Brown, the audience constituted a "picturesque [and] thrilling scene never to be forgotten," as the event unfolded amid "many speeches and much salaaming."[29]

Eventually, following more discussions involving the prospecting rights to be granted to the expedition, a permit arrived from the prince regent as the expedition caravan headed out of Addis Ababa. Basically, they would be making two thrusts into different regions of Harrar Province. From September 26 until December 22, 1920, a caravan of thirty-six mules and two horses would be exploring the highlands, its equipment including "2 large squatter tents for white men, 1 small squatter tent for cook," and "3 bell tents for men."[30]

As the company approached the provincial capital of Harrar, they got a scare. As Moon relates in the journal: "The Major, George and I heard and in another moment saw a bunch of Abyssinians coming quickly toward us in what appeared at the time a very threatening manner. We all got ready for the onslaught of a band of ruffians. They, however, passed with a great deal of noise and shouting, but with no other signs of hostility. We subsequently found out that they were going home after a drinking bout in Harrar as the mascal feast [a Coptic holiday held at the conclusion of the rainy period] was just over and they were merely saying cheerio or their equivalent."[31]

Once in Harrar, they met the local governor, who arranged for a parade of his forces. The event created quite an impression, as Moon relates:

> We had waited ten minutes when there was a lot of shouting, and a bunch of mounted troops dashed across the square shouting and waving spears, at the same time making their horses buck and rear causing themselves to look very fierce. Next came the camel corps; these were Somalis in very bright uniforms and red turbans and waistbands. Then came trumpeters and a crowd of soldiers in no formation at all, more resembling a mob than a company of troops. The Dejasmatch [local governor] came next with his lady mounted on mules . . . the din was indescribable; but we got a very good picture [photo, presumably].[32]

Thanks to the permit from Prince Tafari, arrangements for additional help and supplies fell into place and the mule caravan was soon off for the hinterlands. Soon after they left Harrar, they were joined by a guide provided by the governor. His function, however, seemed to involve more than simple geographic guidance: "The Dejasmatch's guide caught up with us; a very fierce looking chap with a very big sword, a gun, pistol and cartridge belts. His name was Otto Demisi, but he would answer to Dinasterly, Tinasterly, or Tinhats ... in fact, almost anything. He was very fond of rum or sardines, and very keen on going with the Doctor to see what he was really doing. The Doctor says he would be delighted to have him, so he arranged to go with him the next day."[33] Geological and paleontological reconnaissance apparently did not fascinate Demisi as much as it did Brown, however. The next day, Moon relates,

> On our arrival at Dokgou we were met by the governor and a crowd of his friends. . . . This it appeared was the work of Otto Demisi who had gone with the Doctor. He was not however struck with this kind of amusement, and being satisfied that the Doctor was quite harmless, and was not laying mines, or in any other way working for the destruction of the country as he first thought, he gave up going with him and devoted his time to frightening the various governors that we came in contact with, and getting as much food and fodder out of them as possible.[34]

The potential for peril surrounded the group always, not only in the form of ethnic and political conflict but in terms of more natural threats as well, as Brown relates:

> Jackals howled on the hillsides every night and hyaenas prowled about. One of the chief entertainments after dark was to give flashlights to the soldiers, who at a signal flashed them into the darkness, and when gleaming eyes reflected the light the white men would fire. One morning, there were four hyaenas lying a short distance from camp as a result of this practice. . . .
> At Dogou, the Dijazmatch Vake, Governor of Gara Mulata, passed us, leading his army against a raiding party of the Mad Mullah's Somalis who had burned several villages, killing the men and carrying off the women and cattle. Vake made a successful trip and brought back a large number of prisoners, among them the Mad Mullah's father-in-law who had been leading the raid. It fell to our lot to give medical treatment to several of his soldiers.[35]

Between 1900 and 1920, Mohammed Abdullah Hassan, known by the British and other foes as "the Mad Mullah," had led a two-decade jihad to rid Somalia of infidels, including the colonial forces of Britain and Italy as well as Abyssinia. An eloquent poet and orator, Hassan made the hajj to Mecca as a young man, where he joined the conservative, fundamentalist sect of Mohammed Salih. Upon his return to northern Somalia in 1895, he attempted to unite the disparate clans of the region, and in 1899 he launched

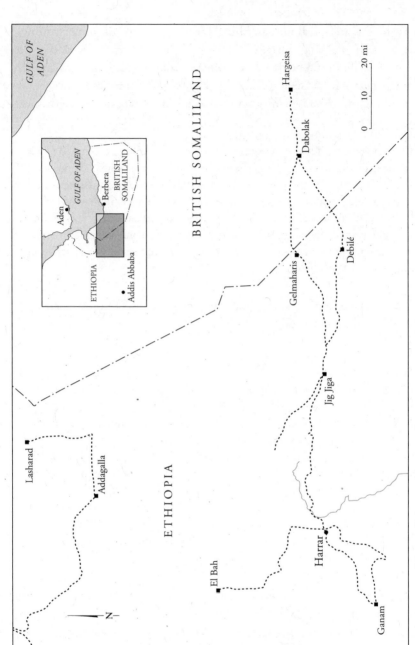

MAP 5. Route of the Dudley Expedition to Abyssinia in 1920

FIGURE 26. Unidentified criminals hanged in an Abyssinian town in 1920, during Brown's Dudley Expedition for the Anglo American Oil Company (5:4 B11, American Museum of Natural History Library)

his drive to throw off the shackles of the Christian, colonial powers. Over the next twenty years, which were characterized by brutal battles and ephemeral treaties, his fortunes waxed and waned as his warrior ranks expanded after victories and deflated after defeats. His ability to rouse Somalis' spirits with fierce and flowing oratory was balanced by the people's distaste for his harsh religious strictures. Finally, in early 1920, the British unleashed an air assault, backed by their Camel Corps, against his fortresses at Medish, Jid Ali, and Taleh, which led to a rout of Hassan's army. Although Hassan was not killed, he fled across the border into Abyssinia (which is why the Dudley caravan encountered bands of his fighters) and died of influenza in February 1921. Today, Hassan is widely revered in Somalia as a fierce freedom fighter and icon for Somali nationalism.[36]

As the fighting that Brown mentioned had occurred only about ten miles from the caravan's camp, they decided to delay their departure by a day to avoid meeting up with Hassan's militia.[37] As is common with crews of this size, disagreements occasionally erupted, as on October 21:

That evening some of the men got drunk and started fighting; one hit another one on the head with his rifle and raised an enormous bump. Several of the men jumped on him but before he could be secured he picked up a rock and bashed another fellow's head open. Then one of the Shunkallas, a very big man, got him roped. He was kept so all night with two other men tied to him to see that he did not escape....
Next morning Joe held a court of inquiry on the offenders of the night before. They were good friends again, and would not give much information, so they were fined a small sum to be deducted from their pay, with which treatment they were quite pleased.[38]

In early November, the expedition returned to Harrar to resupply. Hearsay abounded about their activities, including the notion that Brown's mission was to construct a new electrical power plant, to drain a nearby lake and drill for oil, or to confront Hassan's band of rebels.[39] Back on the trail, the caravan continued to have brushes with danger, as on the evening of November 15, when the camp was stalked by a Somali robber, leading the wary crew to fire a few warning shots at him.[40]

By early December, most of Brown's survey in the highlands of Harrar Province was completed. He now divided his company, with the goal of leading a smaller camel caravan to survey a few lowland regions near British Somaliland. Accompanied by his head assistant, Hassan, seventeen riflemen, five camels, and two saddle mules, Brown was the only Westerner on this leg of the journey, and he knew that since his contingent included both Muslims and Coptic Christians, diplomacy in dealing with his men was essential. Gazelles, in herds ranging from a dozen to several hundred, afforded Brown's crew with ample meat—a good thing too, because the crew would not eat food that had been touched by either Brown or someone from the other sect. Brown's hunting protocol reflected these restrictions. When game was shot, Brown would immediately announce whether the animal was for his Muslim crew members or Christian. Then, as it dropped, "two members of that faith would rush forward to cut the throat before it died."[41] If it stopped kicking before the throat could be cut, no one of either sect would eat it.

On December 4, the team was reminded that the country into which they were headed was home to the most feared of the continent's denizens: "At eight o'clock met a large camel caravan of Somalis without guns, and a few spears. A lion jumped into their zariba [makeshift wooden enclosure or corral] last night and grabbed a sleeping man, but was driven off after biting the man terribly in the breast. He was carried off to the village of Djibeli, but will probably not recover."[42] Brown's closest run-in with these symbols of natural royalty came on December 13:

> While camp was being pitched in an old zariba, I strolled out and saw a lion, but when the gun bearer arrived, had lost sight of it. Followed up and saw tracks of two lions when I reached the spot. . . . I instructed the guards to call me if the lions came during the night. Was wakened at twelve by the guard who said he had seen the lions. After some time watching I used the flashlight, but could discover nothing. Went back to bed. The relief guard reported next morning that he had seen the lions but failed to call me. On examination, we found tracks of the two lions within ten feet of the tents, but saw nothing of the lions.[43]

On New Year's Day of 1921, Brown's party, accompanied by heavily armed guards, set out for territory controlled by the Donkali tribe.[44] Not only do these warriors have a reputation for fierceness, but the region is one of the most forbidding places on Earth. Many of his crew refused to go, including some of his guides. As they progressed, Brown's contingent passed dozens of deserted homes constructed of lava in the typical Donkali fashion; their abandonment indicated that they were in land that was disputed by the Donkali and a rival tribe, the Issa. One day as they were returning to camp, a group of Issa warriors approached with the intent to attack, mistaking them for Donkali. But when the Issa realized that the party was led by a Westerner, they altered their strategy and demanded baksheesh, a sort of healthy bribe common then as today throughout the Middle East and northern Africa.[45]

Brown also learned from his fellow crew members of a particularly loathsome cultural ritual practiced by the Issa. In an unpublished note written for his memoirs, Brown writes:

> Among the Issa Somali tribe of northeastern Africa there is a practice not generally known.
> Mothers nurse their babies from the breast and if the baby is a girl when weaned the mother aunts and female relatives gather. One holds the child and the mother with a pair of scissors cuts off the child's clitoris. As explained to me this does not impair her capacity to bear children when married but it keeps her from seeking other men for she has no amorous feeling during intercourse.[46]

FIGURE 27. Group of Issa warriors who at first mistook Brown's caravan for an enemy tribe in disputed Donkali territory during the Dudley Expedition to Abyssinia in 1920–21 (5:4 B11, American Museum of Natural History Library)

The barbarous practice of female circumcision continues to this day, not only in this region but also in ethnic communities of Western Europe.

A week after his run-in with the Issa, Brown decided to do a little hunting:

> With the chief [presumably Hassan] and five picked men went down the canyon to a deep waterhole.... We soon saw hippo tracks... and followed them to the pool. Presently, I heard splashes and grunts from the rushes, and just spied one before he submerged, but could not get a shot. We stayed there for five hours, but he never came up. The natives say they will remain submerged all day.
>
> This was a beautiful pond of water 200 ft. long by 30 ft. wide fringed with cat-tail rushes and tall plumed grass. I spent the rest of the morning here fishing, and longing for dynamite. Caught seven large catfish. With fish hooks and line I was a sight of great interest to a large number of natives, who collected along the banks and watched. They had never before seen fish caught in this manner.[47]

In terms of expedition goals—to find new lands laced with rich deposits of oil and fabulous fossils—the trip was a disappointment. Although Brown discovered a rich diversity of fossilized marine animals without backbones, including new species of corals, urchins, clams, and snails, he found no dinosaurs or even fossil mammals. In fact, this was his only expedition that failed to find troves of vertebrate fossils.

Little did he know, however, how close he came to discovering one of the most important fossil fields ever. On the last leg of his camel caravan through the northern reaches of Harrar Province, he passed easily within a hundred miles of the Middle Awash drainage at Hadar, which in 1974 yielded the spectacular skeleton of the then earliest known human, Lucy (*Australopithecus afarensis*). Today, these sediments continue to yield an array of important fossils—of not only hominids but also the other mammals and vertebrates that inhabited this Plio-Pleistocene landscape. The discovery of Lucy and other hominid fossils from Hadar greatly influenced our modern theories of human evolution, and moved the cradle for human origins from Asia (as Osborn and Matthew believed) firmly into Africa.

Although Brown presented "unfavorable conclusions" to the Anglo-American Oil Company regarding the likelihood of easily attainable reservoirs of oil, he certainly laid the geological groundwork for further exploration. In 1923, the company announced that it had secured exclusive rights to prospect in the northern half of Harrar Province, the aim being to find an oil field near the Mediterranean, thus alleviating the need to ship large quantities of oil from fields in the Gulf of Mexico, Venezuela, and India. In the 1930s, however, disputes over oil rights in the region triggered an "undeclared war" between Mussolini's Italy, to which Rockefeller was allied, and Haile Selassie's Abyssinia. In the negotiations that followed the conflict, Selassie was offered peace if he would surrender the oil-rich Fafan Valley in Harrar Province to the Italians. Although Selassie rejected that proposal, the Ethiopian oil fields were eventually controlled by the Rockefeller interests through the Sinclair Oil Corporation. Interestingly, Sinclair would play a major role in Brown's fieldwork in the western United States during the 1930s, although there was apparently no direct connection between that work and Brown's work in Abyssinia.[48]

As Brown wrapped up his expedition, a vivacious, aristocratic twenty-one-year-old from New York, Lilian McLaughlin, ostensibly on a world tour with her aunt, was traveling through Egypt and awaiting Barnum's arrival. The pair had met several years previously in New York before Barnum departed for England to begin the Abyssinian project. Fourteen years younger than Barnum, she was, as revealed in photos, a slightly rotund woman, with a round face adorned with dark eyes, short hair, and an infectious smile. Frances reveals that Lilian was personable but often opinionated and strong-willed; raised a Roman Catholic, moreover, she was "city-born and convent-bred." She was also naturally flirtatious, and her gaze was not exclusively fixed on Barnum any more than his was on her. Nonetheless, Lilian "had decided back in New York that Barnum was the husband she wanted, and, if he would not come after her, she would go after him, even if it meant crossing a couple of continents to do so."[49]

Lilian may have harbored an ulterior motive in her pursuit of Barnum. Fancying herself a budding writer, she intended to serialize her experiences and adventures in magazine articles and possibly even books, and Barnum could certainly provide a ticket for exotic exploration. Although we know of no diary that she kept, the AMNH archives preserve a prolific correspondence that she wrote, primarily to her two sisters, while overseas. In reading her letters, one senses that the words and thoughts simply streamed out of her nimble mind, such that her hand could barely keep up. Her penmanship is an indecipherable scrawl, but later she took the time to type out the texts, revealing torrents of flowing descriptions and narratives, punctuated by daggers of derisive gossip, occasional racial epithets, and nuggets of heartfelt advice.

As 1920 came to a close, a lonely Lilian wrote her family from Cairo on New Year's Eve, lamenting that she had not yet heard from Barnum, even though she was sure he was on his way.[50] But a week later she had and seemed determined to make the most of her opportunities while she waited: "I am in the throes of British entertainments and must say I have fallen for the British officers—not any one of them, but they are all fine chaps—good looking and real white men."[51]

Lilian had still not heard from Barnum by January 18: "I cannot understand what has happened to him or where he is, or why he has not let me hear from him."[52] The reason was that the fieldwork was not finished until January 19. A month later, Lilian had made her way to Suez, and Barnum had finally written. Lilian wrote her family that he had been delayed and that she was unsure of his plans. But the uncertainty did not seem to cause much anxiety, largely because she had "been having such a wonderful time in Egypt. Everyone knows 'the little American' and there are many British officers here who have fallen very hard for little Tish [Lilian]. I am wild about the Englishmen. They are adorable. It has just been one round of dances and teas and dinners."[53] Eventually, however, the couple was reunited.

Meanwhile, unbeknownst to Barnum, many of his letters sent from the depths of the Abyssinian desert had not made it through to Europe or the United States. On May 18, Charles W. Brown, Marion's father who was taking care of Frances in Oxford, anxiously wrote Brown's colleagues at the museum: "We have not heard from Barnum since the last of March, and we are getting a little concerned about him again. Has anything been heard at the Museum from him since that time?"[54] Matthew responded that he had just been drafting a letter to Marion's father to ask whether Brown had contacted the family, because the museum itself had received "no word whatsoever from him since last December. . . . We would be very much obliged to you for any information you can give us as to his whereabouts."[55] "It was during this long period of silence [between March and May]," Frances later wrote, "that the American Museum notified Barnum's father-in-law that the expedition must be presumed lost or destroyed. Always before, Barnum's silences had been punctuated by the arrival at the museum of carloads of fossils, but this had not been that kind of an expedition. When the oil company had given up hope for its explorers' survival, the museum felt that it must pass this sad information along."[56]

Finally, though, word made it through to Brown's family; just three days after Matthew's response, Charles Brown wrote: "I have received information that Barnum Brown has been in London."[57] Osborn was especially cheered by Brown's reappearance, writing: "After giving you up for lost in the Abyssinia deserts, I was greatly relieved to hear of your cable to your daughter that you had arrived safely in London."[58]

Now, Brown turned his attention to a new worry: Would his recent exploits derail plans for an expedition to Asia, especially given that thorny legal action? In an unusually direct appeal to Osborn, Brown wrote on June 3, 1921:

> The work in Africa took much longer than I anticipated, largely due to the attitude of the Abyssinian government. This has upset my plans, and unfortunately brings personal affairs into consideration. I want your advice.
>
> You may recall that a blackmail case was brought against me last year. My counsel, Edward Stetson Griffings, 140 Nassau Street, New York, wrote me that this case was indefinite; it might possibly come up this spring, but probably not until fall. Now it has moved forward rapidly. . . .

Under the circumstances what shall I do? Go to India as planned for a limited survey subject to recall by court, or go home and finish this case before going to India? It has occurred to me that it might be possible for you to have the time for this case extended if you think it advisable.

It will of course work to my disadvantage if known that I have been connected with an Oil Company, and I hope you may be governed accordingly.[59]

By the end of June, Barnum and Lilian were in Turkey. While Brown was doing reconnaissance in preparation for fossil collecting in the region, he was also clearly carrying out work for the U.S. government. As he observed in a letter to Osborn on September 1: "The Embassy has been very grateful for the data I secured for our State Department in Turkey, and want more from Greece."[60] Those "data" are discussed, curiously, in a 1941 popular article about Brown published in *Natural History:*

The Museum work [fossil collecting] was interrupted when his two other "bosses" clamored for his services more or less simultaneously. The oil company wanted [Brown] to interview Mustapha Kemal with a view to making some important commercial arrangements. At the same time, the United States Government asked him to see if he couldn't find out why the Greeks were being systematically peppered with British-made ammunition despite the fact that Great Britain openly backed them against the Turkish adversary.

With one eye cocked for signs of contraband explosives, the other alert to any natural history items that might come to light along the way, and his mind busy with the problem of properly approaching the redoubtable Mustapha on that oil matter, Barnum Brown made for Ankara, undaunted by a few Turkish shells that geysered uncomfortably close to the American destroyer that carried him across the Dardanelles. Asked how he fared on this complex mission, Doctor Brown replied that the oil transaction was successfully consummated. And the ammunition leak? "I secured the information they wanted." And did he collect anything for the Museum in Turkey? "No," he said testily, "I couldn't find the time."[61]

(Mustafa Kemal Atatürk was the military and revolutionary leader that led the political and military campaign referred to as the Greco-Turkish War of 1919–1922, also called the War in Asia Minor or the Greek campaign of the Turkish War of Independence, which led to the establishment in 1923 of the Republic of Turkey.)

Meanwhile, plans for the next leg of Brown's international odyssey, this time to India and Burma, were being finalized, and Lilian wrote from Istanbul: "Well, it is all decided about India, and we will go from here to London, just stay there long enough to make out the report..., and we are on our way.... The President of the Museum is now awaiting B in London, so the details I do not know.... This case of B's will of course hasten our trip a little, and it is a shame.... If my papers are in London we will be married before we go to India. They surely should be there by this time."[62]

As later correspondence clarifies, the papers for which Lilian was waiting were her divorce papers. But Barnum approached the prospect of a new marriage with considerably more trepidation than Lilian; indeed, given the legal case as well as other complications, he seemed intent on hedging his bet and delaying the marriage. A true test of wills loomed on the horizon.

TEN

Jewels from the Orient: Raj India
(1921–1923)

LIKE PARTS OF A PALEONTOLOGICAL pinball machine, Brown, together with Lilian and her aunt, careened off over the Mediterranean, through the Suez Canal, and across the Indian Ocean to Bombay. Yet there was a method underlying this madness. Brown was now hot on the trail of more mammals for Osborn and Matthew, the ultimate goal being to discover fossils of Cenozoic primates that would support Osborn's contention that Asia was the cradle of human origins. Brown's agenda was ambitious, with an eye to collecting from outcrops of terrestrial Eocene, Miocene, Pliocene, and Pleistocene sediments exposed along the foothills of the Himalayas from what is now India, Pakistan, and Myanmar, then returning to Asia Minor and finally prospecting on the Greek island of Samos.[1]

Brown's successor as curator at the museum, Edwin Colbert (who married Matthew's daughter Margaret), laid out Brown's thinking in a monograph about the Indian expedition, written about a decade after Brown completed his fieldwork. Osborn and Matthew, he wrote, dispatched Brown to the subcontinent "in order that the Museum might have representative faunal collections from these areas, so elaborately studied by various European paleontologists." In part, securing such a collection would solve some financial and logistical issues for researchers in the United States: "American students of American evolution [will] now be able to gain a comprehensive understanding of the Upper Tertiary faunas of India, and without the necessity of a long trip to London or Calcutta, they will be able to compare these Asiatic faunas, a method of study that undoubtedly will throw a great deal of illumination on vexing problems of phylogeny, of mammalian migrations and of international stratigraphic correlations."[2]

Brown's efforts in India and Pakistan were aided significantly by one of the European paleontologists to whom Colbert alluded. Guy Pilgrim was a British geologist and paleontologist who joined the Geological Survey of India in 1902 and for the next three decades roamed the Tertiary exposures of the subcontinent, Persia, and Arabia in search of vertebrate fossils.[3] He was essentially following in the footsteps of Hugh Falconer, a Scottish geologist, paleontologist, and botanist who in 1830 signed on as an assistant surgeon with the British East India Company. Wandering the Siwalik Hills of what is now India and Pakistan, often in the company of Proby T. Cautley, a British engineer who garnered fame for building the Ganga Canal, Falconer discovered the first fossils of Cenozoic reptiles and mammals from the region. His research on the Tertiary geology and paleontology of the Siwaliks led him to the conclusion that although species stay virtually the same over long periods of time, when a new species does evolve, it occurs rapidly—a pattern similar to the punctuated equilibria posited in 1972 by Niles Eldredge and Stephen Jay Gould.[4] Falconer's interpretation of the pace of evolutionary change put him at odds with Charles Darwin, whose theory of evolution postulated a more continuous and gradual pace of change. Nonetheless, Falconer and Darwin remained close friends, and Darwin valued Falconer's work and discoveries from distant regions of the globe.

While combing the Siwaliks in the 1830s, Falconer may have collected the first-known fossil anthropoid, or ape.[5] In 1878 an anthropoid palate, described by Richard Lydekker a year later, was discovered in the Siwalik sediments of the Potwar Plateau of modern-day Pakistan.[6] But it was Pilgrim who found the first significant fossil primate specimens in the Potwar Siwaliks, which he described in two papers published in 1910 and 1915.[7] These, in addition to the 1891 discovery by Eugene Dubois of "Java Man," now recognized as belonging to the species *Homo erectus,* are what led to Osborn's interest in investigating whether the cradle of human evolution might not lie in Asia.

Osborn's peculiar views on human evolution reflected his neo-Lamarckian views on evolution as a whole. Brian Regal, in his book *Henry Fairfield Osborn: Race and the Search for the Origins of Man,* describes Osborn's perspective on human origins as a fusion of his religious and political beliefs, on the one hand, and his scientific convictions on the other:

[Osborn] believed human evolution was purposeful, had specific direction and was built upon divinely inspired laws of organic change. He argued that the present human races first appeared from human-like ancestors with little direct connection to the primates. The difference was in their level of struggle. The primates lived in environments that were luxurious and placid, while humans lived in a harsher, less forgiving climate. Osborn said that man's "struggle for existence was severe and evoked all the inventive and resourceful faculties he possessed." These early "Dawn Men" began to evolve into more advanced creatures because they used the inherently superior characteristics of their "race plasm" (a pool of hereditary material) to overcome the harsh environment they were born into. The population that had the greatest "race plasm"—and therefore advanced the furthest—became the Aryans, Nordics and Anglo-Saxons.... He claimed... the "drama of the prehistory of man" had begun... in Central Asia millions of years before.[8]

Osborn's concept of the "Dawn Men" was a reaction to the argument made by Darwin in the *Descent of Man* (1871) that the first home of close human relatives had most likely been Africa. As Roger Lewin points out in *Bones of Contention: Controversies in the Search for Human Origins*, Darwin's reasoning, later supported by Thomas Huxley and others, was biogeographically and phylogenetically compelling:

Darwin had based his African prediction on what he saw as a close evolutionary relationship between humans and the African apes. "In each region of the world, the living mammals are closely related to the evolved species of the same region.... It is, therefore, probable that Africa was formerly inhabited by extinct apes closely allied to the gorilla and chimpanzee: and as these two species are now man's nearest allies, it is somewhat more probable that our early progenitors lived on the African continent than elsewhere...."

Although Osborn was the chief public defender of evolution in the United States, it was not Darwin's evolution he was defending.... The whole system, he said, was driven by effort, whose reward was progress and in the end a clear superiority of a few. As a result, an immense gulf separated humankind from the rest of the animal kingdom, and no small gulf divided the "superior" from the "inferior" races of humanity. Naturally, there was absolutely no question about which rung of the ladder he, Osborn, occupied.... [There was]... no place for close relations with a tree-climbing ape.[9]

In addition to Osborn's racial bias, he also invoked environmental conditions to justify his theory. As Lewin notes, quoting Osborn himself:

I suddenly found myself forming an entirely new concept of human origins, namely that the actual as well as ideal environment of the [human] ancestors was not in warm forested lowlands... but in the relatively high, invigorating uplands of a country such as Asia was in the Miocene and Oligocene time— a country totally unfitted for any form of anthropoid ape, a country of meandering streams, sparse forests, intervening plains and meadowlands. Here alone are rapidly moving quadrupedal and bipedal types evolved; here alone is there a premium of rapid observation, on alert and skillful avoidance of enemies; here alone could the ancestors of man find materials and early acquire the art of fashioning flint and other tools.[10]

Thus, Osborn believed that fossils of humanlike "Dawn Men" would be found in the Middle Tertiary sediments of Asia. It was primarily to test this hypothesis that Osborn sent Roy Chapman Andrews on the museum's legendary Central Asiatic Expeditions (1922–30) to search for hominid and other fossils. And while Andrews and Walter Granger, Brown's former field mate and expert fossil collector, were penetrating the badlands of the Chinese and Mongolian Gobi Desert, Osborn dispatched Brown to see what he could find in India.

Brown and his entourage spent only a few days in Bombay in early December before heading east toward Calcutta, where they arrived on December 16. Before getting down to Osborn's business, however, Brown couldn't resist a brief diversion for dinosaurs. His prospecting at Jubbulpore in the Central Provinces only yielded "badly broken and disassociated specimens, but I did secure several parts of carnivorous skulls resembling Allosaurus," along with a tail plate of an ankylosaur.[11] Though only fragments, these specimens, including the carnivorous *Indosuchus*, would later provide important evidence that the ancient fauna of India shares its origins with Africa and the other southern continents.

Upon arriving in Calcutta, Lilian wrote her family on Christmas Day 1921 to express her wonderment at "the ceaseless jabber and cries of a passing throng of half naked natives who make it very hard to keep my mind on my job." Other distractions included the arrival of the Prince of Wales, as well as horse races, "dances, shows, . . . and everyone is binged and happy except B and I who remain horribly sober, and so everything gets on our nerves. . . . Everything is so different, and weird . . . an absolutely unchristian world with millions of human beings we call 'heathans,' but who are in reality more religious than most Christians."[12]

It wasn't just the locals who caught Lilian's eye. Though intent on getting married, she was, she wrote on New Year's Day, again distracted by the "corking fine Britishers . . . [who] make life wonderful for a good looking girl." After attending a "punk movie" with her husband-to-be, she bemoaned the boisterous night life, with "horns tooting, and guns cracking, and people

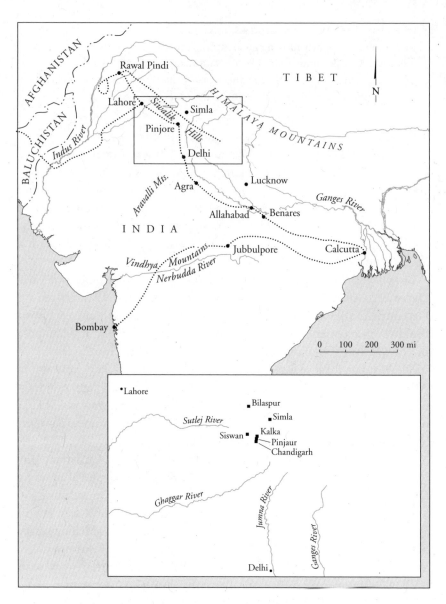

MAP 6. Brown's route through Raj India during the 1921–23 expedition, showing the main Siwalik collecting areas (inset)

acting hilarious like." But the real problem seemed to be that "B and I are horribly sober and I am lonesome as hell."[13] Lilian's loneliness may have resulted in part from Barnum's continued hedging about their marriage date, for a few days later she wrote her sisters:

> Darlings, I have the papers at last, the final decree, which is absolutely necessary to me here, and I intended to get married now in Calcutta, but since I am free and can marry again any minute, after waiting five years about for this time to come, B has decided that he must write home to father Brown and announce our engagement and ask him whether he thinks he should get married again or not, and if so why so, and if not, why not. He does not think it fair to Frances to be married without her knowing it, etc. and when I told him he had five years in which to write home about me, he said he wasn't sure whether I would get my divorce or some such excuse. In London he knew perfectly well it was only a question of weeks till I received my final papers, which he kept urging me to get as quickly as I could. He could have written then, and had the answer now, but he did not, nor is he very anxious to write now. We had all plans made for our marriage here, and he let me make them, and for the past year has led me to believe we would be married as soon as we got to India. I know B well enough to know that this father Brown affair is only an excuse. I love B, and I never could have stuck to him all these years, especially after the Texas affair, the keenest hurt of which was that when it came to the showdown he stood by the other woman at my expense, and to her made me the excuse for the lies and the dealings that got him into that wretched affair. B loves me too I know, and he raises hell if I look at another man, but he carries on affairs with other women of which I have proof. There is a woman in the Museum who has been writing to him for years. When we were home he used to read me her letters, which were "Dear Pal" stuff, and once he told me he had asked her not to write him any more. The other day by accident I found a 20 page love letter, dated November 4th, 1921. It was in among some postcards and I thought of course it was an old letter, knowing her writing. Now I know perfectly well that he has no intention of marrying that woman any more than he had of marrying Texas, but he likes to string women along year after year. Texas nipped it in the bud after three weeks, otherwise he would still have been writing her too. I don't blame her either. Well, so it goes darlings. Ask me not when we will be married, because you know as much about it as I do.[14]

We have found no documents that identify either woman mentioned by Lilian, but the woman from the museum may be the one who filed the breach of promise case against Brown, given that Brown's New York lawyer handled the case. As to why Barnum read the love letters to Lilian, one might suggest arrogance or vanity, along with a desire to impress or anger Lilian, but the truth remains unknown.

Meanwhile, Brown prepared to chase down Pilgrim, who was in the field southwest of Simla, the summer capital of the British Raj, collecting in a sequence of Siwalik strata. Pilgrim had recognized that this 20,000-foot-thick section of sediments was divided into three basic units, which he termed the Lower, Middle, and Upper Siwaliks. That thickness may be an exaggeration, at least in any one field area, but the Siwaliks do comprise an exceptionally thick sequence of strata. That is actually not surprising, given the tremendous forces generated when the Indian subcontinent slammed headlong into southern Asia between 40 and 50 million years ago, initiating the uplift of the Himalayas, with the Siwaliks forming that range's southern margin. Today, these same three Siwalik units are referred to as informal lithostratigraphic subdivisions of the Siwalik Group, with the Kamlial and Chinji faunal zones included in the Lower Siwaliks, the Nagri and Dhok Pathan zones in the Middle Siwaliks, and the Tatrot, Pinjor, and Boulder Conglomerate zones in the Upper Siwaliks.[15] In terms of age, the lower Siwaliks generally represent a period from about 18 to 11 million years ago, the middle from about 11 to 4 million years ago, and the upper from about 5 to 0.5 million years. However, these rock units are often time-transgressive—that is, they are of slightly different ages at different localities—and other named units are sometimes used for layers in the Upper Siwaliks. Geographically, the Kamlial, Chinji, Nagri, Dhok Pathan, and Tatrot are all closely situated in Pakistan, but the Pinjor is far removed from the others over in India.[16] The Siwalik outcrops, exposed along the drainages of the Indus and Ganges Rivers, are now basically interpreted to represent stream and floodplain deposits that were tilted, faulted, and uplifted as the result of plate tectonic mechanisms in the early Tertiary. In general, the Siwalik strata have yielded a host of mammal fossils that include primates, rodents, bears, hyenas, horses, rhinos, pigs, hippos, camels, deer, giraffes, and elephants. Additionally, important reptile remains have been collected, including relatives of the living gharial and, arguably, the largest terrestrial tortoises ever to have lived.

Brown wrote to Matthew on January 10, 1922, that he planned to "go from [Pilgrim's] camp west to the Salt Range in the Punjab where upper and lower Siwaliks are most fossiliferous."[17] Two weeks later and halfway

FIGURE 28. Outcrop of Upper Siwalik strata exposed near Siswan, India, photographed during Brown's AMNH expedition in 1921–23 (5:4 B10, American Museum of Natural History Library)

around the world, Osborn wrote that he had received a letter from Pilgrim in mid-December saying that Brown had yet to show up in India. In a tone that evinces both concern and displeasure, Osborn scolds: "I am again mystified about your movements. . . . It is extremely important that you should make it a rule to write me not less often than once a week."[18] Osborn goes on to reveal that Brown's expedition was being funded by a gift from Mrs. Henry Clay Frick, the widow of Andrew Carnegie's former partner in the steel industry.

Unaware of Osborn's anxiety, Brown left Lilian in late January in Simla, where they had traveled by train. (Since they were not married at this time, and since Lilian's aunt was still with them, they separated during his stints in the field.) Brown set off for Pilgrim's camp by mule over a challenging trail that went "up 300 feet then down 400 then up a ways then down some more continuously, winding round the faces of cliffs and mountainsides always but very picturesque."[19] It took five days to track down Pilgrim, who was twenty-one miles up the Jumna (or Yamuna) River from Bilaspur. After dealing with suspicious officials and broken-down mules along the way, Brown seemed to have forgotten whatever tensions existed between him and Lilian, leaving him to lament: "I am lonesome for my darling and wonder how she keeps her feet warm without hubby these cold nights."[20]

After spending two weeks with Pilgrim, Brown returned to Simla and promptly wrote Matthew about Pilgrim's cordial reception and the trip's results:

> It was a revelation to him to see me develop prospects. Practically all work done heretofore has been surface collecting, and most of that by native peasants.
> ... Pilgrim's method is to go into the exposures, camp in a nearby Bungalow or tent, and send out word to farmers and shepherds that he will pay for the fossils they bring to him.... I must say that he pins them down to exact locality.... But the chief error is that they see and collect only those parts that have been exposed, while embedded portions are entirely lost.
> ... I secured one good Hipparion [horse] skull and jaws broken into very small pieces.[21]

Brown went on to outline his plans to collect in the Upper Siwaliks near Kalka, then proceed south to the type locality of the Upper Siwaliks at Ghaggar—where those beds were formally described in print—on to the Bughti Hills in Baluchistan, then to Hasnot, Chinji, and Dhok Pathan in the Salt Range, and finally to Burma and Java.

Success quickly followed at Chandi near Kalka, where, in what Brown describes as the lower part of the Upper Siwaliks, he encountered a richly fossiliferous zone that yielded "a perfect uncrushed skull ... of Hippopotamus; a hippo palate with good teeth [now *Hexaprotodon*]; camel skull lacking part in front of canines; a young Elephas [elephant, perhaps *Stegolophodon*] jaw and several minor specimens."[22]

Meanwhile, social unrest was on the rise in India, with some cities beset by riots and railway strikes protesting British rule. While Brown was in the field, Lilian got a firsthand look at one protest in mid-February from her bungalow in Kalka. Writing in the middle of the night to her family, she said: "there's something doing, and I'm in the middle of it ALONE." A mob of locals had awakened her with their commotion, so she "crawled to the window in the living room to watch the passing show. All night long they went up and down the road singing weird funeral dirges to the ominous beat of drums. They would always stop in front of the bungalow and hold forth for about fifteen minutes, then go on.... My ears are keened to strange crys,

FIGURE 29. Brown and assistants with skull of "Platelephas" (AMNH 19818) collected along Amlee Creek near Siswan, India, 1921–23 (5:4 B10, American Museum of Natural History Library)

my eyes search the dark corners, and I am wondering what the devil is going to happen next."[23] After a repeat performance the following night, Lilian hired a horse and evacuated to Chandi to join Barnum.

Given the strikes, Brown, advised to leave the area, hurried on to Jacobadad before heading to Sibi in Baluchistan. As is true still today in this tribal region of what is now southwestern Pakistan near the Afghan border, trouble between the tribes and the governments often bubbled to the surface, as Brown reported to Matthew: "Baluchistan is different from other parts of India, the tribal chiefs having full say of all matters in their own territory subject to appeal. It was necessary for me to see the chief of the Bugtis.... He is very loath to grant permission for me to visit his section. Last year an American oil geologist was killed out there and the Bugti tribe was compelled to pay 50,000 rupees to the American Consul for the widow and four of the murderers were hanged." Despite further appeals and Brown's willingness to hire a military escort, he was forced to delay that leg of the expedition to the Bugti Hills southeast of Quetta and return to Jacobabad, before heading on to Hasnot, Chinji, and Dhok Pathan. Frustrated, he complained: "I hate to be balked like this when I know perfectly well I can take care of myself."[24]

Lilian, however, was relieved, writing home: "I am RATHER glad it is off. It would be better for B to lose his fossils, than for me to lose my 'Topiwaller.'" (*Walla* is the common Urdu word for man, and one meaning of *topi* is "pith helmet"; so "Topiwaller" may be a name for a colonial Westerner.) Despite these sentiments, her eye continued to wander; the locals, she remarked, were "great big men, tall and straight, with complexions like coffee and cream, soft black silken locks that hang in curls about their shoulders, and gleaming white teeth that would put any Pepsodent ad to shame."[25]

Problems continued to bedevil Barnum as he tried to get to his next destination, for in Chakwal, "the bubonic plague was rampant, and the town mostly deserted." Although his cook and interpreter had abandoned him, he wrote, he was finally able to assemble a camel caravan for his onward journey.[26]

Meanwhile, Lilian, apparently unable to wait for the honeymoon she hoped to have there, took off on her own for Kashmir. There she encountered an official who tried to get her drunk on whiskey before asking her to sleep with him; however, she made it clear that "if there is anything I LOATHE it is whiskey . . . in fact I was getting madder by the minute. . . . The hours wore on and still he paid no attention to my insults. . . . I told him to go to hell, and that the American Consul would make it hotter than that for him."[27] Ironically, the official turned out to be the chief of police, who had been given the assignment of investigating Lilian to see if she was doing anything illicit; when she refused to drink or sleep with him, he concluded she was legit. Lilian stayed three weeks in and around Srinagar, roaming around the romantic streams and lakes in her gondola-like rowboat, while Barnum labored near Rawalpindi. She remarked to her family, "This will probably be the last letter that I shall write as Lilian Mac," because she and Barnum planned to get married soon after Easter and return to Kashmir for a honeymoon on a houseboat.[28]

By April 23, Brown was expressing to Osborn frustration not only with the local officials but also with the fossiliferously sparse outcrops of the Siwaliks: "In some ways this enormous series of Siwalik strata are discouraging. Thousands of feet of alternating reddish clays [and] thick bands of gray sandstone [with] occasional zones where fossils are numerous but usually broken and rarely associated."[29] He had prospected around Haritalyangar, Chandi, Hasnot, and Dhok Pathan. Yet despite the paucity of articulated bones, he had found some important specimens. Around Hasnot, he "secured several skulls and dozens of jaws, many fine exhibition specimens... [including] a perfect skull of Samotherium?, [a] fine skull of a large Rhinoceros, several antelope skulls, two lower jaws of Palaeopithecus (Middle Siwaliks), and a skull of Sivatherium, Upper Siwaliks." At Dhok Pathan, even though he was pleased to be able to stay in government *daks,* or bungalows, conditions on the outcrops were brutal, with the

> average daily temperature now... considerably over 100 degrees between 9 A.M. and 5 P.M.... I work without shelter from six until nine and then it is necessary to put up the tent. By two P.M. it is a little inferno.... Only ants and flies remain active.
>
> The present quarry is most important because of definite association of species and also because it is the first time I have seen associated skeletal material.... I have a skull and partial skeleton of Hipparion with many jaws and separate bones in sight.[30]

Having returned from Kashmir, which she described as "a beautiful place!," Lilian grudgingly set up housekeeping in Rawalpindi, where they planned to marry. "There are residence laws in India," she explained to her family in late April, "and this is the only place we can claim residence for six months. We won't have been here that long, but we will say we have anyway." These problems, along with the heat, weighed on Lilian and her aunt, leading Lilian to lament: "maybe I shall be married [about June 1st], and MAYBE not. Lou says the sooner the quicker, for she is fed up with India, and is only staying on my account."[31]

In mid-May, Brown summarized his collections for Matthew: "There are hundreds of jaws and parts of jaws, largely antelope. Primates, Carnivora and Rodents are extremely rare in the Middle Siwaliks. Mastodons, Antelopes, and Hipparion are equally common. Here along the Sowan [= Soan] River in the vicinity of Dhok Pathan Mastodon and Hipparion are most abundant. At Hasnot at the same horizon Antelope were most abundant. The fossil horizon at the two places is in the same stratigraphic position that is above the middle of the Middle Siwaliks." Brown was especially excited about two fossil elephant specimens he had collected, though he was not sure how to transport them to the railroad. One was a "magnificent mastodon skull [of] *M. punjabiensis*?" and the other a different and as yet unidentified species. From Dhok Pathan, he said, he planned to move on to the Lower Siwaliks at Chinji about June 1. He then went on to describe his grand agenda: "I want to start work on Samos Island if possible then trace that series of beds into Asia Minor. No work has been done there heretofore on account of the Turks. When I visited Angora [sic: Ankara] and met Mustapha Kemel and his Cabinet they said they would give me every possible help in doing geological work in Anatolia. That was in June last year. Since then that area opposite Samos has been evacuated by the Italians [who occupied the region after World War I] and is again held by the Turks with whom I got on famously."[32]

On another front, Brown had continued to lobby for access to the Bugti Hills, but on June 1 Osborn received a letter from the British embassy in Washington, D.C., indicating that the Viceroy of India had

> ascertained that the chief of the Bugti Tribe is very reluctant, from fear for the safety of Mr. Barnum Brown, to allow him to enter this wild frontier hill tract. This tract is visited by transborder men from independent tribal territory and Afganistan over whom the chief has no control, and who are often of a fanatical and dangerous nature....
>
> In these circumstances, the Government of India feel that they are unable to assist Mr. Barnum Brown in his projected visit to the Bugti Hills and consider it inadvisable for him to attempt to go.[33]

Otherwise, however, Matthew and Osborn were ecstatic, and on June 15 Matthew crowed: "I... congratulate you on a really magnificent collection.... I hope you will have something, even though fragmentary, in the way of primates, which we can use to throw at Mr. Bryan"—an apparent reference to William Jennings Bryan and his anti-evolutionary rants.[34] Animosity between Bryan and Osborn over their contrasting views of Darwinian evolution had burst into public view the previous spring on the editorial pages of the *New York Times*. The paper had invited Bryan to write a piece "excoriating what he saw as the heretical belief of Darwinism" in late February. They then asked Osborn to respond, which he did in early March, lecturing Bryan on horse evolution and stating that the study of evolution did not necessarily conflict with religious beliefs. To see this for oneself, Osborn argued, all one had to do was "drop your books and narrow overspecialized thinking and go out into nature and observe it first hand."[35]

The running battle between Osborn and Bryan eventually culminated in the Scopes trial of 1925, in which Tennessee teacher John Scopes was tried and convicted of teaching evolution to his students in violation of a new state law. Bryan led the prosecution team. Osborn, although he disagreed with the choice of agnostic Clarence Darrow as the lead defense counsel, met with Scopes in New York and provided funding for him. Although Scopes was convicted, the sensational trial (which Osborn did not attend) was seen as a humiliating defeat for Bryan, who struggled under Darrow's questioning about literal interpretations of biblical scripture. Bryan died five days later in his sleep.

Meanwhile, Brown had tackled the dilemma of transporting the proboscidean skulls to the rail line at Chakwal. It took "one week," he told Matthew, "to build a passable road for carts out of the badlands, and then required four bullocks and twenty-one men to move each Mastodon skull." Then, it took two weeks to travel "sixty-five miles, thirty-five of which was without road." Running low on funds after having spent 823 pounds (a little over $4,000 at the time), Brown estimated that to continue operations in India, then go on to Burma and Java, would require $2,500 more.[36]

In a letter to Osborn from India dated mid-June, Brown mentions both his legal case and his impending marriage to Lilian in successive paragraphs. The trial had been postponed several times already due to his travels, and Brown was worried whether the judge would grant more time for him to complete his fieldwork.

> I am uneasy about that case against me which the attorney writes has already been put back several times. I would not like to have judgment entered without a hearing. Will you please see how the court feels about more delay and advise me if I should continue program as mapped out.
>
> Another matter I wish you to know which I have not announced to anyone else.
>
> I shall marry Miss McLaughlin of New York before returning home. No date has been fixed as yet. She and her aunt are traveling in Europe and will arrive in India shortly.
>
> She is not a scientist but I hope she can aid in me being a better one than I have been for several years past.[37]

A couple of sentences are confusing. For one thing, there is no evidence in Lilian's correspondence that she and her aunt had left India at this time for Europe. The meaning of the last sentence, moreover, is unclear, and no further explanation is offered. Perhaps he meant that she would help take notes to better document his field activities, or perhaps it was a reference to the blackmail scandal and an intimation of humility.

At the end of June, in any case, Lilian wrote to her family with the long-anticipated news: "Yes, darlings, it's all over, even the shouting. After waiting all this time it happened all of a sudden, and one morning I woke up MARRIED." The ceremony took place on June 20, after Barnum and Lilian chased down the commissioner, who was on vacation in Murree, and got him to sign the papers, which they then posted for four days on the courthouse door. And so Lilian finally found herself "duly 'hitched up' to Brownie." They left that afternoon for a weeklong honeymoon on a houseboat in "the most beautiful spot on earth," Kashmir. From there, Brown returned to his fieldwork near Rawalpindi, while Lilian prepared to send her aunt Lou back home. She looked forward to settling down after the trip, telling her sisters that home would be in New York "because B would not give up the Museum for all the millions in the world, for which I thoroughly cuss him out."[38] Lilian does not state what she would have preferred he do. Possibly she wanted him to take a higher-paying job at another museum, which in earlier correspondence with Osborn he said he'd been offered, or she may have wanted him to take a more lucrative position with an oil company or mining firm.

In mid-July, unaware that the wedding had taken place, Charles Brown wrote to Matthew saying that his family hadn't heard from Barnum since January and requested that the museum pass along any news of his safety and whereabouts.[39] The department secretary quickly fired off a note in response, followed in two weeks by another, letting the Brown family know that the museum had received letters dated May 14 and June 21, which indicated that Brown was well and working near Dhok Pathan.[40]

With Brown back in the field, Lilian made the best of things in Murree by partying with the British officers: "Last night three of the sexually unemployed went to an amateur theatrical given by one of the regiments.... After an hour journey up and up to the top of the hill, we arrived at Captain Heatley's bachelor quarters, and fell on the booze, of which he had a goodly array of whiskey, port, sherry, beer 'n everything. Thus fortified, we went to the show, which seen through very rosy glasses couldent help but be gleeful."[41]

On July 22, Lilian wrote to her sisters saying she hadn't heard from Barnum since the 8th and was growing worried. She further lamented about having no one at hand to be intimate with; "I wish I could have twins or something, but so far, nothing doing???" Things were still up in the air with regard to future fieldwork as well. She expected Brown to continue working in the Siwaliks until September, but whether they would then go to Burma, Java, or the Bugti Hills was unclear. The prospect of Brown venturing into the latter region, though unlikely, still concerned her greatly, "because they are death on the white face there.... When the gentle Baluchistanis capture a white man they turn him over to their women to pass the time away. If they don't skin him alive and send their heads back to the regiment by special bearer, as they had done to some of the poor devils on the frontier, they cut off their 'works' and send them back to camp minus ammunition, gun 'n everything???" The soldiers, she said, even carried vials of poison for use in avoiding this fate in the event of being captured. She ends the epistle with a plea that her sisters pass along any gossip about the "love affairs of the family.... At present I have none but my old darling Brownie. We fight like heck, but it's mostly his fault, and he says it's mine, and we make up."[42]

By early August, Brown had returned to Rawalpindi and reported to Matthew and Osborn that he had finished his work in the Lower Siwaliks, having sent "83 cases of fossils, and 2 cases of alcoholic specimens to Calcutta... to await shipment to America." He now intended to return to Chandi for a month or six weeks to finish collecting in the type locality of the Upper Siwaliks. His take from the Lower Siwaliks, he wrote, included "six cases of specimens, the finest of which is a complete skull of a baby tusker Mastodon about a year old, dentition complete; several good Mastodon jaws; two Rhinoceros skulls, different genera; Hyaena skull; Hyotherium (?) skull; a pair of Anthropoid jaws, and many jaws not classified, antelope, camel, Bos." Again, Brown said that he had never seen a fossil field where material was so fragmentary. Moreover, since he was "the first white man who has ever collected for himself in these fields," he had encountered resentment on the part of the natives, who were used to doing the collecting for foreign paleontologists. One of his paid collectors had even complained that if Brown collected everything "there would be nothing to sell the next officer." The expense report Brown submitted for lodging, meals, hauling teams, and transportation totaled about 74 pounds (somewhat less than $400).[43]

On August 8, Osborn, having just received Brown's letter of June 17, wrote to express his pleasure at Brown's collecting success and to suggest that, after finishing work in the Siwaliks, Brown should focus his efforts on the Bugti Hills and Burma. Osborn was planning to travel to China in order to assess the progress of Andrews's Central Asiatic Expeditions and hoped to be able to meet Brown in Burma along the way. In response to Brown's request for an update on the legal case, Osborn responded: "Regarding your personal affairs, I have not heard recently from your attorney but will try to get into communication with him next Monday, if I can find his address; the fact that I have not heard from him would indicate that matters are progressing favorably."[44] A few days later, Matthew wrote even more effusively, complimenting Brown on his work and lauding the Indian excursion as "your greatest scientific achievement yet"—the words, clearly, of a mammalian paleontologist rather than a dinosaurian expert. Matthew seconded Osborn's advice that collections from the Bugti Hills and Burma should round out the endeavor and went on to stress the importance of collecting at Samos in order to "have the materials for an adequate research on the Indian faunas," since the ages of the fossils from Samos corresponded with the age of some Siwalik faunas.[45]

Meanwhile, Brown's attention was being diverted to other geological prospects. Instead of going to Simla from Rawalpindi, Brown had responded to the challenge of an associate, a Mr. Anderson; as related by Lilian: "'You're not going to Simla today, you're going to Kashmir.' 'The deuce I am,' says B, 'the train leaves in a half hour, and I'm on it or bust.' 'Well, look at this and decide,' says Mr. Anderson, handing [Brown] a huge chunk of stone that was mostly lead. As it happened, B is looking for lead deposits here, and it looked interesting."[46] Brown indeed reconsidered and left for a week of lead prospecting in Kashmir.

Lilian was tiring of life on the road and lamented to her sisters that she often suffered bouts of homesickness: "It would be much more enjoyable to take it in small doses, a country at a time, and go home between drinks. But we can't afford to do that. I think I will like married life if I ever get a chance to settle down to it. So far [over almost two months], I have only been with B about one week since our marriage."[47] Although they were now legally married, they were also making plans to have a formal church wedding when they eventually arrived back in Calcutta, a rite they apparently could not arrange in Murree because of Lilian's divorce.

In mid-October, Brown wrote Matthew to inform him that he had almost suffered a wild goose chase in the Jammu Hills of southern Kashmir, but that the excursion was redeemed by a remarkable discovery—some more ammunition to fling at Mr. Bryan. The fossil he found was "a primate of importance... the anterior end of a mandibular rami, broken off after the [front of the lower jaw, which] includes the complete symphysis." Brown recognized it as a "large Simian type" but was unable to identify its genus. Additionally, apparently near Chandigarh, Brown reported collecting "many teeth of Elephas and Stegodon; a Hippopotamus skull; Sus Giganteous skull; a skull of a Bovid that I do not recognise; a skull of a deer with a fine pair of antlers; [and] two large turtles." Brown ends by saying that he had not given up hope of collecting in the Bugti Hills of Baluchistan, and that to press that quest he would meet with the viceroy in Simla, though he also commented that there had been "continuous raiding and ... fighting on that part of the frontier all summer."[48]

Now that they were married, Lilian could accompany Barnum to the field, a circumstance that pleased her. Before, she had felt isolated and lonely when Barnum was off. Then too, life and travel in India as a single woman carried some risk, as in Kalka, where she had spent two sleepless nights alone, except for a mongrel pup she had adopted, as rioters and protesters clamored outside her lodging. But now in Chandigarh, she could set up housekeeping in a government bungalow and manage Barnum's domestic affairs while he was off collecting. As she described her situation, "B and I have decided to be together from now on.... I shall go right into camp with him.... This life is ideal, and I am enjoying it much more than when I [was] knocking about alone in hotels."[49]

Both Barnum and Lilian, however, were tiring of dealing with the locals. "For the Camera and the story book," Lilian wrote, "this country is fascinating, but to live among these black birds grows more galling to the white man every day.... East is East and West is West ... thank God they'll never meet." One incident that prompted her vitriol was what Lilian took to be an assault that occurred when she was shopping for supplies with Barnum. Someone grabbed her by the arm and turned her around, but it was the shopkeeper, not Barnum. In defense, she struck him with her umbrella and called to her husband, who rallied to her aid by making "a beeline for the devil, knocked off his turban (which is like throwing down the gauntlet) and beat him up good and proper."[50]

Back in New York, Osborn and Matthew were thrilled with Brown's results to date.[51] Brown, meanwhile, having failed to persuade the viceroy to let him venture into the Bugti Hills due to continued raiding by Afghan tribes, began to focus on wrapping up his work in India and heading to Burma—but only after one last month of work in the Upper Siwaliks between the "Sutlej River and the border of Nepal State."[52] His efforts were rewarded by the discovery of "the greater part of a camel skeleton" with a

FIGURE 30. Barnum (far left) and Lilian Brown (on horse) traveling by caravan through Punjab, India, during the 1921–23 expedition (5:4 B10, American Museum of Natural History Library)

fragmentary skull and complete lower jaws, along with a "fairly good skull and jaws of an alligator." He also collected one of the immense turtle specimens he'd mentioned in earlier correspondence, which, although severely broken up, included a good deal of the carapace and plastron "with some limb bones [and weighing] approximately eight hundred pounds. The femur is as large as that of a Rhinoceros."[53]

In late November, Lilian reported that they were still at their last camp near Kalka, but that Barnum was preparing to send in the horses and close up operations.[54] They then planned a three-week journey by train along the foothills of the Himalayas on their way back to Calcutta, where they would regroup and sail to Burma.

In early December, Osborn wrote to congratulate Brown on his discovery of the primate jaw, though he longingly wished that "it had included the molars" and said he hoped it would spur Brown on to search for more primates. Osborn had not been able to go to Asia, but, unaware that permission had already been denied, still hoped that Brown could gain access to the Bugti Hills. Beyond that, Osborn felt it essential that

Brown go to Burma to examine some Eocene beds on which Pilgrim had just published, and from there head to Samos before returning home.[55] Matthew, writing a week later, concurred and added that he and Osborn agreed that a trip to Java held a lower priority, since it had already been extensively prospected. Matthew did not concur about the Bugti Hills, however, confiding: "I will be rather glad if you have to call it off, for I don't like the idea of your going into such a disturbed country while there are so many things to be done in regions that are peaceable."[56] If any hope remained on the matter, it must have been laid to rest when Brown received a letter dated December 8 from the Foreign and Political Department of India once again denying him permission to venture into the Bugti region.[57]

In mid-December, Matthew wrote to Childs Frick, the sponsor of Brown's Indian work, to relate that the estimated value of specimens collected totaled $11,870, whereas expenses to date, including an estimate for shipping, amounted to $11,890.[58] The accounting seems to represent a statement to the donor that, although expensive, the expedition had produced results that justified the costs.

In retrospect, it's interesting to note that Roy Chapman Andrews, supported by Osborn, raised about $600,000 (about $7 million in today's currency) to fund the Central Asiatic Expeditions between 1922 and 1930.[59] Although that venture was tremendously successful in discovering new Cretaceous dinosaurs as well as a wealth of new Cretaceous and Tertiary mammals, it found no primate fossils. In contrast, Brown's expedition to the Siwaliks cost only about $150,000 in today's dollars and succeeded in securing a significant sample of *Sivapithecus* specimens for the museum, as well as a host of other Tertiary vertebrates.

The list of paleontological jewels included two mastodon skulls, two hippo skulls, three rhino skulls, skulls of *Giraffokeryx, Bramatherium,* and *Chalicotherium,* a pig skull and braincase, seven antelope skulls, four *Hipparion* skulls, a camel skull, a gavial skull, a bovid skull, a cervid skull, and two large turtles. Such success kept the museum's engine for generating field funds humming as the holidays approached, and at the end of December Osborn wrote Brown with the news that "the generous benefactor who has hitherto supported your work in India has contributed funds to enable us to carry it on another year." Osborn continued to hope that Brown could secure permission for work in the Bugti Hills, he continued, but in the meantime it was imperative to "clean-up while you are on the ground with no rivals" by visiting Burma and then Samos. What with Brown's finds and the collections coming in from Andrews's expedition in Mongolia and China, Osborn foresaw that the museum could boast having "the finest collection of Asiatic mammals in the world."[60] Three specimens that Brown collected on the Indian leg of his overseas odyssey are now on display at the museum: skeletons of the crocodilian *Gavialis* and the giant tortoise *Geochelone* in the Hall of Vertebrate Origins, and two mammoth teeth in the Hall of Advanced Mammals.

In 1926, following the initial preparation of Brown's Siwalik specimens, Matthew went to India to study Pilgrim's collections in preparation for writing a monograph on the Siwalik faunas.[61] He published a preliminary study in 1929 but was unable to complete his monograph before his untimely death in 1930 at age fifty-nine. Colbert picked up where Matthew left off and completed the monograph, in which he credited Brown for the collection, describing it in glowing terms: "The American Museum now houses one of the outstanding collections of fossil mammals from the Siwaliks; a collection that ranks not only with the classic assemblage of fossils gathered together early in the last century by Hugh Falconer and Proby T. Cautley, and now contained in the British Museum in London, but also with the Geological Survey of India collection in Calcutta, accumulated and described by Richard Lydekker and by Guy E. Pilgrim."[62]

In the decades since Brown explored the Siwaliks, many of his specimens have been scientifically described and identified, and phylogenetic and taxonomic revisions continue to be made. Even by today's standards, Brown's collection produced a diverse and significant set of holotypes, which include a species of bamboo rat (*Rhizomyides punjabiensis*), a porcupine (*Sivacanthion complicatus*), a marten (*Martes lydekkeri*), an aardvark (*Amhiorycteropus browni*), a rhinoceros (*Gaindatherium browni*), a peccary (*Pecarichoerus orientalis*), a deer (*Cervus punjabiensis*), four relatives of elephants (*Zygolophodon metachinjiensis, Protanancus chinjiensis, Gomphotherium browni,* and *Paratetralophodon hasnotensis*), seven species of antelopes (*Selenoportax vexillarius, Tragoportax salmontanus, Tragocerus browni, Sivaceros gradiens, Strepsiportax gluten,* and *Strepsiportax chinjiensis*), and two species of gazelles, (*Gazella lydekkeri* and *Antilope subtorta*).[63]

As the Indian leg of his transcontinental trek neared an end, Brown continued to be troubled by extraneous events. He apparently inquired about the legal case, although no record of his queries has been found. But in mid-January, Osborn wrote Brown with a caution: "If you wish to write me confidentially, mark your letter 'Personal,' and it will not be opened except by myself."[64]

Barnum and Lilian spent Christmas Eve and most of Christmas Day packing fossils in Chandigarh.[65] He then went to Rawalpindi to try and collect his fee for the mineral prospecting and report he had done in Kashmir. While there, he wrote Lilian that he'd had trouble tracking Anderson down but finally caught up with him at the railway office, at which time "I gave him straight talk and of course he wiggled and squirmed, thought I was working through friendship, etc. I told him he would have to pay my fee or I would bring suit against him and the firm."[66] The outcome is not discussed in further letters.

Meanwhile, Lilian headed to Patiala for a fling with Indian royalty. Decked out in a red, white, and blue satin dress, topped off with a "statue of Liberty" crown, she joined friends at the palace of a Sikh maharajah for New Year's Eve. Being the only American at the gala, she caught the maharajah's eye: "I went into dinner with the Prime Minister, and sat on the Maharajah's left . . . and danced two dances with his Royal Highness. . . . He is six feet three and built in proportion. . . . I liked him very much and he put himself out to be sociable with young America. . . . His palace is beautiful in a way, but his real homes are with his Maharanes (queens). He has six of these little things, one for every night of the week except Sunday. Now he is looking for a Sunday wife."[67] Although Lilian does not mention this in her letter, Barnum's daughter, Frances, reveals that the maharajah "not only entertained her royally but even allowed her to spend some time with the women in his harem, something no Westerner had ever been permitted to do previously."[68] Summing up her impressions, Lilian wrote her sisters that "when an Oriental takes a fancy to a woman, he immediately thinks of her from the standpoint of breeding, while our men are hoping to heavens [her] good points lie in the opposite direction!"[69]

With the fossils packed, Barnum and Lilian arrived in Calcutta in mid-January 1923, where, as Lilian confided to her sisters:

> I've had a little excitement. . . . Just when things got unbearable I decided to relieve the monotony by having my wisdom teeth out, and on the same day B took a notion to have a church wedding. So on the bright and cheery

FIGURE 31. Portrait of Lilian Brown dressed in a sari, ca. 1922 (AMNH Vertebrate Paleontology Archive, 2:6 B7 F2)

morning of February first we Rolls Royced over to the Thoburn Methodist Church where a very nice American Padre married us according to the rites of the Christian religion.... I feel that I am really married until death or someone else do us part. This is the wedding B will announce. It took place at ten A.M. At twelve I was in the dentist chair, so life was made up of one thrill after another.[70]

Apparently content that, having discovered a treasure trove of fossils and squeezed in two weddings as well, they had accomplished all they could in India, the newly remarried newlyweds hopped a ship headed for Rangoon.

ELEVEN

Perils and Pearls
Up the Irrawaddy: Burma

(1923)

IN HIS LAST LETTER TO Osborn and Matthew from Calcutta, Brown laid out his plans for work in Burma with characteristic confidence. He had consulted with a Mr. Stamp, who had previously collected there and provided Brown with detailed information about his fossil sites. The country contained exposures of latest Middle Eocene sediments, about 37 million years old,[1] as well as Mio-Pliocene beds about 7 to 3 million years old that represented an eastward extension of the Siwaliks. G. D. P. Cotter from the Geological Survey of India had also prospected in the Eocene outcrops in the southern region of Burma's Pondaung Hills in 1913.[2] But neither Cotter nor Stamp had published their finds, so very little was known about the faunas that had inhabited the area in the geologic past. Although Stamp was "sanguine" about the prospects for gathering a large collection, Brown begged to differ, telling his superiors that "I anticipate that I will get better results than have been obtained heretofore in Burma." Brown expected to arrive in Rangoon on February 7, from where he would "begin work first near Yenang Yaung on the Irrawaddy, thence up the river about fifty miles, going inland about seventy miles where the best Eocene exposures occur. After this work I will proceed to Mandalay to examine the Upper Siwalik exposures." He estimated that the expedition would take two months and cost $1,500 (about $17,600 in today's currency) for travel and shipment of specimens.[3]

By late February, Brown and Lilian had reached Yenangyaung, which Lilian explained to her family meant "creek of bad smelling waters better known as the Burma Oil Field." Brown was again mixing paleontology with his geological consulting, but Lilian, ensconced either in her bungalow on the riverbank or at the nearby American Club, was enjoying the life of luxury that the oil industry provided its foreign agents, especially given that prohibition was in full swing back home. As she described the scene to her sisters:

> First my almond eyed Burmese chef de bar, in his pink silk bathing cap, tight silk skirt and short white linen jacket will shimmy amongst you with some cocktails the like of which has not tickled my palate anywhere else in the world.... You may drink to the whole world, and let the whole world know you are drinking, just like the olden, golden days of not so long ago. After the cocktails, we will have Cheroots ... dainty tobacco morcels a foot long and an inch and a half in diameter, [which] are the women's choicest tidbit. (doctor and nurse in attendance in case anyone passes out after the first round)[4]

The oil field, however, was a stark contrast, as workers labored in the "forest" of 1,200 wells. Each well was dug by hand, with laborers working in three-hour shifts, and the walls were stabilized with wooden shoring as the excavation deepened. When oil was struck, as Lilian described,

> they bring it up in diver's shifts of three hours each. The diver is naked except for a G-string, a helmet made of a kerosene tin, and an air hose coiled around his body. He is let down into the well (those I saw were 270 feet deep) from a beam across the top, and the air was pumped to him through a large machine like a coffee grinder.... The oil is drawn up from the bottom in kerosene tins attached by a rope to the beam at the top. This rope is lowered and drawn up by a corps of eight or ten Burmese girls.[5]

Despite the females' subservient status at the rigs, Lilian was amused by the generally matriarchal nature of Burmese society, in which the "women rule the roost," through handling all the family business, smoking the "biggest cigars you ever saw," and ogling as many men "as their hearts desire and their heads think profitable." Such customs led to relatively informal marriages, she observed, which amounted to "a legal affair binding the two parties. When they wish to separate another legal ceremony frees them from each other. Each takes about five minutes to perform."[6]

By early March, Brown had worked his way up the Irrawaddy, stopping whenever promising outcrops came into view and he could convince the captain to pause. Essentially, as Brown explained in a 1925 *Natural History* article, Burma is composed of three geologic provinces: the Shan Plateau in the east, which is an extension of the Yunnan Plateau in China and

FIGURE 32. Workers preparing to retrieve oil by hand from a well at the oil field near Yenangyaung, Burma, as photographed during Brown's expedition in 1923 (5:4 BIO, American Museum of Natural History Library)

formed primarily of Paleozoic rocks; the north-south trending valley of the Irrawaddy River in the center of the country, which once contained a narrow marine gulf that has been filled by sediments of Eocene-Pleistocene age washed down from the Himalayas; and the then little-studied Arakan-Yoma Range to the west.[7] Of particular interest to Brown were the late Middle Eocene, Pondaung sandstones and clays, thought to be equivalent in age to Uintan sediments in the United States, and the younger Mio-Pliocene sandstones and clays of the Irrawaddy Series, which are similar in age and fossil content to part of the Siwaliks in India and Pakistan.

From Pokokku on March 5, Brown reported to Matthew that the pickings were rather slim, "mostly fragments valuable only for identification of fauna." Nonetheless, he had collected a "fairly good Hippopotamus skull from the upper Irrawaddys at Yenanyaung," which he thought was equivalent to a Dhok Pathan species of the Siwaliks, as well as a *Stegodon* tooth, though its fragmentary skull was not well enough preserved to collect. All in all, he lamented that fossils were "few and far between."

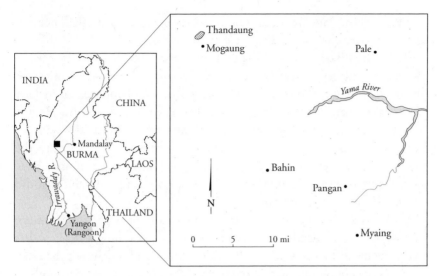

MAP 7. Brown's fossil site near Mogaung, Burma, where he collected in 1923

From Pokokku, he wrote, he planned to set out cross-country by bullock cart for a 150-mile jaunt to the Pondaung Hills, a journey that would take him through Myaing, Kyetmauk, Kine, and Chinbiyit to Monywa on the Chindwin River. There he could connect with a rail line. This excursion would require that he be incommunicado for a period of time, but as he explained:

> Myaing is where Cotter found most of his Irrawaddy fossils, and Kyetmauk and Kine are localities that Stamp tells me are the best for Eocene fossils.
>
> If this trip does not turn out successful (Survey people are not at all optimistic) I shall go on up the Irrawaddy to Mandalay where a native Barrister has found a considerable number of jaws and bones of Stegodon and Bos in the Irrawaddy series.[8]

He hoped to complete his work in Burma in late May and sail for Greece on May 26, if fossils weren't too numerous and things went according to plan—which, as it turned out, they would not.

That same month, Frances's guardian, Charles Brown, wrote to Matthew that Brown's daughter, now fifteen, was planning to travel with one of her teachers to New York City and was anxious to see her father's collections at the museum. Matthew graciously replied that he would take "great pleasure [in meeting] the daughter of my old friend and associate and I shall not delegate to anyone else the privilege of showing her around a bit."[9]

Meanwhile, Barnum and Lilian left the Irrawaddy, heading by cart to the remote village of Gyat to set up a base camp. While Brown searched for fossils, Lilian made friends with the locals, who happened to include an orphaned baby elephant, which she christened Bimbo Brown. He was "hardly larger than a St. Bernard dog, . . . [and] constantly at odds with his feet or his trunk which were always getting in the way."[10] To help pass the time, Lilian adopted the pint-sized pachyderm, which like any human baby often required extra attention, including 3 A.M. feedings—a habit that did not sit well with a weary Barnum.

At Gyat, Barnum spent his days taking long walks or rides, trying to trace the fossil-bearing beds through the dense jungle. One evening when he didn't return at the normal time, Lilian became alarmed and sought out the town elder in hopes of mounting a search party. When she said that Barnum had set off that morning toward a village called Thittadaung, the elder scowled and said that bandits from that town occasionally ambushed Westerners, stealing all their belongings, including their clothes. He refused to mount a search party, however, and Lilian returned to her house quite distraught. Her housemaid tried to calm her by saying, "Madame please not to worry. . . . Dacoits [bandits] not making troubles with mens anymore; just womans. Tying naked to trees and holding for ransom." This intelligence failed to soothe her anxiety. "It was bad enough to have Barnum wandering through the jungles without his pants," she later recounted, "but naked girls bound to trees along his path . . . that was a ticklish situation." As dawn broke, the hoof-falls of Barnum's horse resounded along the road into town. Relieved, Lilian dryly inquired: "With whom did you spend the night, Mr. Bones?" Fidgeting with his pipe, Barnum explained:

> Well, I'll tell you. . . . It was growing dark and I was heading for camp through the hills when suddenly, only a few feet away, I saw a ball of light big around as my thumb. Naturally, I thought it was a huge firefly. . . . But this light was different. . . . I dismounted and pressed the brush aside till I was directly over it. Then I struck a match. There in full view, Pixie, was a luminous spider, the large oval abdomen glowing in the dark. . . .
>
> Fearing it might be poisonous, I wrapped a handkerchief around my hand and made a quick grab. The cloth caught on a twig and my treasure scurried away. I watched the light vanish in the brush and spent the rest of the night looking for it. . . .

It would have made big news in the scientific world.... No one has ever caught a luminous spider, [and] only one other white man, to my knowledge, has ever seen one.[11]

Soon, Brown had finished his work in the area and began to make plans to move on, which raised the question of what to do with Bimbo. Lilian wanted to take her pet with them, but Barnum refused. Pressing her cause with calculated logic, Lilian exclaimed: "Suppose you had been presented with a live baby dinosaur. Would you leave it? We have a cart full of fossil elephant bones—a lot of them old, cracked and busted—yet you have seven fits if someone so much as breathes in their direction." Barnum suggested that Lilian ask the town elders what would be best for Bimbo. To his relief, they counseled that baby elephants didn't survive long outside their native habitats, so it would be better to leave the tot with them. Grudgingly, Lilian accepted their advice.[12]

Toward the end of April, Barnum and Lilian emerged from their nearly two-month-long trek through the Burmese jungles. It had been a somewhat frustrating quest; due to lack of roads, Brown was largely unable to collect in the western exposures of the Irrawaddy Series and had to limit his prospecting to the Eocene exposures of the Pondaung sediments. Only one unit within the Pondaung Formation, a fifty-foot-thick sequence of purple clays, had proved to be worth investigating for vertebrate fossils, and Brown had doggedly tracked it throughout the backcountry. Although dense vegetation impeded his quest for these beds, which he described as "like following a rainbow," he was able to trace the deposits for seventy-five miles. However, he told Osborn and Matthew, he "would frequently ride and walk ten hours without seeing a fossil, for they occur in localized areas with many barren patches in between." In the end, he reported, "The collection is small, but better than all collections made heretofore, and I believe contains all [previously] described species; the choice specimens are a complete skull and lower jaws of a small Anthracothere; lower jaws of Metamynodon; upper and lower dentition of Anthrocohyus, and half of a lower jaw of Stegodon with good tooth from the Irrawaddy series.... This is a curious fauna in which Carnivores, Insectivores and Rodents are absent."[13]

Brown's accounting of his collection omits one spectacular discovery.[14] Near the village of Mogaung, Brown encountered an outcrop of the purple clay that was littered with pieces of bones and teeth. One specimen, about the size of a half-dollar, was a fragment of a lower jaw containing three teeth.

Though unable to identify it, he collected it and packed it up to be shipped back with the rest of the fossils. Only later, when E. H. Colbert examined the specimen during his study of Brown's Siwalik collection, was the jaw fragment recognized to belong to a primate, and not just any primate. In 1937, Colbert named a new genus of primates based on the specimen: *Amphipithecus mogaungensis,* meaning "near-ape from Mogaung." Along with another primate fossil found earlier by Cotter, which Pilgrim named *Pondaungia cotteri* in 1927, these specimens now represent the earliest-known remains of the group called anthropoids, which contains monkeys and apes, including humans.[15] Thus, once again, Brown had hit the primate jackpot for Osborn, even though he didn't realize it at the time.

Given the general paleontological paucity of the Eocene outcrops, Brown intended to set off for the region around Mandalay before returning to Rangoon and sailing for Samos. A five-week-long gap in Brown's correspondence reflects more than his pursuit of specimens near Mandalay, however. In fact, he almost died. At the start of June, he informed Matthew that "I am just out of hospital where I have been confined two weeks with a severe case of malaria; temperature of 106 plus for four days."[16] According to Frances, Lilian, as much as the medical staff, was responsible for saving his life: "Apparently [Barnum] had been overdoing the work and had depleted his resistance and physical strength.... Lilian was competent but frantic and brought a doctor into the camp to do what he could. When Barnum's temperature reached 106 degrees, the doctor said the case was hopeless. How Lilian saved him should have been written up in medical journals. It must have been a combination of twenty-four-hour-a day very skillful nursing and very determined love. Whatever it was, the fever broke, and ever so gradually the patient began the slow road back, very literally from death's door."[17]

Lilian's account provides more details. Upon returning from an excavation, Barnum had seemed unusually fatigued. She put him to bed and sought out the British doctor, who told Lilian that Barnum had contracted black malaria, the worst form of the disease. The doctor prescribed sixty-grain doses of liquid quinine to be taken six times a day, an unusually large dosage, then told Lilian: "It's kill or cure, you know ... may pull him through ... no nurses available ... you'll have to go it alone.... Feel up to it Mrs. Brown?" For five days and nights, the illness brutalized Brown in daily cycles as Lilian packed him in ice to try to keep the fever down: "The ghastly routine went something like this: Six o'clock, fever rising ... 103 degrees ... 104 ... icepacks ... *pray.* Then the long watch through the night ... delirium ...

fever... 105... 106.2... 106.3... life hanging in the balance of a half-degree. *Pray.* Then drenching sweats as the fever declined, and then—the long sleep so close to death that I often thought my darling had slipped away, only faint moisture on a mirror held to his lips showing that he still lived." Through the flames of fevered delirium, Barnum relived his past exploits in weak whispers that Lilian could barely make out: following the scrapers on the family farm in search of his first fossils, studying at the University of Kansas with Williston and his buddy Elmer Riggs, discovering the first dinosaur specimen for AMNH at Como Bluff, being shipwrecked on the Patagonian coast. On the fifth night, the fever broke, and the doctor declared Brown stable enough to be moved to the hospital. As he came to and recognized Lilian, Brown whispered, "Just tired, Pixie. Just tired—that's—all. Day's rest—fix me up." The immense doses of quinine left him temporarily deaf, and he weighed less than one hundred pounds.[18]

In the aftermath, the doctor ordered Brown to recuperate at a facility in the hills at Maymyo for about a month, because the microbes had "destroyed a large percent of the red blood corpuscles so that I am so weak."[19] Lilian described his condition as being "better but very thin. He puts me in mind of a man walking on stilts, but the stilts happen to be his legs."[20] Of course, Brown was most irritated at the loss of time for collecting. The illness had "scrambled all my plans."[21]

Before contracting the disease, however, Brown had managed to prospect the sandstones in the upper Irrawaddy Series near Mingun, where, although fossils were once again rare, he had found "a good palate with teeth, and a pair of lower jaws of another individual, both Stegodon."[22] Although he had hoped to find more primates, that quest was unsuccessful. With the delay caused by his illness, Brown now planned to sail for Samos later in July.

As yet unaware of Brown's plight, Osborn wrote in mid-June to congratulate his collector on "the very valuable list of fossils constituting the Burmese collection of 1923." Osborn also told him that the shipment of fossils from the Siwaliks had safely arrived in New York. To celebrate Brown's successes in India, Osborn informed Brown that he had "decided to place your collection in large special cases [in the exhibition halls] and not mingle them with other collections, so you will have a monument of your energy and devotion."[23]

As Brown slowly recovered, Lilian spent most of her time marveling at the Buddhist monks and pagodas around Mandalay: "I like to wander through the little side streets... early in the morning when the pongyis (monks) are collecting their donations, mostly food. They wear a bright yellow costume somewhat Roman style, ... and they look like a swarm of yellow jackets going in and out of the hive." She was particularly taken by a climb up Mandalay Hill to the Arakan or Mahamuni Pagoda, where she enjoyed the immense brass statue of Buddha—"that is the foundation is brass, but it is covered with solid gold leaf which the worshippers apply themselves."[24] This nearly two-thousand-year-old statue is thought to have magical powers and is still visited by thousands of visitors today.

At the end of July, Matthew received Brown's letter about his bout with malaria, and he responded characteristically, with acute concern and pointed caution:

> I do not like the idea, not a bit, of your being exposed to such risks. No results are worth it, for there is always plenty to be done in areas where it is reasonably safe and healthful. Of course this is a matter that has to lie with your judgment, but I earnestly hope that you will play [it] safe and not let yourself be persuaded into taking undue risks by any enthusiasm either on your own part or on ours. No scientific results that could be attained by taking them are worth the loss of your life or health, personally or professionally, to your friends and colleagues.

Matthew had also apparently received Barnum and Lilian's announcement of their marriage, about which Matthew expressed great satisfaction, instructing Barnum to "felicitate Mrs. Brown upon a fine husband whose fine character and sterling good qualities we know well from long and close acquaintance."[25]

By August 1, Barnum and Lilian had set sail from Colombo across the Indian Ocean to the Red Sea, where Brown wrote Matthew a summary of his exploits. On his way back to Rangoon from Mandalay, he had prospected further in the Irrawaddy beds near Mandalay and downriver at Kabanee, where he found "great numbers of fragments... but only one specimen... worth taking, the Hippopotamus jaw." In all, he reported, he had three cases of fossils from Burma, which were being shipped from Rangoon. However, he had kept separate the primate material from India to ship from Alexandria, because "I could not risk sending such rare things by freight, nor is the mail service of India or Burma to be trusted; once lost, gone forever." He included a financial statement for the Burmese expedition, which totaled $4,930 (about $58,000 today), along with a specimen list that detailed twenty-one fossils, including parts of two fossil hippos, three mastodons, one horse, an alligator, three anthracotheres, and four *Metamynodon* specimens.[26]

Subsequent fieldwork and research carried out by Russel Ciochon and Donald Savage in Myanmar in the late 1970s resulted in the discovery of several more complete specimens of *Amphipithecus* and *Pondaungia,* as well as one specimen that may represent a new genus of anthropoids. Our earliest apelike cousins shared the Eocene landscape of the region now represented by the Pondaung Hills with a host of other intriguing vertebrate relatives, including hippo- and piglike anthrotheres, rhinolike brontotheres, diminutive deerlike artiodactyls, and various rodents, lizards, turtles, fish, and crocodiles. They lived, wrote Ciochon, along "a medium-size river that drained seaward toward the Burmese Gulf, which in the past was located much farther north. Along the banks of this river, which was partially covered by a forest canopy, anthracotheres, brontotheres, and small artiodactyls came to drink. Turtles, crocodiles, and fish swam in the river. In the trees above, the ancestors of the higher primates romped."[27] Brown's discoveries had sketched out the outlines for this compelling portrait. For a small fraction of the sum invested in the Central Asiatic Expeditions, Brown had opened a new window on the earliest phase of higher primate evolution, one that eventually led to humans and other apes.

Now that he had wrapped up his Burmese excursion, however, he was lacking some critical collecting gear, as he lamented to Matthew: "The last trip I made down the Irrawaddy, I had the misfortune of having my coat and knapsack stolen, losing pick, chisels and geological hammer. Send me two picks with handles, a geological hammer, and a bottle of ambroid thinner."[28] And so, as they ricocheted back through the Red Sea, the intrepid pair prepared their assault on the small Mediterranean isle of Samos.

TWELVE

Samos: Isle of Intrigue
(1923–1925)

IN 1921, BEFORE LEAVING WITH LILIAN and her aunt for India, Brown had laid the groundwork for his sojourn in Samos by conducting reconnaissance. Earlier collecting on the island, especially by British and Germans, had established that fossils from the Miocene exposures on Samos would be key to interpreting the evolutionary and environmental history of that period from Europe eastward to the localities in Asia from which Brown had just returned.

Explanations for the fossils found on Samos extend back to around 200 B.C., when the Greek geographer Euphorion identified them as the remains of menacing creatures called Neades. Other ancient commentators interpreted them as belonging to Amazons, a tribe of warrior women, who had died in battle, with the reddish stain on the bones—which in fact came from the sediment in which they were buried—being explained as their blood.[1] Whatever their true identity, the bones were highly revered, and some were even displayed in the isle's famous Temple of Heraion.

In modern times, Charles I. M. Major conducted the first scientific excavations for fossils on Samos in 1887 and 1889.[2] Other Europeans soon followed. Today, collections of mammal fossils from Samos are found in museums across England, Germany, Switzerland, Austria, Hungary, Italy, and Greece, as well as in several U.S. cities.

Brown was the first North American to collect on the island. Of course, Osborn and Matthew were hopeful that he would discover fossils of anthropoid primates that would shed more light on human evolution, as well as fossils that would provide new insights into other mammalian lineages. Matthew, in particular, had a special interest in the Samos fauna as a result of his paleoclimatic and paleobiogeographic research. In a 1915 paper, "Climate and Evolution," Matthew had argued that the positions of continents and ocean basins had remained relatively constant throughout geologic time and that migrations between continents had occurred across strips of land that joined or nearly joined adjacent continents, such as the Bering land bridge and the Isthmus of Panama.[3] Although plate tectonic theory has rendered Matthew's paradigm obsolete, much of his underlying work involving the description and dating of fossil species remains highly valuable today. Matthew was also correct in realizing that Samos was an important intercontinental link. As Nikos Solounias and Uwe Ring note in their paper "Samos Island," "The fauna is located between three continents and has species that can be related mostly to those of central Africa (rainforest and savanna) and secondarily to those of Asia. The fauna is least similar to those of central Europe."[4] Brown's work on Samos was designed to help establish the critical database that would make such comparisons possible.

Brown arrived in Athens with Lilian on August 18, 1923, and immediately began trying to secure the requisite collecting permits. Although optimistic of eventual success, he described the tribulations as involving "the favorite Eastern pastime, diplomacy, which is intricate and devious." The most problematic barrier was a Greek paleontologist named Theodore Skouphos, whom Brown described as "the gentleman who caused Smith Woodward so much trouble, and who was also instrumental in having the law passed that prohibits exportation of fossils except those not wanted here." Fortunately for Brown, at this time a political tussle was under way between "royalist" and "revolutionary" factions within the Greek government. For the moment, the revolutionists had the upper hand, and, conveniently, the wife of one of the revolutionary generals, Mary White Tsipouras, was an American southern belle who had lived in New York. As Brown succinctly put it, Mrs. Tsipouras—whom he met courtesy of the American chargé d'affaires—was well acquainted with the museum and anxious to help out, which was fortunate because success "will be due chiefly to her influence."[5]

Together with the chargé d'affaires and Mrs. Tsipouras, Brown met with the minister of education, whose duties included the granting of collecting permits. Although Skouphos would have to approve, the minister was favorably disposed toward Brown's request, especially since American relief agencies had "saved so many Greeks from the Turks and starvation during the past year."[6] Skouphos was out of town at that time, however, so the permit could not be finalized until his return.

At the same time, Barnum was fighting on another front as well, as Lilian intimated to her sister Brad: "I am now in the throes of a desperate fight with B. He's a darling and I love him same as ever, but he has an idea that I'm going to trail around with him in these God-forsaken ports-of-call until I'm old and feeble and will have to be shipped home instead of enjoying the trip of my own accord."[7] This fight had been sparked when Lilian expressed a desire to go home to the States for Christmas, about which "B was first mad, then sad, and has now resorted to weeping and wailing."[8] Barnum's fear, she explained, was that he might "lose" Lilian if she went and visited her sister, a Hollywood screenwriter of growing reputation, in California. Part of his apprehension may have stemmed from his having lost his first wife, Marion, to illness fifteen years before. But more immediately, he loathed the prospect of losing his second wife to someone from the "Hollywood crowd."

Lilian shared with her sister a love of writing and clearly wanted to transform her experiences on this round-the-world jaunt with Barnum into books. Her motivations were both creative and financial. Once she and Barnum returned home, she confided to Brad, she expected she would need to find work; although his salary was sufficient to support them in the field, living in New York would be another matter. And "this is what B is afraid of—that I'll get into some work that will take me away from him."[9]

This issue would flare up throughout the couple's tenure in Greece, as would Lilian's concern about her relationship with Barnum's daughter. Frances had just written her father, apparently complaining about her present situation and laying out her plans for schooling. As Lilian told her sister Madge,

> She is 15 years old and has been brought up with her mother's parents.... Of course she has had no hectic youth like most girls have, and she has grown up in a little town of 2000 people, all old aristocrats. She has just finished her second year of high school and wants to go to boarding school in Albany this fall.... It is a very nice school, also a prep for college, so she should go and get out amongst other girls. The fee is 1000 a year. B gives 1000 a year for her support anyway and she might as well be where she is happy. She did not like the idea of her daddy getting married, but she'll get used to it when we know each other better.[10]

By September 10, the permit had been finalized, granting Brown permission to collect under the same conditions that had been afforded to other paleontologists from Great Britain, France, Austria and Germany.[11] The museum archives contain copies of the draft permit, which stated that all laws would be obeyed and that the purpose of collecting was to "classify, establish relationships, and determine migration routes of Tertiary animal life extending from China through Burma, India and Europe to the United States." Excavations would be overseen by a representative of the Greek government, and "of the finds, those which would not be considered necessary for the filling... of gaps in [Greece's] Palaeontological Museum will be placed at your disposition for exportation to America."[12]

Brown reported to Matthew that issuance of the permit had required a vote by the Greek Cabinet of Ministers. The cause for such top-level action was a dispute between Brown and Skouphos, who had demanded that the Athens museum be permitted, at Skouphos's discretion, to take half the specimens that Brown collected. Brown refused, arguing that the specimens would need to be placed in plaster jackets in the field for their protection, making it impossible for Skouphos to evaluate them without opening each jacket. According to Brown, Skouphos grew "insulting and said I could take his terms or leave them.... I asked him where I could see a copy of the law pertaining to fossils, and he replied that it was not necessary for me to see the law, as he was the living law."[13] With Mrs. Tsipouras running interference for him, Brown appealed to Mr. Plasteras, the "Revolutionary President" of Greece, who suggested that Brown submit a formal request to the Ministry of Public Instruction that could be voted on at the next cabinet meeting; this Brown did, resulting in the issuance of the permit—without the approval of Skouphos. Brown was careful to provide copies of all documents and notes of conversations to the U.S. legation. Brown implored Matthew and Osborn to commend both Mrs. Tsipouras and the minister of culture and public instruction, suggesting that Osborn grant Mrs. Tsipouras a life membership to the American Museum of Natural History. Having settled these diplomatic matters, Brown was set to depart for Samos on September 19.

On that same day, Matthew wrote Brown that he had received his cable about securing the permit and that he had begun to unpack Brown's collection from India, which included a rare "creodont," *Dissopsalis,* as well as another specimen that might be another primate. Matthew stated that the advanced nature of the *Hipparion* skulls and teeth from the Middle Siwaliks and Lower Chinji—as advanced, he said, as anything from the Republican River beds—might necessitate that workers at AMNH revise their earlier correlation scheme and "put back our so-called Lower Pliocene into the Miocene, returning to the old Cope correlation." The Republican River beds are now recognized to be Late Miocene rock layers in southwestern

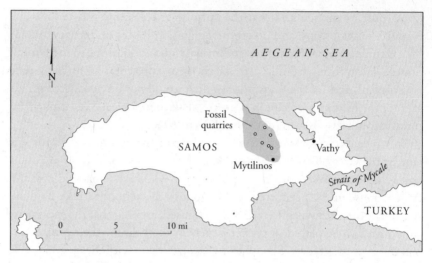

MAP 8. Greek isle of Samos, showing the quarries where Brown collected in 1923–24

Nebraska. Matthew also informed Brown that Osborn had safely made it to China, having averted potential disaster by leaving Yokohama just twelve hours before a massive earthquake struck.[14]

About a week later, Matthew gleefully wrote to Brown that the box of other primate fossils from India had arrived safely, and "Gregory leaped upon them with joy when I showed them to him."[15] William King Gregory had been a classmate of Matthew's under Osborn at Columbia, and in 1900 Osborn hired him—much as he had previously hired Matthew—as an assistant at AMNH. Gregory's primary task was to organize the departmental library (now named in honor of Osborn), and he worked with the departmental artists as well. Also like Matthew, Gregory did much of the legwork for Osborn's major research projects, including papers on sauropods and a monograph on titanotheres.[16] Through such efforts, Gregory became an exceptional comparative anatomist, as well as an influential researcher in functional morphology. Eventually Gregory, at Osborn's suggestion, directed those considerable skills toward research involving the evolution of primates, especially the relationship between humans and anthropoids, a pursuit that dominated his paleontological efforts after 1913. By 1916, Gregory's research had led him to conclude that gorillas and chimpanzees had evolved from species of *Dryopithecus,* and humans had evolved from *Sivapithecus,* another Miocene anthropoid from the Siwaliks. This research in turn led Gregory to argue that chimpanzees and gorillas are the closest living relatives of humans, a conclusion that put him in opposition to Pilgrim and Osborn, who felt humans were at best distantly related to those apes.[17]

Despite their differences (including Gregory's dismissal of Osborn's theory that humans originated through "Dawn Man"), Osborn supported Gregory's right to disagree, and Gregory became the museum's ace on questions on primate evolution.[18] It therefore made perfect sense that Matthew should propose to Gregory and his collaborator, Milo Hellman, a dentist in New York, that they begin to analyze Brown's primate fossils from the Siwaliks: "I have advised Gregory and Hellman," he told Brown, "to go ahead with a thorough biometric study . . . leaving the matter of their zoological description until I hear from you and Osborn [who was off to Mongolia with Andrews]. . . . They are the first real contribution made by our Asiatic expeditions that has a bearing on the ancestry of man. We don't want to get them into the lime-light until we are quite sure what that bearing is, but we do want to get thorough and elaborate research upon them promptly under way."[19] By early October, Matthew had also begun to act, through the museum's administration, on Brown's suggested commendations for Mrs. Tsipouras and the minister of culture and public instruction.

By that time, Barnum and Lilian had set up camp on Samos outside the town of Vathy, near the village of Mytilinos. Lilian told her sisters that the "wonderful" accommodations consisted of two eighteen-by-twenty-foot tents, with one serving as their bedroom and the other as a sitting and dining room, which allowed her to build a suitable if somewhat impromptu nest:

> B has even had to build the furniture. We have a good bed, and B has built a sort of wardrobe for our clothes, and I have a long table for my dressing table. . . . The tent is linen, and I will cover everything with rose cretonne. The floor will be covered with a homespun matting. . . . The living room I will drape in blue, and the boxes of provisions will all be covered for seats. B will make an up to date buffet, long and low, and I have a lot of semi nice things to make the place look cozy. I am even going to put little curtains in the windows . . . cute![20]

A week later, Brown sent word to Matthew that the excavations had commenced on October 4, and to date the crew had etched a sixty-foot-long, ten-foot-deep wound in the hillside as they removed the overburden above the bone-bearing bed, which was later named Quarry 1. The crew consisted of local Greeks, many of whom had arrived on the island as refugees in the

FIGURE 33. Brown's crew excavating Quarry no. 1 near the town of Vathy on the Greek island of Samos during the AMNH expedition in 1923 (AMNH Vertebrate Paleontology Archive, 8:2 B9 S1)

previous two years as a result of hostilities between Greece and its archenemy, Turkey.[21] Although most had been prosperous farmers in Turkey, they had had to flee for their lives because of the conflict. Conditions for them were now bleak, and work was hard to come by. Never one to miss an opportunity, Brown hired eighteen men to excavate his quarry and six women to haul out the dirt in baskets. For this labor, Brown offered "the attractive sum of 35 drachmes (70 cents) per day for the men and 20 drachmes per day for the girls." In a memoir about their stay on Samos, Lilian later recalled that, in true Greek form, the quarry turned into a veritable Olympic venue, with "the girls trying to cue up as many empty baskets as possible against the boys who handled the shovels. The baskets, of course, were kept continuously filled; an empty one invited so many jeers."[22]

Quarry 1 was located adjacent to an earlier quarry excavated by the Austrian ambassador to Samos Karl Acker, who was also a wine merchant and amateur paleontologist. In 1924, Samos was not formally a part of Greece, but rather a protectorate. In mid-October Brown told Matthew that Acker, who was living in nearby Vathy, had just sent him a "threatening letter ... commanding me to desist working on his land, and thus save him the trouble of invoking the Greek tribunal."[23] Brown, however, had made an arrangement with the owner of the land; he therefore suspected that Acker's threat was instigated in part by Skouphos. He had taken pains, he said, to keep the U.S. chargé d'affaires and Greek officials in Samos apprised of the situation.[24] Brown eventually planned to open a second quarry a short distance away.

In early November, Matthew wrote expressing confidence that Brown would resolve his problems.[25] Osborn and Andrews had returned from Asia, he said, and Osborn was especially pleased with Brown's primate and proboscidean fossils from India and Burma, exclaiming to Matthew: "Why, that is the finest Siwalik Mastodon skull that has ever been found!" In all, Matthew told Osborn, Brown's proboscidean collection from the Lower, Middle, and Upper Siwaliks contained five or six skulls belonging to *"Mastodon," Stegodon,* and *Elephas*.[26] Osborn and Matthew also wanted Brown to write a popular article for *Natural History* magazine about his expedition to India, which would accompany one by Andrews describing the treasure trove of fossils the Central Asiatic Expeditions (CAE) were garnering in China and Mongolia.[27]

On December 4, Brown wrote to Matthew that his crew's excavations had progressed to within eight inches of the upper bone layer throughout Quarry 1, which was now 60 feet long, 25 feet wide, and between 35–45 feet deep.[28] Unfortunately, folding and faulting within the hillside had fractured many of the fossils, but Brown had jacketed and removed several of the larger specimens, especially those threatened by falling rock along the back wall. Two layers within the quarry, separated by about 1.5 feet of sediment, were producing bone; both appeared to represent a "wash accumulation," presumably bones concentrated by currents in the ancient streams of the area. These deposits are now considered to represent the Mytilini Formation, a Late Miocene rock unit thought to be 7.2–7.0 million years old based on radioisotopic age estimates and paleomagnetic characterizations of rocks in this and other regions.[29] Solounias and Ring provide an age estimate of "7.6–6.9 ma for the Main Bone Beds and 6.7–6.5 for the Marker Tuffs above them."[30] Brown's best specimens included about six skulls representing two genera of horses, skulls from two genera of rhinos, a good deal of deer material, isolated bird bones, and a five-inch-long skull of a small, rarely preserved carnivore. These results buoyed Brown's confidence in the ultimate success of his Samos enterprise, and although he could not promise anything as spectacular as the collections discovered by his colleagues in Central Asia, he assured Matthew of "another record collection unsurpassed by anything previously secured here."[31]

Osborn, meanwhile, had instructed his administrative staff at the museum to prepare the commendations for Mrs. Tsipouras and the government minister that Brown had requested. In addition to confirming that action to Brown, Osborn exclaimed:

> This is certainly the golden age of the American Museum. Everything is coming in from every part of the world in superb quality and quantity.
>
> I am greatly increasing the preparation force, so as to handle the magnificent fossils from Asia.... As soon as these are worked up Doctor Matthew is arranging to put the whole force on your Siwalik fossils, especially the proboscideans, which are superb.[32]

On Christmas Day, Lilian wrote her sisters once again of being homesick. Nonetheless, she was determined to celebrate the holiday by preparing a Christmas dinner as close to the traditional American feast as she could, featuring Greek chickens and a variety of canned goods, including soup, sweet potatoes, plum pudding, pineapple, nuts, and raisins. The lack of more fresh ingredients led Lilian to quip with an air of sarcastic resignation, "In fact, the whole dinner is CANNED before it ever starts." She finished the letter by confessing that she would "now drown my troubles in 'La Samienne' the delicious wine of Samos."[33]

She would need it, because on Christmas Eve, the camp received a double punch from Mother Nature when the isle was lashed by the most violent storm in many years. In a letter to her sisters, Lilian described her terror as hail, rain, thunder, and lightning pelted their flimsy shelter. As if that weren't abuse enough, an earthquake rocked the region as the strobelike lightning illuminated "pines bent in two" and the winds tore the tents from their moorings. "All that remained," she said, "was for the graves to give forth their dead."[34] For Barnum's part, he described the events to Osborn simply as "the worst gale I have experienced since my shipwreck off Tierra del Fuego."[35]

In the same letter, Brown updated the finds from Quarry 1, listing a series of vertebrae with limb bones of a small deer, a complete skull and jaws of an edentate (a group that includes sloths and anteaters), some incomplete limb bones of a proboscidean, and partial skulls of *Samotherium*. He also requested literature on what had previously been collected from the area, anticipating problems getting the collection out of the country. "The Greek officials," he said, "are calling for a specific list of my collection which so far I have evaded. I must know what genera and species have been published in order to send them a list. I cannot expect any help from Athens; on the contrary all kinds of hindrance. I dislike to work in this underhanded way, but it is the only method understandable to the Greek."[36]

By mid-February, the rains had still not completely let up. Lilian, moreover, was having difficulty remaining tolerant of the locals, who often showed up unannounced in family groups as large as ten or fifteen, expecting to be wined and dined. "There were times when neither B nor I had a chair to sit on," she complained. Finally, when she caught some guests stealing food out of the boxes of provisions, she threw a fit and put locks on all the boxes. Barnum understood her displeasure but was more philosophical, reminding his wife that "we have to live among them, so we must put up with them."[37] A few days later, Lilian added that the first days of spring had finally arrived, albeit with "flys fleeing and fleas flying."[38]

On February 14, Barnum reported to Osborn and Matthew that he was working when he could and had "added two beautifully preserved carnivore skulls with jaws, hyaenas I think." However, he had been forced to jacket them before they were fully exposed because

> there are so many nosey people around all the time that I can scarcely move without surveillance. I do not even let my interpreter know what I find, for none are to be trusted. Many are in communication with Athens.
>
> I am very sorry we have to work with such people. . . . No one could say whether my material duplicates theirs or not, but I am very careful to label specimens so they will come within the Athens Museum collection [i.e., be identified as specimens already in the Athens collection]. The one thought in Athens is to levy all the good specimens from any collection taken out of Greece; "getting something for nothing" which is the motto of the east.[39]

By now, Brown had begun work on the Soumenas place (Quarries 5 and L), which appeared to be lower in the stratigraphic sequence. Although preliminary, his collecting there had already yielded "two pairs of skulls, a horse and an edentate." He had also completed the draft for his *Natural History* article, which he sent off with photos.[40] Osborn responded that he and Matthew would take that draft and develop it into two articles, one on the India expedition and one on the trip to Burma. In addition, Osborn reported that he now had eight preparators working on Brown's Siwalik proboscidean specimens, which he called "magnificent."[41]

In early March, Lilian wrote to her sister Madge that the limited diet was becoming oppressive, bemoaning that "I haven't had a bite of meat for two weeks—simply none to be had. . . . Also, altho we live on the water, the fish are too mean to bite, so B and I have beans and potatoes and eggs at 5 cents each and canned stuff."[42] She further reported that the museum had offered to send them to China to join the CAE, but Lilian refused, unless she could first go home for a spell after four years on the road.

Osborn wrote again in mid-March to congratulate Brown on his success thus far, as well as to express the hope that he would eventually find better proboscidean material. He sympathized with the interference Brown was facing from Skouphos and his allies: "I regret that you are working under such miserable and half-hearted support from the Greek Government. You will be interested to know that this is what the Americans are being treated to everywhere [a reference to problems the CAE was encountering in China and Mongolia], and you are justified in taking the steps you describe to avoid interference." Osborn ended by saying that J. P. Morgan had sailed for the Mediterranean and had hoped to visit Brown's operation, but Osborn requested that he refrain from doing so "because it would simply stimulate the avaricious tendencies of the Greeks."[43]

At about this time, Brown sent a set of maps to the museum documenting the localities in India and Burma that he had explored. He had intended to keep them with him to ensure that they did not get lost in the mail, but mice had eaten through the box in which they were stored and nibbled around the edges. Brown also noted a lack of response to his request for a list of previously collected fossil taxa from Samos, imploring his supervisors: "It is important that I should have this information, because if I get the collection through without division, it will be chiefly by camouflage."[44]

On Easter, Lilian wrote her sisters saying that the steady stream of freeloading visitors had ebbed in the wake of her earlier tantrum and that with the coming of spring they had finally been able to secure some meat. She also noted that although Barnum tentatively planned to join the CAE, she intended to go home for a year before joining him in China.[45]

Brown's operation in Quarry 1 was back in full swing by the first week in April, and on the 4th he filed an encouraging report to Osborn. Among the specimens that had come to light were several "perfect" horse skulls, three well-preserved skulls of *Samotherium,* and skulls representing at least two species of rhinoceros and two or more genera of carnivores. However, no primates or proboscidean skulls had yet been located.[46]

To take a break from camp life, Lilian occasionally ventured to Vathy, where she could shop and hobnob with the only other Americans on the island. She described these sojourns in typically humorous terms:

> By way of diversion, I ... crank up the family mule and ride over eight miles of damn all trail to THE city. Vathy [is] a quaint little town rolling from the water's edge up the hillside.... I seem to be living in some prehistoric existence or in a movie set which might be a fishing village.... I shop for camp in the crooked little stores along the crooked little [streets] among the crooked little people, after which I swap scandals with the two other Americans on the island ... tobacco merchants. Here I get mildly binged sampling their newest wines none of which are less than 40 years old. I have to hand it to Samos for wines and cigarettes [of which I am very fond].... I stop all night in THE hotel, next morning I crank up the ole mule, and two asses instead of one wander back to camp.[47]

From New York, Matthew sent a list of the fossil genera and species that he had gleaned from the scientific literature about collections from Samos and Pikermi, another fossil site on the southeastern tip of the Greek mainland. He could not determine what exactly was in the Athens Museum, he said, but indicated that he would write colleagues at the British Museum to try to find out. Matthew ended by telling Brown that he and Osborn felt it would be best if Brown returned to New York after completing his work in Greece, then return later to conduct further work in Asia Minor, if he wanted.[48]

Meanwhile, Brown continued to lay the groundwork for fulfilling the terms of the agreement with the Greek government by sending an interim report to the science teacher in Vathy, Mr. Therianos, which included an updated list of taxa collected so far. Brown emphasized that nothing had been packed in cases for transport and that, when his work was finished, he would haul it all to a storeroom in Vathy where the specimens to be turned over to the Athens Museum could be selected before the rest were shipped to New York.[49]

Back at camp, Lilian had fired all her domestic helpers except one local girl who spoke English. And Barnum was suffering from a relapse of malaria, relatively mild but still debilitating: "This time there was no need of urging him to stay in bed and take his beverage of quinine, which he did like a good little boy. Nevertheless, he was in a state of pajamas from Friday night until Monday night, when he got up a little while." Lilian's care for her patient, she observed, required individualized food preparation and "undivided personal attention." In the middle of Barnum's bout, another local family showed up for an extended visit, which taxed Lilian to her limits. Even once they left and Barnum had recovered, Lilian felt trapped: "[I] can't go [any] place because the dam fossils are stored here, and if I go away and B is away they will be stolen along with everything else.... I am never going to camp with B again. I came here thinking I would have a wonderful opportunity to round up my writings on India, and start my little book of letters from all over the world, but alas and alack the petty affairs of every day eat up my hours and my life."[50]

As the wrap-up phase of the Samos venture began to unfold in the middle of June, even Lilian was anticipating problems with the Greek officials when they returned to Vathy "to box and ship the fossils and go thru the agony of government formalities, which... will mean a big FIGHT." Citing the law that required foreign collectors to leave half of their bounty in Greece, Lilian, echoing her husband, bristled: "In this way they labor not,... but they sit back waiting for the other fellow's choice tidbits on which they pounce." With her frustrations reaching the boiling point on several fronts, she fled the camp for some down time in Athens.[51]

In New York, meanwhile, Osborn jumped into action on Brown's behalf by drafting a letter to Secretary of State Charles Evans Hughes. He indicated that although the museum had a legitimate permit to collect fossils on Samos, "unfortunately the officials have been changed since our permit was granted and the local inspector may give us trouble." Osborn implored Hughes to cable his chargé d'affaires in Athens and "request his continued help in the shipment of this collection to America." He closed by arguing that these fossils had little commercial value, indeed their only value was to AMNH, and that Brown deserved support in view of the fact that he had nearly died in the "Far East" during this arduous expedition.[52]

In other news from New York, W. K. Gregory wrote Brown that he and Hellman had drafted a preliminary scientific description of the primate fossils Brown had discovered in India and hoped that, after commenting on their work, Brown would "do us the honor to permit the use of your name as senior author." Brown's colleagues had identified the fossil jaws as belonging to *Dryopithecus,* and they argued that the *Dryopithecus* "group (including the European and Siwalik species) form[s] an ideal common ancestor for the whole man-anthropoid series. It is really amazing how almost every little wrinkle of the enamel, or every fundamental character of the crown and roots of any of the cheek teeth of *Dryopithecus* may be found in some human teeth, so that it is extremely difficult to find constant differences other than proportions."[53]

In early August, back on Samos, Lilian wrote her sisters that "AT LAST we have finished in this God Forsaken neck of the woods" and Barnum was preparing to move the fossils into Vathy.[54] Once safely stored, a process that they estimated would take a month, they would return to Athens to get the export permit. In mid-August, thanks to Osborn's missive to Hughes, Brown received a letter from the U.S. legation in Athens offering "to lend you appropriate assistance in the shipment to the United States of paleontological specimens." The legation had again enlisted the support of the museum's new member, Mrs. Tsipouras, and they wanted Brown to specify what difficulties he envisioned so they could "take such steps as will settle the matter satisfactorily to you and the Museum."[55]

Brown, meanwhile, was enmeshed in his long-anticipated scuffle with Skouphos and his agent, who had sailed to Samos to make his selections from Brown's collections for the Athens museum. As Brown subsequently related to Osborn and Matthew, it took forty mule pack-loads to transport the blocks and jackets from the quarries to the road, where they could then be loaded onto vehicles bound for Vathy. After the last load had been stored, the government inspector came to present Brown with

> a belated list of selections made by Skoufos from the many lists I had from time to time sent in.
> He had chosen practically two-thirds of my collection not according to the agreement, that is such material that would fill gaps in the Athens collection, but random selections most of which duplicated things they already have. Of blocks, he had selected all but three out of the seventeen.
> Naturally I was furious and refused to comply, locked up the warehouse and came to Athens.
> In conversation with our new Minister Mr. Laughlin, and Charge d'Affaires Mr. Atherton, it was decided that Mrs. Tsipouras should again be our General in chief.[56]

At the end of August, Brown traveled to Athens and, backed by Mrs. Tsipouras, called on the minister of culture and public instruction once again to resolve the dispute. He informed the minister:

> A large part of this [fossil] material was found in inextricable masses which had to be taken up in blocks.

A selection of this lot has been made by Director Skoufos in a manner that practically destroys the value of the collection to the American Museum and to science.

To my own knowledge, and by admission of Professor Skoufos, his selection of individual bones simply duplicates specimens already to be found in the Athens Museum, while his selection of block masses would entirely destroy the value of these fossils to either Institution; that is, a part of some of the bones would be in America, and a part in Greece.

I have tried to reach an understanding with Professor Skoufos, but found it impossible.

I now appeal to you to issue a permit at your earliest convenience for me to ship these bones to the American Museum of Natural History, New York.[57]

Sensing that his hand was the more powerful, the card player in Brown had led him to raise the bet and go all in. And it worked: on September 1, Brown received two notes from the Greek government stating that he was authorized to ship his entire collection of Samos fossils to AMNH, "as a token of appreciation of the many benefits received [by the Greek government and its people] from the United States."[58]

While Brown was in Athens, Lilian, unbeknownst to him, devised a plan to smuggle some of the choicest specimens, including *Samotherium,* out of Greece on a boat owned by a friendly local captain. Soon, the most precious of the crates were floating three miles off the coast outside territorial waters. As Frances later recounted, Barnum met Lilian at the dock upon his return and triumphantly exclaimed: "'Dearie, I've got it [the export permit]!' ... Realizing from Lilian's looks and way of speaking that something was amiss, Barnum finally got the story out of her. After rocking with laughter at the very thought of Samotherium being smuggled out of Greece, he told her to get word to the skipper to bring the treasure back right away because they would be leaving ... as soon as they could ... [and needed to] get the fossils packed and off."[59]

On September 10, Brown wrote Osborn and Matthew saying that Mrs. Tsipouras's influence and intervention had proved essential. "This is the first collection ever to leave Greece without being looted," he observed. Although Skouphos had again become insulting and tried to block the fossils' export, Brown and his colleagues had met separately with all eleven ministers; although some initially resisted Brown's request, eventually Mrs. Tsipouras managed to deliver a unanimous vote of the council in his favor. As one special consideration, Brown had agreed to provide specimens or casts from

AMNH showing the evolution of horse feet from early forms with several toes to modern forms with one; one or more fossil horse skulls with jaws; a fossil camel skull and/or jaws; and a fossil rhino skull and/or jaws. A dinosaur egg was added to sweeten the deal. In all, Brown's Samos collection comprised "a good representation of the entire fossil fauna," with the exception of primates and proboscideans.[60] Today, three of Brown's specimens are on display in the museum's Hall of Advanced Mammals: the porcupine, *Hystrix;* the hyena, *Ictitherium;* and the antelope, *Prostrepsiceros.*

With the fate of the collection resolved, Barnum wrapped up the packing operation and saw the shipment off to Holland, from whence it would make its way to New York and the museum. His plans now included evaluating some fossil sites on the mainland of Greece before heading back to New York himself via the Continent. Lilian, meanwhile, set off for Rhodes, on an adventure of her own. There, as she discreetly wrote to her sister Brad, "I met an honest-to-God-Englishman—fine fellow—we had one damn joyous time for a week, wined and dined and everyone simply wonderful to little me." Later in the letter, she said: "Am going back to Samos tomorrow after one week of reckless fling." At the end, the Englishman added a note of his own: "Glad to meet you! (on paper) and if you are anything half as charming as 'Pixie,' well, you're just what I've been looking for these last 'umteen' years and I shall look forward to meeting you in the flesh. (?) Harold"[61] In her next letter to her sisters, three weeks later, Lilian vented: "I am fed up with B and his lies. I am simply bored to death with him"[62]—though what she meant by "his lies" or being "bored to death" is not at all clear.

In early November, back in Athens, she revisited her fling in another letter: "As I told you I met a fine Englishman on the boat, and a plot immediately thickened. As I was the first American girl who had been to Rhodes for some years I reigned in queenly fashion, with everyone from the Governor down in attendance. With no hubby in the offing one can have a glorious time. I like Rhodes."[63] Her next letter went on to describe her relationship with Barnum and his lack of sympathy for how homesick she had been on the trip, especially her need to see her family:

> B can't understand this lonesomeness at all. In disposition we are as different as day and night. He really hasn't a love in the world but me. By this I don't mean that I completely fill his life by any means, but he hasn't a single tie in the world except his daughter, who he doesn't know, and never longs to see. So when I'm with him he's perfectly content to be at home wherever he hangs his hat, and he can't understand why I'm not perfectly happy batting

around with him. That's the man of it... or rather the Anglo Saxon of it, and unfortunately there is a good deal of Latin in our make-up, and thereon hangeth many tales... or tails however you choose to take it.

Lilian then reiterated her desire to visit her sister in California once she returned to the States, especially so as to jump-start her own writing career. But Barnum was still not happy about her scheme: "Whenever I speak of going to Calif. B weeps and says if I go he'll lose me. What can you do when your sweetie cries, even tho it's time we lost each other anyway."[64]

Brown, meanwhile, having seen the collection off to Holland, exhaled in relief to Matthew: "Again fortune has favored us, for the Greek Ministry resigned yesterday. The next Government may be still more unfavorable to us, but it doesn't matter now that the collection is out of their jurisdiction." He enclosed a list of specimen and case numbers, warning Matthew that the list was "camouflaged and mostly incorrect excepting for quarry data and association of material"; rare specimens were identified as the common taxon *Hipparion*. In all, Brown crowed that it was "a magnificent collection."[65]

Matthew responded, "We are all immensely pleased at the success of your work and even more of your diplomacy." To accommodate the wishes of the Greek government, Matthew had selected for exchange "a good series of specimens along the lines you suggested. Both parties will gain by this arrangement, which is of the essence of good business."[66] The museum had already started preparation of some of the Samos material that Brown had sent previously, and Matthew said he was especially pleased with the fossils of carnivores, rodents, antelopes, and aardvarks, which would nicely complement those from Mongolia, China, and the Siwaliks. Indeed, the fauna from Samos, which Brown helped illuminate, still plays a critical role in studies of mammalian evolution. Today, more than two hundred publications deal with various aspects of the seventy-eight species of mammals and eighteen species of reptiles, birds, and snails that were discovered by Brown and others on Samos.[67] In addition to their biogeographic significance, many fossil species shed light on the origins of closely related modern fauna. Thus, Brown's collections, from Samos as well as colonial India and Burma, helped set the stage for new research on the global evolution of mammals, including our own human lineage. Although Brown found no fossil primates in Samos, his Indian discoveries were apparently already garnering special attention, as Matthew cryptically intimated:[68] "I have been a little hesitant about writing too freely and Osborn has instructed me to hold back any published notes for the present, for reasons that you will understand. He appreciates its value, however, no less than I do, and is in private most enthusiastic about it."

By November 9, Barnum and Lilian were preparing to depart Greece on separate routes to Paris. While Barnum took a detour to Trieste, Vienna, and Berlin to study more museum collections, Lilian made a beeline for France to satiate her desire for comfortable surroundings.[69] In a hand-written note on November 20, she exclaimed to her sisters: "France at last, binged as an owl, and intend staying so—Gods! Civilization, at last!!! Paris tomorrow."[70]

In late December, after a month lounging around the castles and cafes of Western Europe, Barnum and Lilian boarded the *Mauretania* for the final leg of their globe-trotting paleontological odyssey. As Lilian later recounted, they stood transfixed on deck as the skyline of Roaring Twenties New York rose from the horizon to greet them:

> It was New Year's Eve—the midnight hour, to be exact.... We heard the chime of church bells; the distant hub-bub of auto horns... the wailing sirens spiraling up from fire-boats.... Old New York was out to welcome us home.
>
> Home! What a wonderful thought.... We'd both been away for four long years....
>
> Barnum took me in his arms, hugging till it hurt, and we laughed and babbled and shouted Happy New Years to everyone....
>
> Weeks of indulgence followed, sating ourselves with all the things we had longed for while away.... Cocktails at the Colony Club; steaks at Cavanaughs;... dancing at the Plaza.... It was paradise.[71]

The couple took a "luxurious, sprawly suite" on the Upper West Side near the museum, which soon became well known as "Brown's Bailiwick"; there, Lilian and Barnum often entertained guests with curries, pilafs, and other exotic dishes—a "potpourri composed of nearly everything in the kitchen." Barnum stunned Lilian with his homemaking skills, learned long ago in the hills of eastern Kansas and at other stops along the way: "The man [could] cook and bake,... put up jams, and bread-and-butter pickles, and mix the most divine salad dressing you ever tasted."[72]

At the museum, Brown helped direct the preparation of his immense collections, readying the specimens for exhibition and assisting with the associated scientific analyses. For at least this short period, he seemed content to be at home rather than wandering the world in search of fossils. As Lilian summed it up:

When he returned after a hard day's work to loll around all evening in smoking jacket, or linger through a quiet game of bridge, the Browns, globe-trotters no longer, were living the lives of normal human beings—and loving it.

"No tents to pitch!" Barnum would gloat, smiling reminiscently. "No camels to unload! *No monkeys to get in your hair!*"[73]

True for the moment, but of course, it was only a matter of time....

THIRTEEN

Ancient Americans Hunting Bison? Birds as Dinosaurs?

(1925–1931)

IN THE WAKE OF HIS five-year international marathon, the list of field trips that Brown compiled near the end of his life for his unfinished autobiography states simply, "No fieldwork," for the years 1925 and 1926—the only time in his forty-five-year career that he abandoned the field during consecutive years.[1] However, as documents in the Vertebrate Paleontology archives at the museum demonstrate, Brown's characterization was not entirely accurate.

As Brown reintegrated himself into a daily routine, a monumental change loomed. In late May, the museum's acting director, George Sherwood, wrote Brown to inform him that the trustees had decided to appoint Matthew as acting curator in the Department of Geology for the rest of the year and "to relieve him of certain of his duties in the Dept. of Vertebrate Paleontology by assigning them to you."[2] In essence, Brown was appointed acting chairman of his department. His annual report for 1925 indicates that, thanks to an enlarged preparation staff, considerable progress had been made in preparing, cataloguing, and photographing the torrent of specimens arriving from Samos, India, Central Asia, and Snake Creek.[3] Regarding the latter collection, therein lies a tale.

In March 1922, while Brown was on his jaunt through Raj India, a rancher and geologist from Nebraska named Harold Cook had sent Osborn a fossilized tooth that he'd discovered in 1917.[4] After comparing the tooth to those of horses, bears, and peccaries, Cook concluded that it belonged to a primate.[5] The stunned yet gleeful Osborn agreed: "I sat down . . . and I said to myself: 'It looks one hundred per cent anthropoid.'"[6] He also recognized that if it was anthropoid, the tooth would rewrite the history of higher primate evolution.

It took only a month to rush a description of the tooth into print, along with a formal scientific name, *Hesperopithecus haroldcookii*. Although Osborn dismissed the idea that the tooth resembled those of extant apes, the genus name means "first anthropoid ape in America."[7]

One motivation for both the name and Osborn's haste to describe the tooth was his running battle with William Jennings Bryan over Darwinian evolution. What better weapon could there be for Osborn to wield against Bryan than a fossil of early human relatives from his home state of Nebraska? The press even christened the genus with a nickname that must have infuriated Bryan: "Nebraska Man." In a 1922 statement to the National Academy of Sciences, Osborn taunted Bryan directly:

> It has been humorously suggested that the animal should be named Bryopithecus after the most distinguished Primate which the State of Nebraska has thus far produced. It is certainly singular that the discovery is announced six weeks to the day (March 5, 1922) that the author advised William Jennings Bryan to consult a certain passage in the Book of Job, "Speak to the earth and it shall teach thee," and it is a remarkable coincidence that the first earth to speak on this subject is the sandy earth of the Middle Pliocene Snake Creek deposits of western Nebraska.[8]

Many of his colleagues met Osborn's announcement of "Hesperopithecus" with skepticism, especially given the small sample that Osborn had to study. The press, however, had a field day. Osborn's biographer Brian Regal relates:

> Osborn was nervous... that the European press was referring to *Hesperopithecus* (he never called it Nebraska Man) as a human ancestor: something he did not believe it to be. His worst fears were realized when, almost immediately after voicing this concern, the ancient dung hit the fan. The June 24, 1922, edition of the *London Illustrated News* ran an article on *Hesperopithecus* with a large double-page spread illustration showing the creature as an archaic, but clearly humanoid, hunter of ancient game, traipsing about the landscape with a crude stick in its hand.

And despite a disclaimer that Osborn published in *Nature,* the idea continued to gain traction.[9]

The prospect of public acceptance of a direct evolutionary link between an apelike progenitor and humans in fact appalled Osborn. In line with his developing ideas about "Dawn Man," he favored the anti-Darwinian concept that apes and humans had long traveled separate evolutionary paths. He therefore remained cautious regarding how "Hesperopithecus" fit into the human lineage, asserting that the tooth more closely resembled human teeth than any apelike tooth. To Osborn, "Hesperopithecus" represented "a new and independent type of Primate," but more material would be required to determine its relationships.[10]

William King Gregory was assigned the task of analyzing the isolated "Hesperopithecus" teeth, and in 1923 he and his co-researcher, Milo Hellman, published two expanded studies affirming the teeth's identity as primate, having compared them to the teeth of Old World monkeys and apes. In their first paper, they stated that the "Hesperopithecus" tooth "combines characters seen in the molars of the chimpanzee, of *Pithecanthropus,* and of man, but ... it is hardly safe to affirm more than that 'Hesperopithecus' was structurally related to all three."[11] (*Pithecanthropus* was the generic name given by Dutch paleontologist Eugene DuBois to hominid fossils known by the common name Java Man, which are now assigned to *Homo erectus*.) Their second paper downplayed any close evolutionary relationship between "Hesperopithecus" and humans, instead concluding that "the prevailing resemblances of the 'Hesperopithecus' type are with the gorilla-chimpanzee group."[12]

This was essentially where the argument stood when Brown returned from Samos in 1925, and as Osborn had stressed from the start, more complete specimens would be required to resolve the outstanding issues. Matthew, by now deeply enmeshed with his responsibilities in the Geology Department, was unable to return to Nebraska to search for more "Hesperopithecus" fossils.[13] That operation was therefore spearheaded by a long-established collector at the museum, Albert Thomson. In mid-June, Thomson wrote Osborn: "Yesterday,... [Austrian paleontologist] Dr. [Othenio] Abel, Harold Cook and I went over to the famous Hesperopithecus locality. Dr. Abel was fortunate enough to pick up a very beautiful Hesperopithecus upper molar.... I am going to try my best to make a deal with Mr. Ashbrook [another local rancher] for permission to excavate on his place."[14]

Even before Thomson's letter arrived, Brown was thinking that he would make a "short visit" out to Nebraska to help devise "ways and means and plans for excavating."[15] On June 23, Osborn hastily sent a response to Thomson emblazoned as "CONFIDENTIAL." In it he strongly endorsed Thomson's plan of making an arrangement with Ashbrook, and suggested that payment of $150 (about $1,725 in today's currency). He also indicated that he had already dispatched Brown to help out:

> I gave Mr. Brown $250.00, so that he will have the cash on hand in case you have made the bargain.
> Meanwhile, it seems best to guard the secret of the tooth very closely and not to mention the new tooth or give any inkling of the importance we attribute to this quarry, as this would probably lead to his standing firm on his former exorbitant figure of $400. Mr. Brown is a very wise and experienced man to advise with, and I will leave the matter to you and to him.
> In the meantime guard the tooth as if it were the Koh-i-noor diamond, because I consider it priceless. I am now carrying around in my pocket the cast of *Hesperopithecus,* wondering how we made so much of this little, water-worn tooth. The new tooth will put us on the right track, I feel sure.[16]

Osborn had instructed Brown to set up a screening operation to sift through all the matrix from the previous dump piles in which the new tooth had been found as well as any new matrix that was excavated. Osborn had also given Brown a chimpanzee skull so that Thomson would be able to recognize skull fragments other than teeth.

On July 3, Brown reported to Osborn that the new tooth was probably a perfectly unworn second lower molar that belonged to a primate other than "Hesperopithecus." After surveying the Sheep Creek and Snake Creek exposures in the area, Brown and Thomson had "spent the day on hands and knees finding another Primate tooth and several fragments of striking interest. Wednesday we started in and sifted most of the day getting a broken jaw with roots of premolars present and certainly Primate." The work was being done "surreptitiously," Brown said, because an agreement with Ashbrook had yet to be worked out. Ashbrook liked Thomson and was friendly toward Brown; however, Brown discovered that Cook had "been instrumental in getting a highway surveyed through Ashbrook's property, a proceeding that he [Ashbrook] does not like, and we are slightly tainted by contact with the Cooks."[17]

Brown made two trips to Nebraska that summer (June 18–July 11 and July 17–August 21).[18] He was back at the museum by the end of August, when he wrote Thomson to congratulate him on finding more "monkey teeth." Although Brown seemed initially to accept the possibility that the

"Hesperopithecus" fossils were primate teeth and supported Osborn's efforts to collect more complete specimens, doubts had begun to color his view of the supposed primate fossils, as he further confided to Thomson: "In looking over the teeth, I am still very doubtful as to whether they are 'primate' or 'peccary.' It is quite possible that these teeth are an undescribed form of 'peccary.' I think the question will not be settled until you find a jaw containing one or more of these questionable teeth. So good luck to you."[19] Brown's views apparently mirrored those of Gregory, and even Osborn was being more cautious. (Although the sensationally publicized Scopes trial played out in July 1925, "Hesperopithecus" was not mentioned.)[20]

Still unable to reach a deal with Ashbrook, Thomson had moved his operation for screening to an adjacent area called Kilpatrick Hill. On August 27, Thomson excitedly reported: "We discovered yesterday evidence of early man." That "evidence" consisted of three objects: a three-inch-long bone fragment, shaped like an awl with a hole in it; a seven-inch-long fragment of camel bone "which has been hacked with some more or less blunt instrument"; and a fragment of turtle bone, shaped like a "trowel or paddle." Thomson additionally reported the discovery of another dozen "monk teeth"—his slang term for the "Hesperopithecus" fossils.[21]

Four days later, Brown reported back to Thomson that Osborn was "tickled to death and thinks this is the greatest find of the season." However, though "bubbling with excitement" himself, Brown cautioned Thomson to photograph the potential artifacts in situ: "I am no 'doubting Thomas' as you know, but there is always a question when finds of this nature are published.... The awl-like needle seems to me most likely a true Artifact... but the other specimens I have seen more or less duplicated as the result of erosion, etching by plants or gnawing by rodents before the bones were fossilized."[22]

With that, the correspondence for the season ends, and Brown apparently did not go to the field in 1926. Then in 1927, the concept of "Hesperopithecus" collapsed under the weight of new specimens, primarily those collected by Brown and Thomson in 1925.[23] That year, Gregory published a paper in *Science* concluding that the teeth, despite their strong resemblance to anthropoid molars, probably belonged to an extinct peccary, *Prosthennops*.[24] It was a bitter pill to swallow, especially for Osborn, who, due perhaps to his arrogance as well as embarrassment over the error, did not co-author the retraction. Osborn never published on the alleged "Hesperopithecus" artifacts collected by Thomson either, though he had presented a scientific talk about them at the meeting of the American Association for the Advancement of Science in December 1926.

The year 1927 held another disappointment for Osborn and the museum community. As Brown reported in his annual summary for the department, W. D. Matthew "resigned from the scientific staff in June, after thirty-one years service in this department, to become Professor of Paleontology" at the University of California, Berkeley. "This announcement," he continued, "is made with sincere regret, not only of members of this department, but of the entire staff of the Museum."[25] Matthew's relative silence on the affinities of "Hesperopithecus" teeth may well reflect his disagreement with Osborn and Gregory's identification. As those authors noted, "In 1909, W. D. Matthew and Harold Cook had the following to say in describing *Prosthennops:* 'The anterior molars and premolars of this genus of peccaries show a startling resemblance to the teeth of Anthropoidea, and might well be mistaken for them by anyone not familiar with the dentition of Miocene peccaries.'"[26] Given that Matthew was essentially Osborn's second in command, Osborn must have known about this early warning. In the end, Matthew said very little about the identification (or misidentification) of "Hesperopithecus" as a primate; his published comments on the tooth stressed only its stratigraphic position, not its affinities. This disagreement with Osborn, coupled with Matthew's different views about evolution and apparent displeasure at being assigned to restructure the Geology Department, thus limiting his ability to conduct research and fieldwork, probably led to his resignation.[27] Fortunately for the museum, Matthew was succeeded by George Gaylord Simpson, then at Yale, who would become one of the foremost evolutionary biologists and paleomammalogists in the world.

Fieldwork for Brown returned to a more normal pattern in 1927, when he ventured to the Rocky Mountain West—Colorado, New Mexico, Utah, Wyoming, and Montana. Prospecting on the Crow Reservation near Pryor, Montana, in what is now known as the Cloverly Formation, he discovered the skeleton of a "fine little dinosaur about the size of Camptosaurus," which he then covered for later excavation.[28] He also identified several other sites on the reservation and to the south in the Bighorn Basin that he felt merited further scrutiny. The fact that Brown continued to call the ornithopods from these exposures camptosaurs suggests that he still considered the beds to belong to the Morrison Formation. He was well aware that the fauna of dinosaurs found in these beds was quite different from that of the Morrison. J. H. Ostrom notes that the Cloverly was first recognized and named by Nelson

Horatio Darton in 1904.[29] From that year through 1927, no less than twelve scientific papers discussed the Cloverly, and in all those studies the stratigraphic boundary between the Cloverly and the Morrison varied considerably. It is therefore not altogether surprising that Brown seemed to think the outcrops along Beauvais Creek were related to the Morrison.

Brown also collected several "small mammal jaws" from Eocene deposits in the Bear Creek Coal Mine of Montana. The most notable collecting, however, was done in New Mexico, near Folsom, and it again landed Brown in the midst of the debates about prehistoric humans in America. Not only that, but the museum's involvement in this enterprise once again hinged on the staff's relationship with Harold Cook.

As compellingly chronicled by Southern Methodist University anthropologist David Meltzer, controversy surrounding the antiquity of human habitation in North America had ebbed and flowed since the 1860s, when evidence of prehistoric humans was discovered in Europe.[30] In the 1870s, in supposed Pleistocene gravels around Trenton, New Jersey, Charles Abbott uncovered stone artifacts similar to those found in Europe; more such discoveries followed in the 1880s. By the end of that decade, the presence of Paleolithic humans in America was widely accepted by the scientific community. But not all agreed; chief among the skeptics was the prominent anthropologist William Henry Holmes of the Smithsonian Institution, who argued that these alleged Paleolithic artifacts looked primitive, like those from Europe, not because they were ancient but because their more modern makers had not finished working them into their final form. Controversy ignited and more or less raged into the 1920s, with the "Hesperopithecus" debacle representing one front in this scientific war.

Cook, in his efforts to establish the legitimacy of ancient humans in America, had found, in addition to Osborn, a willing partner in Jesse Figgins, the director of the Colorado Museum of Natural History (now the Denver Museum of Nature and Science), where Cook served as an honorary curator of paleontology. Figgins was intent on acquiring skeletons of extinct bison to mount for exhibit. In 1924, he received word from Colorado City, Texas, that a "huge skeleton" of a bison was weathering out of a bank along Lone Wolf Creek. Arrangements were made for a local artifact collector and rancher to collect it. Projectile points were found in association with the bones, but the points were not photographed or otherwise documented in situ. (Such procedures, though considered obligatory today, were not common at that time.) Nonetheless, Cook and Figgins immediately realized that the find might be used to establish the presence of ancient humans in America and published their account of the discovery in *Science*.[31] But without careful documentation to back up their claim that the points had been found in situ, many, including Holmes and his Smithsonian colleague Aleš Hrdlička, remained vociferously skeptical. Barnum Brown, however, was not so skeptical.[32]

While the debate about Lone Wolf Creek was playing out, Cook and Figgins got wind of another, similar site that had been discovered at Folsom in Colfax County, New Mexico. Sometime after a flood in 1908, George McJunkin, a cowboy on the Crowfoot Ranch, had spotted a bison skeleton weathering out of a streambank.[33] He brought the find to the attention of Carl Schwachheim, a blacksmith in nearby Raton, who in turn informed Fred Howarth, a local banker and fellow natural history buff. In December 1922, shortly after McJunkin died, Schwachheim and Howarth visited the site and photographed the protruding bones. Unable to interest the New Mexico government in excavating the site, the pair traveled to Denver in January 1926, where Cook and Figgins enthusiastically received them. A subsequent shipment of bison bones from Folsom convinced them that the bones represented an extinct species, whereupon Figgins agreed to commit funds for an excavation that summer to be led by Schwachheim and Howarth. In July, the first projectile point was found, but not in situ. Another fragment was found later that fit onto a "sliver of stone that attached to a rib," but again, there was no photographic documentation.[34] Brown encouraged Cook and Figgins to publish their findings in *Natural History*, which they did,[35] and Figgins took the Folsom artifacts to the Smithsonian for examination by Hrdlička and Holmes. Although they remained unconvinced, their reception of Figgins was courteous, and Hrdlička implored Figgins to stop excavations immediately if more points were found in situ so that they could be photographed and examined before being collected.

The excavation resumed in May 1927, but it wasn't until August 29 that Schwachheim found another point in situ. Figgins telegraphed the crew telling them to leave it in place and guard it intently. He then wired institutions around the country to announce the discovery and invite colleagues to come investigate. Brown was already in the region and immediately responded, as did a representative of the Smithsonian. Brown took stratigraphic notes and posed for a photo. All who saw the specimen were convinced that it was indeed in situ and that the specimen and the associated extinct bison skeleton were contemporaneous. In December, Brown and others presented talks on the Folsom discoveries at the meeting of the American Anthropological

FIGURE 34. Brown (right) and Carl Schwachheim sitting next to an in situ Folsom point during Brown's visit to the Folsom site in New Mexico in September 1927 (AMNH Vertebrate Paleontology Archive; Meltzer 2006, fig. 2.12)

Association. Their conclusion was that the earliest Americans had arrived on the continent between 15,000 and 20,000 years ago. Hrdlička, still skeptical, wrote Brown asking for copies of the photos he was using in his talks as well as for his opinion on how the points had come to be associated with the bison bones. Brown confidently responded: "I personally removed the five inches of stratified clay that immediately covered the fifth point, excepting one of the barbs.... There is absolutely no possibility of any introduction of the points subsequent to the natural covering over of the bison skeleton."[36]

Hrdlička never overtly accepted the evidence that Brown and others collected to establish the Paleolithic legitimacy of the Folsom site, but his tacit acceptance is manifested by the fact that he dropped his strident attacks and barely mentioned Folsom afterward.[37] Meanwhile, in his report for the AMNH, Brown crowed that he and his colleagues from the Colorado museum had made "the most startling of the year's discoveries—flint arrow heads of exquisite workmanship and of a culture so far unknown, which have been found... in association with the bones of species of bison now extinct—finds which open a vista into the American past of unexpected and fascinating possibilities. [I] visited one of these arrow head localities and through the generosity of the Denver Museum... brought certain of the specimens to New York."[38]

Brown's reconnaissance around Folsom in conjunction with the Colorado museum allowed him to lead a full-blown excavation in 1928, aided by his longtime sidekick, Peter Kaisen.[39] The quarry encompassed an eighty-by-sixty-foot plot, and it contained several associated skeletons and partial skeletons of the extinct *Bison taylori* (now *Bison bison antiquus*). As had been hoped, more evidence of human predation turned up, principally in the form of eleven arrow points, including a "perfect" one nestled among a cluster of bones, which was collected in situ for exhibition. As if to compensate for the museum's missteps in the recent "Hesperopithecus" fiasco, Brown trumpeted: "Archaeologists throughout the country are convinced that this is the most important work of recent years relating to Prehistoric Man in America. The date [now thought to be about 10,500 years ago based on radiocarbon dating] establishes the presence of man with an extinct species of bison near the close of the Pleistocene while the implements probably represent a distinct culture."[40] Indeed, that culture is now referred to as the Folsom culture. The specimen mentioned above that Brown collected at Folsom is on display at the end of the Fossil Mammal Halls; it comprises a jumbled lower jaw and tibia of *Bison bison antiquus*, complete with a projectile point lying alongside.

To the north near Grand Junction, Colorado, Brown also prospected in "Lower Eocene" strata, where he discovered a "Dakota" duckbill dinosaur and identified a Jurassic skeleton of "*?Brachiosaurus*" with a nearly complete vertebral column and associated limb bones. Brown had also sent a student, Rachel Husband, to follow up on his earlier exploration of the Eagle Mine at Bear Creek; there, she succeeded in screening out an Eocene microfauna that included fishes, turtles, champsosaurs, crocodiles, and mammals. (Although female paleontologists were rare at this time, Brown not only helped develop Husband's career but also worked with two female invertebrate paleontologists—Ethel D. Currie of the Hunterian Museum, University of Glasgow, and Gayle Scott of Texas Christian University, Fort Worth—to describe fossils collected during his expedition to Abyssinia.)

In his summary of recommendations for future work, Brown bemoaned the "contraction" that had occurred in staff and fieldwork in the wake of World War I. In 1921, the department's budget had been $48,000, but by 1928 it had slipped to $45,000 (about $530,000 in today's currency). Without substantial private funding during the last three years, "no fieldwork could have been undertaken." To maintain the prestige of the Vertebrate Paleontology Department, he argued, "a material increase in funds should be provided for... expansion. Dinosaur fieldwork ceased in 1916 and this work should be renewed and diligently continued, for [new] institutions are sending out expeditions in increasing number, whereas fossil localities are becoming less prolific through repeated searching."[41] It would be a tough bill to fill, especially with Black Monday and the stock market crash of 1929 looming just six months over the horizon. Nonetheless, thanks to Brown's natural tenacity, this clarion call would lead to the great collector's last great dinosaur campaign during his formal career at the American Museum of Natural History.

He kicked things off in 1929, with a whirlwind summer reconnaissance under the banner of the Walter Herring Endowment Expedition, a joint effort of AMNH and the University of Pennsylvania, which saw him touch down in Arizona, Nevada, Utah, and Colorado. In all, he was out from mid-May to late July.

Near his old haunt of Cameron, Arizona, Brown discovered a remarkable series of "300 dinosaur tracks... representing four different species of Triassic dinosaurs," which, he noted, will "aid us greatly in determining posture and foot structure of early dinosaurs." Also near Cameron, he collected a "partial skeleton of a primitive reptile at the very base of the Triassic," apparently a new taxon. "Some characters are common to alligators, and other elements in the skull indicate affinity with the dinosaurs."[42] Today, this animal is known as *Hesperosuchus*, a relative of living crocodilians. One specimen that Brown collected on this leg of the journey, a skull of the early relative of amphibians called *Buettneria*, is still on display in the museum's Hall of Vertebrate Origins.

In the Jurassic of southern Utah, Brown went on to locate specimens representing *Apatosaurus* and *Diplodocus*.[43] Both were marked and covered for future excavation. A later trip to the same area in November and December identified two more prospects, another apatosaur and a carnivorous form, which he thought might be new, related to *Allosaurus*. None of these specimens was ever recovered, however.

Brown also pursued a geographically expanded program of archeological and paleontological fieldwork related to the Folsom project, conducting a survey of caves in northern Arizona and throughout Utah in search of human artifacts and Pleistocene fossils. In this effort he was corresponding with Clarke Wissler of the museum's Anthropology Department.[44] From Brown's point of view, the evidence was good enough for the museum to initiate excavations in several caves, including a large one near Lovelock, Nevada, where he found artifacts that included "nets, decoys and bones"—items suggesting that early inhabitants "lived almost exclusively on birds," which would have lived in habitats associated with the Ice Age Lake Lahontan. Other caves Brown investigated were Grimes Cave near Stillwater, Nevada; one near Battle Mountain, Nevada; and some along the Green River near Vernal, Utah, which had yielded "exquisite bird and rabbit traps."

Brown especially implored Wissler to visit the area around Yuma, Arizona, where local collectors had amassed a "truly enormous collection of arrows [points, presumably], scrapers, knives, and metates" from sandy desert blowouts, including a dozen Folsom points. Then as now, the kill site at Folsom was thought to represent a hunt conducted by a group of Paleoindians that was on the move.[45] The region would have been inhospitable during the winter due to frigid temperatures and a lack of abundant food resources when grass became scarce and the bison migrated in search of more favorable habitats. Despite a thorough program of reconnaissance, no evidence of a permanent settlement near Folsom was discovered during the excavations in the 1920s, nor has one been uncovered since. This raised the question of where the hunters' more permanent settlements might have been. In summarizing his thoughts about this mystery to Wissler, Brown counseled: "I strongly advise that you visit Yuma for three or four days. . . . This area may lead to the Folsom homeland but the sandhills were not permanent camps—no wood, poor shelter, hard winters, and infrequent water supply."[46]

Brown's reports for 1930 reveal that Sidney Colgate funded the season—which followed the stock market crash—with a $2,000 grant (about $24,000 in modern currency). Such grants from private benefactors would increasingly become the norm as museum funds grew more limited during this financially difficult decade. Brown's agenda for the season reflected a continued interest in the archeology of the Southwest, where he excavated an ancient dwelling just east of Cameron for Wissler. Three five-foot-high walls exemplifying "typical period 3 pueblo workmanship" still stood atop a sandstone cliff. Brown envisioned the dwelling as "a circular wood structure, probably similar to a modern Hogan, 10 feet in diameter at floor level without evidence of a stone basement." He called it the "Neo House," *neo* being the Navajo word for "blowing wind."[47]

Within the structure, Brown found a metate partially buried in "a square corn meal bin formed by four flat stones standing on end." Excavation of the bin further yielded a "beautiful obsidian" point. Sixteen pots and a "mano" were arranged on the floor, just "as the householder would have placed them at the end of the day." Brown characterizes the pottery as representing three types: "Black on white, black on red, and built up coiled forms. Two of the designs seen are identical with those from Pueblo Bonito [a massive and architecturally complex dwelling built by the Anasazi between A.D. 1000 and 1100 in Chaco Canyon, New Mexico]." Other artifacts recovered included a string of shell beads and pottery pendants. Because the Navajo regard the Anasazi as their ancient relatives, Brown considered the "Neo House" and its artifacts as representing "Mother Neo's bequest" to both the Navajo and science. In addition to their research value, the specimens from the house would, he said, make an excellent "single house exhibit."[48]

Brown's paleontological work focused on the nearby Triassic exposures around Cameron, where he hoped to recover more fragments of the "Stammosaurus" specimen he had discovered in 1929. Aided by L. I. Price of the University of Oklahoma, he screened fifteen tons of debris on the quarry hillside, then hauled it ten miles to a spot where it could be washed, dried, and picked through. The result was a collection of fragments "that covered the bottom of a cigar box one-half inch deep." Although this sounds like a meager reward for three weeks of hard labor, Brown was pleased, reporting that the effort "was well worth the expenditure as it will enable us to make a paper restoration of this rare beast considered ancestral to the Dinosaurs and Phytosaurs."[49] In addition to this material, Brown discovered a large phytosaur skull and a partial skeleton of a small labyrinthodont.

After swinging through Los Angeles to arrange an exchange of fossil material with the Los Angeles County Museum from their Rancho La Brea collection, Brown headed home through eastern Montana, where he examined the Hell Creek beds near Glendive. Of special interest was a *Triceratops* skull that had been found by a local, which Brown thought might be "a new species with [an] extraordinary short wide crest and sutural separation of the central frill element."[50]

By the end of the season, Brown had wandered around the West from July 24 to October 9, spending about $2,700 (roughly $32,500 in today's currency). But his annual report describes another key acquisition for the museum, a skeleton of the sauropod *Barosaurus,* and Brown didn't shovel one pound of dirt to acquire it. Brown's note indicates:

A *Mesohippus bardi* skeleton was assembled, restored and sent to the University of Utah as a partial exchange for the Jurassic dinosaur (*Barosaurus*)....

This department has [thus] acquired the major part of a large Jurassic Sauropod dinosaur skeleton [of] *Barosaurus,* the body from the University of Utah and part of the tail from the Carnegie Museum. The neck of this specimen was collected by the United States National Museum. The entire skeleton was approximately 74 feet in length.[51]

Eventually, Brown would wheel and deal for the neck, and although he did not mount it himself due to spatial constraints, a cast of that specimen, rearing up on its hind legs to protect its baby from an attacking allosaur, now greets AMNH visitors as they enter the Roosevelt Memorial Hall off Central Park West.

The fuse for Brown's 1931 foray was lit back at the start of his career, when he roamed the ranges of southeastern Wyoming from near Como Bluff up to the Crow Reservation south of Billings, Montana. The 1931 season, which was funded by the sale of a horned dinosaur skeleton, *Monoclonius,* to Yale, saw him heading back to the reservation with his longtime assistant Peter Kaisen.[52] He started his trip in May, first making a detour to California for a scientific meeting, followed by a swing through the Folsom site in New Mexico, where he discovered another point in the dump pile for the 1928 excavation. He then continued on to the Triassic exposures near Cameron, Arizona. There, he discovered an "almost perfect skeleton including external plates in position of a Middle Triassic [now Early Jurassic] reptile . . . —a form that is directly ancestral to the alligators."[53] This is surely a reference to *Protosuchus,* a very gracile, lightly armored, and almost catlike primitive crocodile, one exemplar of which is still on display in the Hall of Vertebrate Origins.

From Arizona, Brown headed north to Vernal, Utah, in early July to meet with officials from Congress and the National Park Service and discuss plans for a museum at Dinosaur National Monument. Funded by Andrew Carnegie, Earl Douglass had discovered this rich site packed with Jurassic dinosaur fossils in 1909, and he continued to collect there for the Carnegie Museum until 1922, when, following Carnegie's death, his funding disappeared. The U.S. government established the national monument in 1915. From Billings, Brown reported to Osborn that Congressman Don B. Colton had agreed to request that $200,000 (around $2.6 million in today's currency) be set aside in the next federal budget to "complete the Monument."[54] Brown's annual report summarizes the project:

> Plans were approved whereby twelve sections of land were set apart as a National [Monument] and the Government is to spend (not to exceed)

$200,000 in exposing a sandstone face 190 feet long by 30 feet high on which Jurassic dinosaur bones are embedded; build custodian houses, road, and when work is completed erect a museum building over the monument [i.e. outcrop].

... an American Museum force would undertake to relief the bones, identify them and interpret the occurrence adding models and copies of restorations in return for the privilege of removing certain desirable skeletons and surplus material that would obscure the main exhibit.

The work is estimated to require three years and an expenditure by the Museum of $50,000 thereby obtaining skeletal remains not available elsewhere.[55]

In his letter to Osborn, Brown said that he had held out for developing an exposure "nearly three times the dimension [originally envisioned] in order to give us greater latitude in examination, selections and preservation of material to be actually housed in the Museum [AMNH]."[56]

Brown set out on July 6 from Salt Lake City to hook up with Kaisen and crew members G. Edward Lewis, a student at Yale, and Darwin Harbicht, a collector, in Billings. From there they headed to the Cashen Ranch and its exposures of what is now recognized to be the Early Cretaceous Cloverly Formation along Beauvais Creek. Although field correspondence from this season is paltry, we do have one letter from Brown from the end of July, when he wrote Granger, now heading up the department back in New York, saying that he had injured an ankle or knee: "While I am laying up for repairs I'll surprise you by writing a letter. I'm beginning to walk a little less like a sprained horse now, expect to be on my way again Sunday."[57]

Brown's initial goal at the Cashen Ranch was to excavate the ornithopod skeleton discovered and covered over for protection in 1927. In taking up that skeleton, Brown noted in his final report, the crew ran across five others within a quarter mile of the Cashen Ranch house: "Of these, four are of the genus Camptosaurus [*Tenontosaurus*] of different sizes— Closely associated with one is a small carnivorous dinosaur but encased in lime [which is] difficult to prepare."[58] Brown's reports of this expedition make it clear that he did not yet realize that these strata and specimens belong to the Cloverly Formation. Rather, Brown assigned the exposures along Beauvais Creek to the Lakota Formation, which he considered to be Early Cretaceous in age. Because rock units containing vertebrate fossils of Early Cretaceous age were rare relative to those containing fossils of Late Jurassic or Late Cretaceous age, he was determined to assemble as comprehensive a collection as he could of this important new fauna. Curiously, even though he realized that the Beauvais Creek fossils were Early Cretaceous in age, he still considered the ornithopods to belong to the genus *Camptosaurus*. It would be almost four decades before John Ostrom recognized that these skeletons represented a new genus, *Tenontosaurus*. Ostrom, following Ralph Moberly, also stabilized the stratigraphic boundaries of the Cloverly.[59] By the time Ostrom completed his stratigraphic study, the Cloverly had indeed been confirmed to be Early Cretaceous in age (Aptian-Albian) based on biostratigraphic correlations of ostracods and charophytes.[60] Today, the claystones, sandstones, and conglomerates that make up the Cloverly are thought to represent stream and associated floodplain deposits, mixed with volcanic ash, which were deposited under hot climatic conditions in ephemeral swamps and lakes between about 115 and 108 million years ago.[61]

In his July letter to Granger, Brown expanded a bit on the discovery near the Cashen Ranch: "The perfect skeleton is a peach, a new species of Camptosaurus [*Tenontosaurus*] intermediate [in form] between the two species of Jurassic Camptosaurus but with very different teeth.... We have three other incomplete skeletons the same size; an incomplete carnivore skeleton massed with one of the three Camptosaurs but in lime and not so good... — a regular cove of Camptosaur skeletons."[62] Granger replied a month later with congratulations and relief, quipping: "I'm so glad they aren't sauropods"[63]—his relief reflecting the difficulties involved both in collecting such enormous specimens and in storing them, given the limited space at the museum.

Satisfied with their take, the crew next moved in mid-August to other outcrops of the Cloverly north of Pryor, where their efforts were rewarded by another discovery: "a skeleton of Hoplitosaurus [now *Sauropelta*] ... all exposed and very fragmentary but [it] will enable us to determine the chief characters of this plated dinosaur. Vertebrae and plates in the dorsal region were preserved in position some of them rising to a height of 27 inches."[64]

At the end of August, Brown returned to the Folsom region and a cave to the south near Carlsbad, New Mexico, where a Folsom point had been discovered in a fire pit associated with the horn of a musk ox. Brown's role was to collect fossil mammal bones from the layer that produced the point; his finds included "two species of horses, a camel, a musk ox, an extinct bison, an extinct four-horned antelope, Mexican deer, and the California Condor," as well as abundant rodent remains.[65] He arrived back in New York on September 26. In his final report for the season, he gave the skeleton of the armored dinosaur, *Sauropelta,* and the mammal bones associated with the Folsom point top billing. Further analysis, however, soon brought another specimen from the expedition into the paleontological limelight.

At the museum, Brown and his technicians began preparing the material. The *Sauropelta* skeleton, though lacking a skull, turned out to be magnificent, and it still helps anchor the alcove of armored dinosaurs in the renovated Hall of Ornithischian Dinosaurs. But Brown was also intrigued by the skeleton of the small, carnivorous dinosaur he had found at Cashen's Ranch—so intrigued that he began preparing a manuscript (no longer extant) with supporting illustrations on this animal, which he informally named "Daptosaurus." (Brown's label still adorns its drawer at the museum.) The specimen included skull fragments, nearly complete limbs, and a series of vertebrae, enough to suggest that it represented a small- to medium-sized theropod about 2.5 meters long. Its thin, hollow bones indicated that it was an active, lightly built, swift runner. Its stiff tail, supported by long, bony rods, would have served as an effective counterweight for the body when the animal ran at high speeds with its backbone held parallel to the ground. Although the illustrations show that Brown recognized the animal's avian affinities, he never got around to publishing the paper.

"Daptosaurus" languished in the museum's collection until the 1960s, when, shortly before Brown passed away, John Ostrom, a young student searching for a dissertation topic, sought Brown out for advice. Brown showed Ostrom the diminutive skeleton. After Ostrom became a professor at Yale, he returned to the Cloverly and discovered other fossils of the same animal; these allowed him to finish what Brown had started thirty-odd years earlier. In 1969, Ostrom published his monograph on the new theropod, naming it *Deinonychus,* which means "terrible claw."[66]

With more complete material, Ostrom could thoroughly document the avian characteristics that Brown had suspected. In addition to having light, hollow bones, *Deinonychus* had a crescent-shaped bone in its wrist, just as in a bird's wing, which allowed a swiveling motion of its grasping hand, much like the flight stroke of a bird. These and other skeletal features led Ostrom to rekindle the century-old debate about the dinosaurian origins of birds, an idea that had been championed by Thomas Henry Huxley and other supporters of Charles Darwin. In a now famous passage describing the anatomy of the leg and ankle of a young bird, Huxley observed: "[If] found in the fossil state, I know not by what test they could be distinguished from the bones of a Dinosaurian. And if the whole hindquarters from the ilium [hip] to the toes of a half-hatched chicken could be suddenly enlarged, ossified [turned to bone], and fossilized as they are, they would furnish us with the last step of the transition between Birds and Reptiles; for there would be nothing in their characters [characteristics] to prevent us from referring them to the Dinosauria."[67]

Ostrom's observations triggered reexaminations of other small theropod skeletons, such as the *Velociraptor* brought back from the Central Asiatic Expeditions, which shared many features with its North American cousin, *Deinonychus*. A decade or two of sometimes rancorous debate in the late twentieth century eventually led to the conclusion that Huxley had been correct: birds did evolve from some small theropod dinosaur, making birds the only lineage of dinosaurs still living today.

But Ostrom went further. He recognized that *Deinonychus*, with its light skeleton, gracile limbs, serrated teeth, and sharp, oversized claws, was nothing like the slow, lumbering behemoths that typified most paleontologists' view of nonavian dinosaurs. It was a lithe, swift, agile predator, with a relatively large brain. Thus, our modern view of nonavian dinosaurs, as well as our knowledge that birds evolved from dinosaurs, flows directly from Brown's discovery of *Deinonychus*, the skeleton of which is now mounted in a vigorous leaping posture in the museum's Hall of Saurischian Dinosaurs.

Brown's connection to both Ostrom and the *Deinonychus* specimens is intriguing. In a personal communication with Norell, John S. McIntosh, a young protégé of Brown's, recalled:

> On a Friday afternoon during my freshman year at Yale, I was working on the skeleton of the small theropod Coelurus agilis YPM 2010 in the prep lab when I realized that someone had come up behind me. Turning around I found my nose about two inches away from that of the great Barnum Brown. In 1931 he had collected the skeleton of a new medium sized theropod (which he called "Daptosaurus agilis") in the Cloverly formation of Montana.... He invited me to come down to New York to examine and measure that animal. I had not cut a single class up to that time but was on the train bright and early on Monday morning. Arriving at the American Museum, I proceeded to the top floor (unopen to the public!) and was issued into the office, where Rachel Nichols called Barnum Brown. He came in wheeling a cart with two trays, one with his "Daptosaurus" skeleton and the other with a much less complete skeleton of a second new much smaller animal which he called "Megadontosaurus ferox." Brown told me that I could study and measure the specimens at my leisure. As he wheeled the specimens down the hall to

a large room with a long table, Rachel Nichols whispered to me "He doesn't show these to anyone." You can imagine how I felt.[68]

Certainly, Brown recognized that his "Daptosaurus" was a close relative of birds. A few surviving illustrations done by museum illustrators compare the skeletal elements of "Daptosaurus" to avians. Departmental memory indicates that a manuscript was prepared and for many years sat with Brown's other unpublished writings in the department library, though what became of it is not known.[69]

Shortly after arriving at Yale, John Ostrom took the "Daptosaurus" and other small Cloverly and Morrison theropods out on loan, assisted by Brown. An undated, unsigned carbon copy of a handwritten note on an AMNH memo pad exists in the AMNH archive, inscribed in Brown's distinctive handwriting simply as "for Ostrom." The memo contains explicit directions to the Beauvais Creek sites, areas the Yale teams would later prospect when they found the "*Deinonychus* localities." However, in neither of Ostrom's excellent monographs on *Deinonychus* and the Cloverly Formation, in which he named *Tenontosaurus*, is Brown credited for his discovery and insight.[70]

At the dawn of the 1930s, with Brown once again reestablished in operations out west, the next two years would culminate in a crescendo of collecting adventures.

FOURTEEN

Digging—and Flying— for Dinosaurs
Howe Quarry and the Aerial Survey of Western Fossil Beds (1931–1935)

BROWN'S 1931 EXPEDITION WAS AIDED by an innovation in field transportation. Undoubtedly spurred on by the CAE's precedent-setting use of automobiles to explore the Gobi in the previous decade, Brown traded in his horse teams and camel caravans for a seat behind the wheel. But while he extolled the virtues of this new vantage point to his colleague, CAE veteran Granger, he was already looking ahead: "It is simply marvelous how much territory one can cover in a car. I am now longing for a Helicopter plane."[1] In the interim, Brown launched another expedition, from mid-July through late October of 1932, to follow up on prospects in northern Wyoming and southern Montana. He intended to return to the Cloverly deposits near the Cashen Ranch, but he also wanted to focus on a thirty-mile-long patch of exposures in the Bighorn Basin that had not been prospected previously.[2]

With funds still scarce after the stock market crash of 1929, the department's budget could not support the work. Brown again sought outside funding, which was provided by a $1,470 grant from Childs Frick, a noted gentleman paleontologist and prominent patron of the department who was the son of the steel industry baron Henry Clay Frick, along with a $635 allotment from a fund established by banker J. P. Morgan.[3] Total funding amounted to about $31,000 in today's currency.

By early August, Brown, Kaisen, and Harbicht were roaming the southern flank of Pryor Mountain in the Bighorn Basin. They soon discovered the exposed tail sections of two sauropod skeletons that appeared to extend into the hill. To check, they began a preliminary excavation on one, uncovering fourteen tail vertebrae along with hip and hind limb bones including the ilium, tibia, fibula, and foot. The bones were not deformed, leading Brown to boast to Granger that "the discovery looks like a ten strike.... It will be necessary to strip an area 45×60 feet, 5–10 feet deep, four feet of which is a ledge of sandstone. It is located on the homestead of B. M. Howe an old man 82 years old who has agreed to hold it for us and supply two men and four horses to do the stripping in the spring.... It will take three weeks to a month to do the quarry work before any further prospecting can be done and then about two months to excavate the specimens."[4]

Back in New York, Granger sent hearty congratulations, telling Brown that his new discoveries were "just the sort of thing we need to fill up that new hall." He went on to say that, "out of the blue," Osborn had decided to visit Brown's camp with Granger in tow, and they expected to be out between mid-August and mid-September: "So unless you are too elusive, it would look as though I would be in a bone camp with you again after all these years—I think the last time was in 1897."[5] He was referring to the first dinosaur expedition conducted by AMNH to Como Bluff, when both these now famous fossil hunters were raw rookies.

By August 7, Brown and Kaisen were back at the Cashen Ranch north of Pryor Mountain, reaping another harvest of ornithopods:[6]

> We are again back on our old hunting ground and yesterday landed another complete Camptosaurus skeleton [*Tenontosaurus*] that looks like a free mount, a tail of another and a prospect that promises another Hoplitosaurus [*Sauropelta*] skeleton.... This area has certainly been prolific for us and the number of Camptosaurs that must have been here is incalculable.
>
> We have a delightful shady camp among the Box Elder trees close to a clear stream of cold water, no mosquitoes and fairly good roads. If you can get word to Mr. Frick, this would be a fine camp for him to visit....

Granger again responded with congratulations, along with a summary of updated plans, a request for a map to help him and Osborn find the ranch, and a note saying that Frances had visited the museum.[7] Osborn and Granger spent a couple of weeks with Brown and his crew, a visit that pleased Osborn, for he wrote afterward: "I am now full of strength and energy as a result of my fine fossil hunting experience with you, the best of guides and camping companions. I enjoyed every moment of... the two weeks in camp... for which I am deeply appreciative. You have no idea how much this complete change of scene and action has benefited me."[8]

Osborn had had a difficult couple of years.[9] The Great Depression had eroded his family fortune, and as if that weren't enough, the family had been rocked by a personal tragedy. In 1930, the health of Osborn's wife, Loulu, had taken a turn for the worse. She had been "the light of his life" for forty-nine years; indeed, the couple had exchanged weekly, if not daily, love letters throughout their marriage. On August 26, Osborn's light was extinguished as he sat beside Loulu on the bed and "held my own darling's hand—while her soul passed." It was a moment from which he would never completely recover, even though he tried to "drown my loneliness and sorrow in hard work," which at the time consisted of overseeing the completion of his monograph on elephant evolution. Brown, who was conducting fieldwork in Arizona at the time, sent a telegram to express his "sincere sympathy in your great sorrow."[10] By 1932, according to his biographer, Osborn was "teetering on the edge of a mental and physical breakdown."[11] Although a round-the-world cruise with some of his family temporarily buoyed his spirits, he continued to suffer from chronic indigestion and other maladies. These were the burdens that had been momentarily ameliorated in the field.

After dropping Osborn off in Billings, Brown and his crew encountered rain on the way back to camp. "We were stuck... for three days, having to walk back to camp ten miles in the rain."[12] Fortunately, Brown had befriended Mr. and Mrs. George Shea, who ran the Northern Hotel, an institution that still stands. Realizing the problems the rain might cause, the Sheas, whom Brown affectionately referred to as "the Billings Branch of the Museum Service," hopped in a plane to make sure everything at the camp was all right. When they discovered that the crew's vehicle was stuck, they sent a truck to pull it out. The Sheas' rescue mission seems to have whetted Brown's appetite for a plane of his own, because the following Monday he reported: "I took a plane southward, over Wyoming and the Big Horn Mountains, tracing the Cloverly Formation southward on the east flank of the uplift a little beyond Sheridan. I will look that country over next year. I obtained some magnificent aerial photographs of the exposures including a beauty of the Cashen Badlands, which I will publish."[13] Brown now had the bone of aerial reconnaissance gripped firmly between his teeth.

In late September, Brown sent Kaisen, who was suffering from "acute stomach trouble," home on the train, which left only Brown and Harbicht to complete the excavations and jacketing. In the process, they "found an excellent skull and jaws, with head plates and some skin impressions of a large Hoplitosaurus [*Sauropelta*] which it took four days to excavate." They also uncovered the skeleton of a "large plated Dinosaur."[14] The two specimens of *Sauropelta,* along with the *Tenontosaurus* skeleton collected that season, are now on display in the museum's Hall of Ornithischian Dinosaurs.

With fall rapidly approaching, the weather was becoming "unsettled." On October 27, Granger wrote from New York to say that he hoped the weather would hold.[15] That same day, Brown informed Osborn that he had shipped sixteen cases of his "new plated dinosaur," which he thought was ancestral to *Peltosaurus,* "with plates in position.... Although incomplete this, I think, is undoubtedly the outstanding discovery of the year. We secured it under great difficulties as five snow storms came during our work with sixteen inches of snow altogether. I had to erect a tent over the specimen and then we walked from camp to it. We ran out of provisions, as did also the cowboys at the Cashen place, and had to kill a beef."[16]

At the end of the season, Brown wrote an abstract for a talk to be delivered at an upcoming scientific meeting. The short piece documents that Brown had collected fourteen dinosaur skeletons over the past two seasons in the exposures on the Cashen Ranch. He deemed these skeletons representative of "a heretofore little known fauna, intermediate in age between the Jurassic and Upper Cretaceous" and representing several new genera and species.[17] He also, at last, recognized that the exposures from which they came belonged to the Cloverly Formation. The Cloverly fauna is still one of the best known of the Early Cretaceous assemblages.

Before heading home, Brown evaluated reports of a huge, 75-ton meteorite that had allegedly been found in northern Montana, as well as investigating a duckbill skeleton that Harbicht had found near his home in Ingomar and a plesiosaur skeleton near Riverton.[18] In all, records show that Brown shipped about thirty cases of fossils back to the museum.

To Brown's disappointment, there was not a large enough field budget in 1933 to mount a full-blown expedition to excavate the specimens on the Howe Ranch. However, department records show that $1,200 from AMNH field funds (almost $19,000 today) was allotted to support a relatively short season that lasted from August 1 to October 31.[19] Kaisen did not participate, so the role of being Brown's chief assistant fell to Darwin Harbicht. The trip also heralded the return to the field of Lilian after eight years on the sidelines.

On the way out west, Barnum and Lilian stopped in Chicago to take in the World's Fair, devoting two days to the extravaganza, which they both "enjoyed hugely." Worthy of special mention for Brown were the Science and Transportation pavilions.[20] Brown must also have been especially interested in "The World a Million Years Ago," a red-domed pavilion that contained a menagerie of robotic models of dinosaurs and other ancient vertebrates. To our knowledge, this exhibition represented the first attempt to construct robotic dinosaurs. Stepping onto a moving walkway, visitors were conveyed back in time, past a "Prehistoric Ape," saber-tooth cat, mammoth, woolly rhinoceros, and giant ground sloth into the realm of nonavian dinosaurs. The star of the show was a fifty-foot-long *Apatosaurus* (= *Brontosaurus*) with a sixteen-foot-long neck. Its supporting cast included *Triceratops* and a 280-million-year-old fin-backed early relative of mammals called *Dimetrodon*. All wallowed in and around a pit decked out to resemble an ancient swamp. The magazine *Popular Science* described the scene: "Controlled electrically, the mechanical monsters swing their heads, roll their eyes, breathe, snarl, roar, and grunt in realistic fashion. A complicated mass of cogs, wheels, bellows, and silent motors produces the life-like sounds and motions."[21] The models were designed by a New York–based firm, Messmore and Damon, which was famous for its movie sets, and the exhibition was funded by the American oil company Sinclair. Barnum kept all this in mind as he headed west toward the field.

By mid-August, Brown was back on the Crow Reservation near Pryor, prospecting on the ranch of Mike Chester.[22] His main intention was to quarry for more of the "skull specimen" he had collected the previous year. He quickly located several vertebrae associated with the skull. Lilian managed the chores and cooking around the camp.

Meanwhile, in the Bighorn Basin to the south of Pryor Mountain, Harbicht took charge of the operation to remove the overburden of sandstone that covered the sauropod skeletons at Howe Quarry, about twenty-five miles northeast of Greybull.[23] Barker Howe's son, Milo, assembled two teams of horses to scrape and remove the sandstone layer across a quarry about 65 feet by 45 feet. The underlying two-foot-thick clay layer, which contained the fossil bones, was excavated only in a few spots to determine that the skeletons did indeed extend throughout. Brown estimated that the skeletons would be about fifty feet long. All in all, the scraping operation required three weeks.[24]

By early September, Barnum and Lilian had completed their work near Pryor; before leaving, they took in "an Indian Fair for photos of Max Big Man's parade."[25] Brown related that they had completed a survey of the exposures around the circumference of the Pryor and Bighorn Mountains without finding any more areas worth working. They therefore planned to head to the Black Hills to prospect there.

In mid-September, Brown reported from Billings that the Black Hills excursion had covered 1,800 miles. They found bones near Devils Tower, Rapid City, and Buffalo Gap, but otherwise the trip produced disappointingly little.[26]

While Lilian headed off to California, presumably to visit her sister Brad, Brown set off to the north toward Harlowton, Montana.[27] In a unit that was mapped as Kootenai but Brown suspected was a correlative of the Cloverly, he discovered two skeletons of *"Camptosaurs"* [now *Tenontosaurus*]. But the prize was a "new type of carnivorous dinosaur no larger than a jack-rabbit, with vertebrae less than an inch in length, long hind legs and short front legs and with tiny sharp dagger-like teeth. This is an adult specimen and apparently a new [species]." The entire skeleton, he estimated, would be "30 inches in length and 15 inches in height."[28] This is probably *Microvenator*, which at the time of its discovery was found with associated but unreferred teeth. *Microvenator* later turned out to be a primitive, toothless oviraptorosaur, and the teeth were errant *Deinonychus* elements.[29]

After finishing at Harlowton, Brown continued his reconnaissance northward to survey exposures of the Kootenai below Great Falls and around the Belt, Snowy, Judith, and Moccasin Mountains.[30] Altogether, the season, though somewhat short on specimens, had seen a lot of ground covered—but it was nothing compared to the following year.

Upon his return to New York, Brown, having seen the Sinclair exhibit at the World's Fair, contacted the company to see if they might fund an expedition. For years, the company had utilized a cartoon image of *Diplodocus* as its logo, a tradition that endures today in the western United States. So it's not surprising that the company jumped at the chance to help Brown collect fossils of the real thing. Besides Brown was, in part, an old oil man himself. The company's president, Harry Sinclair, became so enamored of Brown's enterprise that he personally funded a 20,000-mile aerial survey for exposures of dinosaur-bearing rock units, the first such survey ever conducted.[31] Brown's financial accounting for the 1934 field season records an allotment of $7,438 from the American Museum-Sinclair Fund for the excavation at Howe Quarry and film for the survey.[32] Fully funded at the equivalent of $113,000 in today's currency, the season thus promised to be a veritable paleontological extravaganza.

Although Brown had hoped that Peter Kaisen would be able to manage the quarrying operation, Kaisen was incapacitated. As Brown wrote to another crew member, G. Edward Lewis of Yale, "Mr. Kaisen's operation [on a leg, apparently] proves to be more serious than we anticipated—in that he will have to remain a month, at least, for light treatments [X-ray] at the hospital, consequently I am sending Carl Sorensen out to be in charge of that quarry."[33] Sorensen was an experienced collector who had previously worked with George Gaylord Simpson.

In late May, Lewis set out for the Howe Ranch with one of the expedition's field vehicles and a new Kodak camera that would be used to photograph the quarry operation.[34] Shortly thereafter, he sent photos of the South Dakota badlands to demonstrate the quality of images Brown could expect. Brown responded that he was delighted with "the remarkable photographic results."[35]

Sorensen met Lewis in Billings and arranged to be escorted to the quarry site by George Shea.[36] By early June, the crew was acquiring supplies, setting up camp, and taking photos to gauge the best light conditions.[37] A few days later, Sorensen reported that Mr. Howe had been paid and was grading the road; the cook tent had been erected, and the crew had begun uncovering the fossils previously exposed. Sorensen also noted that an interested young newcomer named Roland T. Bird had shown up in camp and offered to help out.[38] Bird was the younger brother of legendary AMNH anthropologist Junius Bird. As Bird recounts with humor and eloquence in his paleontological biography *Bones for Barnum Brown*, it was not the first time the young man had followed in Brown's footsteps. Born in New York, Bird had suffered from a bout of rheumatic fever as a child before dropping out of junior high school and beginning a career as an itinerate cowboy out west.[39] In 1932, while riding through Arizona on his Harley Davidson outfitted with a collapsible camp trailer, he had stopped between Holbrook and Pine to set up camp at the base of a desolate butte. Dismounting to stretch his legs, he wandered along the talus slope noting the ripple-marked slabs of rock that littered his path. At one point, "a thin sliver of gold from under a low-hanging cloud conspired with reflection from the edge of the butte to give me a minute's respite. In this bit of time a slab of rock tipped up from the shadows. The dark outline of a crocodile-like mouth printed in stone seemed to reach up in the dim light for my ankle. My heart skipped a beat, sputtered, went on. I stumbled, recovered my balance, and the stone mouth dropped back into its pool of liquid darkness."[40]

Bird managed to haul the block back to his Harley and eventually shipped it to his father, who in turn sent it to Brown at the museum. Bird then continued on to Brown's old haunt near Cameron, where he located a chiseled inscription on an outcrop that read: "Barnum Brown, Am. Museum, New York City."[41] Picking up his mail in Flagstaff, Bird received a newspaper clipping from his father saying that Brown was leaving for Greybull to collect two dinosaur skeletons.[42] And so Bird headed to Wyoming, but when he reached Howe's ranch, Brown had already left.

Bird returned to New York in 1933, and one day he and his father paid a visit to AMNH. Brown received them in his office; there, he showed them a cast of Bird's fossil, a stegocephalian (a primitive sort of amphibian first assigned to the genus *Stanocephalosaurus* and sometimes synonomized with *Parotosaurus*),[43] and gave them a tour of the prep lab and fossil halls, during which Bird met Carl Sorensen.[44] In 1934, Brown sent the young man a letter informing him that Sorensen was leaving for Wyoming and suggesting: "If you care to join the expedition at Howe Quarry...." Bird needed no further encouragement.[45]

By early June, work at the quarry was progressing quickly, as Sorensen, Lewis, and the volunteer Bird uncovered part of the two sauropods' tails, hip regions, and some limbs.[46] Bird, Sorensen told Brown, worked hard and had offered to pay for his own food.[47] Lewis, meanwhile, was busily photographing the quarry.

Brown, still in New York, was preparing to leave for Lehigh University to receive an honorary doctorate before heading out to Wyoming. Photos in the museum's archive show him at the ceremony decked out in his cap and gown and beaming with pride. Although he would never receive a Ph.D. through formal study, he could now relish being known to the world as Dr. Barnum Brown.[48]

Brown hastily responded to Sorensen that Bird was welcome to join the crew and that he should not have to pay for his board.[49] It was the beginning of a professional relationship that would last for the next decade, with Bird essentially replacing the ailing Kaisen as Brown's primary field assistant. Brown also recommended that, at the excavation, Sorensen focus on following the skeletal trail of the two sauropods before attending to other specimens.[50] A few days later, Brown informed Sorensen that Kaisen would not be able to return to work until mid-July, although he was recovering.[51]

Brown left New York on June 22, arriving in camp on July 11 and declared himself pleased with the progress that had been made.[52] In late July, however, he heard from Kaisen that his assistant was still receiving X-ray treatments and had been unable to return to work, though he hoped to be able to join the crew soon.[53]

FIGURE 35. One of Brown's crew members, probably G. Edward Lewis, photographing the bone bed at Howe Quarry, Wyoming, from overhead in a barrel lifted by a gibbet during the 1934 Sinclair/AMNH expedition (2A25671, American Museum of Natural History Library)

Brown spent the next few weeks in camp overseeing the excavations. Lilian arrived after him and made an immediate impression. As Bird recalled: "Mrs. Brown joined our camp . . . a sparkling young woman with brown hair, great hazel eyes, and irrepressible humor. 'What's Brownie been up to, before I came?' she inquired, and raised an admonitory hand as I opened my mouth for whatever inane reply. 'Now R. T. . . . don't tell me more than I want to know.' Mrs. Brown fitted into the picture wondrously."[54]

Clearly, within the decade that had passed since returning from Asia and Greece, the couple had gained the confidence required to openly joke about whatever personal indiscretions lurked in their private lives. Drawing on the experience she had gleaned on previous expeditions, Lilian took charge of

FIGURE 36. Photo of bone bed at Howe Quarry taken from about thirty feet above during the 1934 Sinclair/AMNH Expedition. R. T. Bird's grid system for mapping the bones is visible, along with Brown and another person (top center) and a technician excavating bones (center right). (132803, American Museum of Natural History Library)

managing domestic affairs and keeping records. As she related in an article for *Natural History,* "While the Doctor is in the field hunting big game of other days, my hours are filled from sun-up till sun-down; for a woman in camp must often play various roles of secretary, photographer, cook, camp-wrangler, as well as wife. Perhaps the most important is the secretary's job, for camp records must be kept to date, as an impression once lost is difficult to remember accurately."[55]

Because the crew had hired a cook, Lilian's culinary burden was limited to helping plan the meals, which consisted primarily of canned goods heated over a gasoline stove. An increasingly time-consuming job, however, involved dealing with visitors drawn to the quarry by the attention the project had generated in the media. As Lilian related, "Camp was never lonely, because news of the great discovery had been broadcast by one Hans Adamson, the expedition publicist."[56] Brown had become a celebrity of sorts, and wherever he went, hoards of loyal fans, who dubbed him "Mr. Bones," would wait for his train to arrive and compete for the privilege of ferrying him to his destination. Brown himself described the ongoing open house on the Howe Ranch:

> Over [the rough] road came a constant stream of cars: thousands of visitors from all parts of the United States and a few from Europe. From early morning until late at night they came, sometimes during the night, many repeating their visits as the work progressed. So many visitors lost their way on side roads that signs were placed along the trail, but they were soon taken by souvenir hunters—one day a [fossil] rib came in the mail, sent anonymously by some conscience-stricken visitor.
>
> Although it was inconvenient at times, we were glad to show the quarry to visitors and schools, for never have I uncovered a more interesting deposit of prehistoric remains.[57]

The attention, however, triggered a more serious problem for Brown that would once again entangle him with the law. Because of his work at the site, the Howes were overwhelmed by the torrent of sightseers, some of whom, as Brown recounted in an affidavit,

> asked Howe what he was getting out of the operation and several suggested that as the work was financed by a rich oil company the specimens must be worth a lot of money. Finally I heard in a roundabout way that Barker Howe was considering a suit against me, but the matter was never discussed

between Howe and myself until he placed a sign on his fence corner charging visitors ten cents a head to see the excavation.

I called Mr. Howe to our camp and told him as it was a public institution that we represented I did not feel that it was an ethical procedure to invite the public to see this work and then charge admission. But as the appropriation [presumably from Sinclair] permitted it I would pay him ten cents a head for all visitors to the quarry who registered after that date. This agreement Mr. Howe said was satisfactory, and the following day he asked for $10.00 on account, which I gave him. This was on the 18th of August and in the presence of Mrs. Brown I drew a line through our visitors registry which was kept as accurately as could be made of all visitors until the work was completed.[58]

At the start of September, with Howe apparently appeased and the quarry work moving rapidly forward, Brown left Sorensen and Lilian in charge of things and began a new venture: an aerial survey of the region's geology. During the preceding several years during flights between Billings and Denver, Brown had, he said, taken occasional opportunities to "test the value of airplanes in geological work . . . when no other passengers were present." Summoning his charm to ignite a sense of shared adventure, he would encourage the pilots to "deviate from the scheduled course, giving me an opportunity to study the Mesozoic strata where we were then working on the north flank of the Big Horn Mountains."[59]

No doubt excited by the chance to finally take wing on an extended and customized aerial adventure, Brown, with Harry Sinclair's backing and advice, hired D. A. "Mac" McIntyre of Tulsa to pilot his single-engine Stinson prop across what Brown anticipated would be an 8,000-mile swath of ranges and badlands from Montana to Arizona. As he later recalled, "All good ships have appropriate names. Ours was christened 'Diplodocus' with a picture of the beast painted on the side as a talisman."[60] Of course, that "talisman" also paid homage to the survey's sponsor, Sinclair.

Brown was ultimately able to arrange for the use of two sophisticated cameras, a ten-inch unit rented from the Fairchild Aerial Camera Corporation and a twenty-four-inch Fairchild that Brown borrowed from Harvard. Accommodating the larger camera not only made the Stinson two hundred pounds overweight, but it also required some serious modifications to the fuselage: "One door of the ship was removed and the camera suspended by heavy straps from the frame of the ship. In this way the camera could be

FIGURE 37. Brown's crew standing at Howe Quarry watching the plane in which Brown conducted the aerial survey during the Sinclair/AMNH Expedition in 1934 (AMNH Vertebrate Paleontology Archive; Osborn Library; from Brown 1935a: 9)

extended while shooting and withdrawn into the ship while traveling." This arrangement struck Brown as all the more intimidating once he realized that McIntyre had not obtained an essential piece of equipment: "Knowing we would cover much mountainous country, I had stressed the necessity of having parachutes, and it was with considerable misgiving I learned from the pilot that no parachutes had been provided. He assured me that he did not feel the necessity of having them, but if required, they would be supplied later. After flying with Mac a few days, watching him maneuver his ship, I soon forgot about parachutes, and I firmly believe he could land his ship safely on any moderately flat ground lacking major obstacles."[61] That leap of faith would soon be tested.

Brown and McIntyre, aided by G. Edward Lewis and George Shea, took to the air from Billings, but "a pall of smoke" from forest fires in Idaho and western Montana inhibited their ability to take good photos. Therefore, after crossing the Pryor Mountains and landing at Greybull to photograph the quarry and give the "party of excavators... short trips in the air" to admire their work and the surroundings, Brown and his pilot headed south for clearer skies.[62]

From Rock Springs, Wyoming, they veered southwest toward Salt Lake City.[63] To plot the course, they integrated Brown's knowledge of landmarks on the ground with McIntyre's navigational skills, supplemented by highway maps and maps showing the locations of regular and emergency landing fields.[64] Normally, they flew at an elevation of 3,000–5,000 feet, except when evading thunderstorms or crossing over mountain passes. Newspaper reports heralded their impending arrival at each stop, where Brown's old friends would meet him.[65] While the Stinson was being serviced, Brown would drive out to spots where reported fossils had been found.

On the way back from Salt Lake, where Brown had photographed ancient shorelines of Lake Bonneville, to Dinosaur National Monument, McIntyre mistakenly followed the wrong set of railroad tracks, which led them a hundred miles off course. Low on fuel, they redirected their flight path toward Green River, Wyoming. Unable to locate a landing field, Brown directed McIntyre to "a level stretch of sage brush" just outside town. McIntyre negotiated a perfect three-point landing, but about two hundred yards after touching down, "the left wheel struck a gophered sage brush. With the increased weight shifting to that side, the wheel sank, breaking several spokes. Hobbling along... like a maimed jack-rabbit, we finally came to a stop. We looked at each other without printable comment, for it was all in a day's work."[66]

Once refueled, the aviators headed back toward Dinosaur National Monument by way of Price, Utah.

> A good field was marked at Price,... but to our surprise and consternation the runways were covered with mule teams leveling the field. In spite of our circling and signaling, every man stopped in his tracks to watch our maneuverings. We had to have gas in order to reach Vernal, and land we must! One runway was covered with Russian thistles, but looked as though it had been used. It crossed a stream course that appeared to have been filled.... We came down on the end of the runway at sixty miles an hour, a perfect landing, but to our horror, that ditch was fifteen feet deep.... We leaped into the air twenty feet,... landing on the other side of the ditch on the crippled wheel—smashing it to bits—nothing left but the hub. The ship bounded, staggered, pivoted, and finally came to a stop completely reversed.[67]

Having narrowly averted death, McIntyre worked all night to fashion a new wheel out of the mangled hub, the spokes from several Ford wheels, and parts from two derelict planes. Finally, they were able to photograph the exposures in Dinosaur National Monument before continuing south to the Triassic exposures in the Painted Desert and around Cameron, Arizona. Other areas of focus in the region included Meteor Crater, Petrified Forest, Canyon de Chelly, Chaco Canyon with the ruins of Pueblo Bonito, and the Folsom kill site. Then it was back to Denver, where they photographed the nearby Jurassic exposures of the Morrison Formation before documenting that formation's outcrops at Como Bluff, where Brown had made his first dinosaur discoveries in 1897.

When they returned to Billings at the end of September, the smoke from the forest fires had cleared and Brown was able to take photos of the Cloverly exposures around Cashen Ranch, as well as the Morrison and other exposures around Howe Quarry. The final leg of their aerial reconnaissance took them over the Late Cretaceous exposures of the Judith River and Hell Creek Formations from Great Falls to Ekalaka, then east into South Dakota. Back in Billings, Brown's log and report summarize their six-week endeavor, in which they had logged 180 hours in the air, covering 20,000 miles. Brown estimated:

> New areas discovered by this method in Triassic, Jurassic, Cretaceous, and Fort Union strata will require at least three seasons for our field parties to completely explore.
> New important discoveries ... are:
> 1. A quarry of dinosaur bones ... in the Mesa Verde formation (mid-Cretaceous) ...
> 2. A fine ceratopsian skull and jaws in the Judith River formation.
> 3. A *Mosasaur* skeleton and five foot *Mosasaur* skull in the Pierre....
> 4. A new type of Reptile ... , Middle Triassic.
> 5. An unreported Meteoritic Crater.
> 6. An unreported Ceremonial Way of 12th Century.[68]
> 7. Several new oil domes.[69]

While Brown was in the air, Kaisen had written him saying that his leg was "still on the bum"; he also mentioned that he had heard Brown interviewed twice on the radio and seen articles about his work in the newspaper. Kaisen was still receiving treatments at the hospital and had finally resigned himself to that fact that he would not make it out to the quarry, though he hoped the crew had gotten "something new for me to work on and not too big."[70]

Back at the quarry, meanwhile, Sorensen and his crew had already endured two snowstorms by early October.[71] Lilian recalled one such scene in her

Natural History article: "Ma, our trusty cook, poked her head out of the tent and cried, 'We're snowed in!' We were. For three days and nights the blizzard howled, and when it was at its height, two horsemen appeared on the horizon bearing the sad news that Ma's son had met with a fatal accident... and I was alone to keep camp for the men."[72] Despite the frigid weather, Sorensen reported to Brown that they had managed to take out the skull and neck of one of the large sauropods. Plaster, however, was at a premium, since they were now using a bag a day.[73]

Earlier in the season, Bird had accepted a critical assignment from Brown, which, along with helping to excavate and jacket the specimens, had occupied much of his time. Seeing the tangled mass of bones, Brown had lamented that "something's got to be done about recording relationships" among the bones of the skeletons. "How would it be," Bird suggested, "if I stretched strings across the quarry... say a yard apart... and worked up a quarry chart?" "Fine, R. T.," Brown replied; "get right on it."[74] Bird's meticulously rendered quarry map is still the most valuable document available for interpreting the identity of associated fossil bones collected at Howe Quarry. Photos show that the crew also erected an enormous lifting lever, such as is used to load hay bales on wagons, in order to take overhead shots of the quarry. A crew member would crawl into a barrel attached to the end of the lever and be elevated thirty feet in the air to snap shots of the quarry as work progressed.

On November 12, Brown proudly reported that it had been "a banner year with great results aloft and alow." He estimated that four thousand bones had been collected, a veritable fauna that would fill a whole boxcar with its 145 cases.[75]

In his final report, Brown stated that the fossils represented mostly sauropods and ornithischians, with a few theropod teeth, patches of sauropod skin and gastroliths. "These dinosaurs are Jurassic in age [from the Morrison Formation] and represent a northern fauna three hundred miles north of previously described Sauropods with several genera and at least six species new to science. More than twenty individuals are preserved, the largest approximately fifty feet long, but a *Barosaurus*-like species predominates, including one specimen with a delicate skull and jaws attached."[76]

Lilian had left for California in early November to visit her sister, and Brown intended to follow her as soon as he finished up. Sorensen and Bird shipped the last boxes and headed home on November 20.[77] Bird had clearly passed the audition, for in late December his new boss, now wandering around Nevada, wrote the Vertebrate Paleontology Department: "Hope R. T. Bird has reported for the laboratory work."[78]

At the end of this extraordinary field season, however, one unfortunate issue lingered. Two days after Brown returned from his aerial survey, he stated:

> I received a summons—my first notification of this suit.
> The allegation was to the effect that Barnum Brown entered on the property of Barker Howe with a party of men and excavated twelve skeletons of dinosaurs, which are unique, cannot be replaced, and are to the best information worth $25,000....
> On advice, I put the case in the hands of Mr. Thomas H. Hyde, attorney at law, in Basin, Wyoming.[79]

The case remained unsettled as of July 1936, when Brown wrote a colleague in the museum stating that he had expended $7,132 of the $7,438 allotted from Sinclair's grant, plus about $250 in out-of-pocket expenses, and that he was holding the balance in hopes of settling the suit lodged by Howe.[80] The outcome of Howe's suit is not known.

The story of the Howe Quarry collection does not end there. Because of financial difficulties at the museum in the late 1930s, the retraction of scientific activities due to World War II, and Brown's impending retirement, the specimens went into storage. According to museum curator Gene Gaffney, writing in 1992,

> The History of the Howe Quarry crates is a sad one. Legend has it that the collection was originally stored outside [in a museum courtyard] under canvas tarps. Sometime in the 40s or 50s it caught fire and an unknown quantity (estimated at roughly 50%) of material was destroyed. Subsequent to the fire, and before my arrival in 1965, the collection was moved into that chamber of horrors, the infamous "rifle range" (supposedly called that because guards used it for pistol practice, God forbid). The room has always been water soaked, and the collection was put there out of desperation, hopefully temporarily.[81]

In the early 1990s, AMNH curators and staff cleared the room. Sadly, many of the surviving boxes had rotted through, and rats had destroyed much of the rest of the collection. Some specimens were saved and packed in new boxes. Although only a fraction of the original collection, what does remain is being prepared, albeit slowly, as an important component of Brown's AMNH legacy. A few of the bones are still numbered, allowing their location to be identified on Bird's map. One specimen, the delicate neck and skull of a juvenile *Barosaurus,* is on display in the Orientation Center for the museum's fossil halls, while a cast of that same specimen has been incorporated into the skeleton of the young barosaur hiding behind its towering mother just past the museum's main entrance in the Roosevelt Memorial Hall.

But still the story has not ended. Howe Quarry was reopened in the early 1990s by a Swiss firm capitalizing on the big business of dinosaur collecting, something only on the distant horizon in Brown's day. Using the latest in excavation techniques, this team had started to unearth some fantastic specimens. Their tremendous success, however, led to litigation brought on by the landowners, and at present the project lies stalled, its future uncertain.

FIFTEEN

Toward the Golden Years
The Mystery Track-Maker and the Glen Rose Trackway (1935–1942)

FOLLOWING HIS ADVENTURES OF 1934, Brown remained active in the field; however, a lack of funding continued to bedevil his efforts. Part of the problem was no doubt related to the death of Osborn in 1935. Brown's longtime supervisor had retired from the museum in April 1934, after forty-three years of service. Over the next year and a half he sought refuge in the peaceful surroundings of his estate at Castle Rock on the Hudson River near West Point, north of New York City.[1] He also continued to pursue an ambitious schedule of daily research, pouring his energy into his massive monograph on elephant evolution. His steely work ethic, however, exacerbated his physical ailments, which included heart and circulatory problems reflected in high blood pressure and pulse rates. In November 1935, at the age of seventy-eight, he succumbed to a heart attack while sitting at his desk in his library. After thirty-eight years of both professional and financial support, Brown had lost his most reliable patron.

Despite this loss, Brown once again traveled west, this time to Texas, Oklahoma, and Kansas, for about three weeks at the end of October 1935, encouraged by correspondence regarding the discovery of Permian and Pleistocene fossils. With the museum's collection of dinosaurs now on the verge of bursting at the seams, Brown was turning his attention to the origins of our own mammalian lineage. Continuing his new penchant for aerial reconnaissance, Brown documents that he flew two thousand miles and drove four hundred miles "collecting some good Pleistocene and Permian fossils and much valuable information on other areas that should be worked ... as soon as possible." Especially fruitful were Middle Permian (now assigned to Early Permian) exposures about thirty miles northeast of Seymour, Texas, where the Walker Museum had collected a complete skeleton of the early relative of mammals, *Ophiacodon*, the previous summer. Brown apparently found fragments of that taxon in the gravel lenses there, as well as pieces from *Edaphosaurus, Dimetrodon,* and other early relatives of mammals. Twenty miles east of Seymour in overlying Pleistocene sediments, Brown found fragments of armored glyptodonts and mastodons and a jaw of *Equus giganteus.* In all, the season's work cost a relatively paltry $270.86—just under $4,000 in today's currency.[2]

The 1936 season saw a cross-country tour by car for Brown and Bird, funded by a $300 grant from Frick's endowment fund for vertebrate paleontology. First they drove to Dinosaur National Monument, where "the cut [was] about two-thirds completed, with 62 men under [the] National Park Service working on the excavation. When the cut is finished the American Museum plans to do the relief work on the dinosaur specimens on the wall." From Vernal, the pair headed south to Cameron, Arizona, where prospecting around their old Triassic haunts yielded a "gigantic skull, the largest known Phytosaur, more than four feet long, with the jaws, humerus and several vertebrae." The specimen, now called *Machaeroprosopus,* is on display in the museum's Hall of Vertebrate Origins. For Brown and his colleagues, it was not enough to document only the pageant of vertebrate evolution for the public; they wished to depict the flora of past geologic ages as well. Thus, the pair relocated "a forest of 42 petrified trees of Triassic age, discovered first by Dr. Brown in 1904," and stopped by Petrified Forest National Park, where they obtained a "representative" collection of the flora there. At two other localities, they secured a "fine" collection of *Calamites* specimens, a plant related to modern horsetails, for the Dinosaur Hall.[3]

On the way home, Brown and Bird stopped near Argos, Indiana, to collect a mastodon skeleton. Lilian joined the pair on this leg. Upon her arrival, Bird later recorded, with her characteristic wit she inquired of Bird: "How does Brownie manage to look so fit, when he's so long away from me?" To which Brown warned: "Don't you tell her; don't you dare ever tell her!"[4]

In 1937, the Vertebrate Paleontology Department hoped to double the size of its preparation staff to twelve and hire an assistant curator of fossil reptiles.[5] Brown's goal was to restore staffing to 1916 levels in the prep lab and hire an "understudy" for him to train as a replacement. His eventual curatorial successor, Edwin H. Colbert, was already on staff but was focusing his studies on fossil mammals.

In December 1936, Brown was already planning a two-pronged expedition to the West for the following year, with one team working in the Mesaverde Formation near Rock Springs, Wyoming, and another in the Hell Creek and Judith River exposures in Montana.⁶ The trip to Montana, to be led by Darwin Harbicht, would focus on collecting a duckbill skeleton as well as a partial ceratopsian skeleton and skulls, soon to be covered by the lake behind the massive Fort Peck Dam on the Missouri. More ceratopsian material would be the target in the Judith.⁷ The required funding of almost $4,400 had yet to be secured, although plans were afoot for a nationwide radio broadcast on CBS. No record exists of that expedition having occurred.

The trip to the Mesaverde, to include Brown, Bird, Lewis, and Lilian as well as museum employees Gil Stucker, Erich Schlaikjer, Robert Chaffee, and James Ryan, was designed to smoke out an elusive "Mystery Dinosaur" that had left its enormous footprints in sediments buried deep underground in the States Mine in Colorado. Any fossils from this trip would help fill a gap in the record of dinosaurs between the Late Cretaceous Hell Creek/Judith fauna and the Early Cretaceous Cloverly Formation. Brown hoped to entice Sinclair to sponsor this leg of the expedition with a grant of $7,000, and he also planned three nationwide radio broadcasts on CBS to drum up public interest. Brown hoped to use Union Pacific's heavy equipment to help collect the specimens and its railroad to help transport the fossils. By late May, Sinclair had anted up a check for $4,000—so with the equivalent of about $56,000 in today's dollars, the game was on.⁸

Bird was on his way to the field by early June, while Brown was managing the process to secure the necessary permit to collect in Wyoming, which was issued a month later. But a hitch in the plans arose early on: the Mesaverde wasn't yielding up any of its fossils, as Brown lamented to Hans Adamson, a colleague in public relations at AMNH.⁹ A week later, however, Brown, unflustered, indicated that things were beginning to improve, stating: "We have accumulated more important information regarding this Mesa Verde Cretaceous fauna and flora than has ever been known previously."¹⁰ He also suggested arranging broadcasts from Dinosaur National Monument to supplement the expedition's publicity.

On July 10, Brown reported to Granger, who was holding down the fort in New York, that, despite use of the coal company's enormous shovels, only a partial ceratopsian skull had been recovered in the expedition's earliest efforts; but then, as the prospectors fanned out, they had discovered

> half of a skeleton including a disarticulated skull of an enormous [duckbill]; also a skull and ilium of a Ceratopsian; a fragmentary specimen of a low-plated dinosaur, and a weathered skull of a Ceratopsian presumably related

to Styracosaurus.... They indicate a fauna of Belly River age. We have also secured a very large series of invertebrate fossils, and also a large flora.

... We are still anxiously searching for the mysterious Iguanodont that made the huge footprints....

Bird and I ... arranged for securing a section from a coal mine roof on which are two enormous tracks of the same animal with footprints 34 inches across, 34 inches long with a 15 foot 2 inch stride. This will be a wonderful accession to our Dinosaur exhibit....

Mrs. Brown is now filling her post as Secretary, and helping coordinate our work at camp....

During our first two weeks we were overrun with hundreds of visitors including newspaper men who have sent out many unauthorized statements to the Press, including one stating that we had given up the search; and another that our camp was destroyed by a cloud burst.[11]

Brown expanded on the perilous thunderstorm in his final report to Sinclair. It had occurred when Brown and an assistant named Chaffee were collecting the duckbill, which was located on a high ridge that was extremely difficult to access by vehicle. As thunderheads burgeoned on the horizon, the pair persisted in excavating, paying scant attention to the gathering peril. But their focus changed as sheets of heavy rain pelted the quarry. Seeking shelter in the truck's cab, they watched from the ridge in growing apprehension as "the waters began to roar like Niagara in the creek beds on either side of our ridge.... We realized that the truck could not be moved for several days, and that the only way we could get back to camp was to cross one of the swollen streams and walk.... We made coats out of gunny-sacks ... and in the pitch darkness waded to the stream between us and the town—only to find that the stream was an impassable rushing torrent." Their only avenue of escape was a bridge three miles distant, but after trudging through the slippery mud to the span, they discovered that it had already washed away. Soaked to the bone, they hiked back to the truck where they planned to build a fire for warmth, only to find that they had used all their matches lighting cigarettes. Then "A happy thought came to Chaffee and starting the motor he short-circuited a spark-plug and ignited cigarette papers soaked with gasoline.... We built a fire in plaster pans ... [and] took turns in sleeping and tending the fire until morning. In the meantime, members of our party were much concerned.... Rescuing parties were sent out during the night but they could not find us and their anxiety increased until late the following day when we finally reached camp."[12]

During the deluge, reporters in town had gotten wind that Brown had not returned, and word quickly spread that his party had perished in the cloudburst. A week later, Granger calmly wrote to Brown saying that "various high-strung people were running about the Museum expressing great anxiety following the receipt of a press dispatch announcing disaster to your expedition. I assured them, however, that if there [were] any survivors... they would surely wire in to us and I hoped that they wouldn't worry any more than I did—which was none at all."[13]

On July 23, Adamson told Brown that he would follow up with CBS; however, he observed, "since their basic interest was... the Mystery dinosaur, and so long as that is still in hiding, I am somewhat afraid that we will have to consult them on the whole idea" of direct remote broadcasts.[14] To allay the museum's concerns about the broadcasts, Brown reported in early August that they now "had a fine collection of brand new Dinosaurs from a formation in which none have been previously known by skeletal remains." More importantly, he said, "last week we found a humerus (fore arm bone) of one of the giant dinosaurs that made the huge footprints found in the coal mine." In all, they had twenty-three crates in storage and another twenty yet to pack. Additionally, Brown and Bird had collected the footprints in the States Mine.[15]

Brown's own perspective on the difficult work near Rock Springs and the perilous collection of footprints in the States Mine provides an enlightening glimpse into his approach to his work.[16]

Throughout early 1937, Brown had corresponded with Charles States, owner of a coal mine near Cedaredge in western Colorado, about an enormous track in the ceiling of one of his mine's shafts, which was revealed when the coal seam was mined out and the bottom of the rock layer overlying the seam was exposed.[17] Before the season started, a second track had been uncovered. The tracks were in a sandstone layer, thought to be about a foot thick, that cut through the middle of the fourteen-foot-thick coal seam. During a visit to assess the prospects for collecting the footprints, Brown and Bird, with the help of States and his miners, had "formulated plans to cut down these enormous tracks with the rock between them—which [would] remove any doubt as to the length of the stride." Brown estimated the stride to be fifteen feet, suggesting that the Mystery Dinosaur "towered to a height of approximately thirty-five feet."[18] Later reanalysis of the specimens revealed intermediate tracks, making the actual stride a mere 7.5 feet.

Excavating the massive blocks of sandstone with their tracks would be a delicate and dangerous operation, especially when in "drilling through the roof it was found that the sandstone was much thicker than estimated above the second track." Brown described the operation:

> It soon became evident that we had a job in keeping with the size of the tracks, ... to cut a section of the sandstone roof seventeen feet long by five feet wide and forty inches thick, weighing approximately thirty tons. Nine experienced miners were hired ... and assigned to eight hour shifts so that the work proceeded without interruption night and day. Additional props were put under the block and some ... men mined out the vein of coal above and chiseled off the upper layers of sandstone, while others drilled through from below at half inch intervals—outlining the block, while our men plastered the roof to insure the safety of the tracks. Eventually, ... the block was cut into three sections, for the original block could not pass out of the mine. Supporting timbers were then placed in the chamber above and the blocks were lowered successively with a five ton block and tackle.
>
> Even after the block was sectioned some parts of the mine had to be re-timbered so that the sections could pass through. Finally, out of the mine they were still further reduced until the three weighed only eight tons ... , after which they were hauled two hundred miles by truck and loaded into the [railway] car.[19]

While Bird was collecting the tracks, the miners told him about a treacherous caved-in chamber in the nearby Green Valley Mine that housed a spectacular array of palm frond fossils. Bird convinced the miners to escort him into the abandoned "tropical room"—much against their better judgment. But the allure of more magnificent specimens was too attractive for Bird to resist, even though the imprinted rock hung "precariously ... with tons of loose rock above it." Retimbering the room in order to excavate the treasure was, in Bird's mind, a small price to pay.[20] Correspondence from Bird expanded on the dangers and difficulties of getting the specimen out of the abandoned mine by "track and cars": It involved a "seemingly endless task of moving the lanterns on, bringing up the specimen, [and] rolling rocks out of the way—not to mention the use of planks and rollers over those ungodly fallen rooms. When done, it left one with the odd feeling that he'd not only accomplished the improbable, but had survived it too. . . . But I wouldn't have missed it for the world."[21] The trackway formed by the "Mystery Dinosaur" now adorns the transition between the Saurischian and Ornithischian Dinosaur Halls at the AMNH. Contemporary research on these specimens

FIGURE 38. R. T. Bird drilling out a hadrosaur track in the roof of the States Mine in Colorado during the 1937 Sinclair/AMNH expedition (131845, American Museum of Natural History Library)

has indicated that they are the tracks of hadrosaurs, which were not so nearly as large as Brown believed.

Back at the Rock Springs camp, the cook, Ma, had to leave in early August, as she had at Howe in 1934, and Lilian suggested that she not return that summer. What with the comings and goings not only of the crew but of members of the press as well, Lilian counseled Ma, "things will be very unsettled here, and the work [of collecting fossils] is really a man's job."[22] And so Lilian assumed the cooking chores as well as the secretarial tasks.

In late August, Brown was still insistent that live radio broadcasts be conducted from Rock Springs and Dinosaur National Monument, and the museum was lobbying CBS to make those arrangements.[23] Much to Brown's disappointment, CBS was shortly forced to cancel the remote broadcasts due to financial shortfalls.[24] Brown's main concern was that "one of the inducements offered to our sponsors [Sinclair and Union Pacific] was the publicity they would receive in broadcasting and press reports."[25] Accordingly, now Brown planned to feed formerly embargoed information about the expedition's discoveries to the press and try to arrange a broadcast from Dinosaur National Monument through another network, following up with a CBS broadcast upon his return to New York. Meanwhile, Brown reported to Granger that they had discovered the nearly complete skeleton of a small, new duckbill and that more than forty cases of fossils were now in storage, most of which represented new taxa of animals, both vertebrates and invertebrates, and plants.[26]

Roy Chapman Andrews, with whom Brown had long shared a collegial if competitive relationship under Osborn, had taken over as the museum head in 1935. Brown's relationship with Andrews—in particular, whether they actually liked each other or could not stand each other—remains something of an enigma. For instance, the Central Asiatic Expeditions, the largest and most famous expeditions in the museum's history, which were organized by Andrews under Osborn's watchful supervision, were also some of the most expensive expeditions ever undertaken. Andrews was accompanied by Granger as head paleontologist, as well as a number of museum technical staff. Although Brown may have been asked to come too, if any such offer was made it was apparently half-hearted. Why wasn't the museum's most capable fossil hunter conscripted for service? While there has been much institutional speculation on the subject, one thing is certain: the divergent styles of Andrews and Brown would have jeopardized the success of any joint expedition.

While Andrews managed activities in New York, Granger was planning a visit to Brown's camp. Brown, who noted in a letter of August 23 that the first snow had just fallen, looked forward to welcoming his old compatriot, joking: "We have two or three idle picks, and plenty of grub.... Better bring summer and winter clothing, for the nights are cold, and boisterous. Any moment you may find yourself transferred to a neighboring peak... just a friendly hurricane."[27]

In early September, Brown reported that another incomplete theropod skeleton about the size of *Albertosaurus* had been collected, making the haul now fifty-four cases.[28] Brown busied himself arranging for their shipment, which he reasoned might be cheaper if they went by train from Rock Springs to Los Angeles, then by steamship to New York. Brown also asked whether Chaffee and Stucker might be hired by the museum as preparators. Granger replied that although Andrews would need to weigh in when he returned from vacation, the museum administration was hopeful that the two technicians could be hired for $50 a month. He also said that he still hoped to swing by Rock Springs but that his visit would be short.[29] Early September brought the news, too, that the museum's PR department wanted Brown to appear on a new weekly CBS show called *New Horizons,* developed in conjunction with the museum, when he returned to New York in October or November.[30]

Late in September, Brown shifted gears, collecting Eocene fossils of titanotheres, creodonts, and turtles on the Cow Creek sheep ranch southeast of Bitter Creek in Wyoming, funded by a $1,500 grant from Frick (equivalent to about $21,000 in today's currency).[31] Intent on obtaining aerial views of his favorite outcrops around Rock Springs, he also worked to arrange another flight during which he intended to take shots from lower altitudes, between 1,500–2,000 feet, as well as higher altitudes, up to 14,000 feet, at a cost of $18 (equivalent to about $250) an hour.[32] Such surveys were not intended to locate fossils directly; the goal, rather, was to discover new exposures of fossil-bearing formations that could be prospected through ground-based surveys after the flight. By the end of October, sixty-one crates of fossils weighing over 25,000 pounds were wending their way east on the Union Pacific lines.[33]

Brown returned to New York in early November, where he trumpeted the triumphs of the season to an adoring nation over the CBS network. According to press releases and reports, the major achievements included "reptile monsters" from the 80-million-year-old Mesaverde, including the tracks and bones from the Mystery Dinosaur, the skeleton of a "weird dinosaur which apparently resembled a gargantuan horned toad," and an enormous duckbill skeleton that "may eclipse the record of the ferocious

Tyrannosaurus Rex which towers to a height of 18 feet." Filling out the treasures was "a truly fantastic petrified jungle consisting of palms, figs, poplars, willows," and other plant fossils, including a "myriad of ferns."[34] No one could claim that Brown had failed to deliver for his corporate sponsors. The endeavor had cost $7,733 (almost $110,000 in today's currency), of which Sinclair had contributed $7,000.[35]

With Sinclair back in the fold, Brown began planning another foray to the West in the spring of 1938, complete with a round of publicity that he hoped would include the kind of remote broadcasts that had fallen through the previous year.[36] Brown intended to work in the Mesaverde near Rawlings, Wyoming; the Morrison near Sundance, South Dakota; the Cloverly near Greybull, Wyoming, and Harlowton, Montana; and the soon-to-be inundated Hell Creek near Fort Peck, Montana, which he had failed to salvage the preceding year.[37] Estimated costs for the expedition were $6,000, and the crew would include himself, Bird, Chaffee, Schlaikjer, and G. B. Guadagni. Brown submitted his proposal to Sinclair in late April;[38] by early August, however, he was still waiting for an answer. In the meantime, he had approached Frick, with the consent of Andrews, and received a $400 grant, which he used to dispatch Bird to begin collecting and reconnaissance work in Montana, Wyoming, Utah, and Arizona.[39] By August 21, Bird had rendezvoused with George Shea in anticipation of heading out to the Cloverly exposures near the Cashen Ranch.[40] Shortly thereafter, news arrived that Sinclair would not be able to fund the year's field season due to financial constraints.[41] And so the ever resourceful Brown took a consulting contract and prepared to leave for Alberta to investigate oil prospects.

Meanwhile, Bird and Shea hit paydirt at Cashen's. In excavating a series of tail vertebrae, they had uncovered a dinosaur skull associated with another vertebral column, along with the sacrum and ribs.[42] Brown was delighted. When he asked what taxon the skeleton might represent,[43] Bird responded that it was a "Tenantosaurus type" (now called *Tenontosaurus*), its bones well preserved but lacking the feet and forelimbs.[44] Brown congratulated Bird, noting that the tenontosaur skull material would prove essential in distinguishing that genus from *Camptosaurus*.[45] Within a couple of days, Bird had packed the fossils in three crates and shipped them off to New York.[46]

By early October, Bird had traversed the Morrison and Cloverly near Greybull, finding little except an apparent ankylosaur pelvis; he had also stopped by Dinosaur National Monument to photograph the progress there.[47] He then returned to his old haunt near Cameron, where by the end of the month he had discovered two new fossil plant localities and a small phytosaur skull.[48] But just over the horizon lay one of his most important discoveries ever.

On November 20, Bird wrote Hans Adamson to inform him about a just-released Associated Press story.[49] And in a six-page epistle to Brown from Glen Rose, Texas, fittingly written on Thanksgiving day, he detailed the discovery that had triggered the story. While traveling through Gallup, New Mexico, on his way back to New York, Bird had gotten wind of fossil dinosaur tracks—some of them made by three-toed theropods—that adorned a trading post. The trader was not available, but Bird learned that he had bought the tracks in Glen Rose.

> The trail led directly to one . . . Ernest Adams—who denies all and claims all in the same breath. . . .
> Of course, I showed him the two pictures [of tracks]. . . . He didn't seem surprised for it has been a custom lately for people to take up tracks from the river bed and sell them for ten or fifteen dollars. . . .
> But Adams failed to give me [the] slightest bit of information. . . .
> During my prospecting trips up and down the [Paluxy] river . . . some of the finest and most gigantic [tracks] I have ever seen turned up on a ledge.
> Before I had cleaned half of the mud out of them I was convinced I had discovered the trail of a huge sauropod. Depressions like washtubs, some of them six or eight inches deep. [illustrations included]
> [The river runs] rather swiftly . . . [tumbling] blocks about that weigh tons. Some of the best prints were in the most precarious condition. . . . It was impossible to take any of them up alone . . . but I had a sack of plaster. . . . So I decided to take a cast of two of them and have at least the record. . . .
> The fore foot is wider than it is long, being 24×20 inches in size. . . .
> The hind foot measures a full yard in length . . . and is 26 inches wide at its greatest point. There are three large claws arranged [in] typical sauropod fashion and still another mark suggesting a small nail. The mud bulges up in front of the foot as it does with all the tracks—rather spectacularly—proving these great beasts did, occasionally at least, move about where the buoyancy of their native element, water, was partially if not entirely lacking. That small carnivore tracks are found at the same level would indicate this also.
> As for the trail itself, it measures about six feet across. . . . I saw no conclusive evidence of the tail having been dragged. . . . The strides . . . run from 11 feet 5 inches to 12 feet 10 inches.[50]

In a follow-up letter to Adamson, Bird summed up his moment of discovery: "There come rare moments in the lives of all of us when we see things we do not actually believe."[51] Bird thought that a coffer dam could be built to divert the river; then the tracks could be exposed and cleaned out, more accurate measurements made, and the trackway mapped. Actually collecting the specimen, however, seemed beyond possibility, due in particular to the expense.

In a press release, Adamson hit some of the high points of sauropods, including the idea, proved by the tracks, that a sauropod "could, and did, desert his aquatic environment when the spirit moved him." He also noted that the trail could be followed for nearly one hundred yards in the riverbed. In characterizing the ancient setting, the release stated: "The trail was made across a wide and level mud flat, presumably above tidewater at the time (as is indicated by the sharp clarity of the tracks) and [the sauropod tracks] were directly associated with both large and small carnivorous dinosaurs—known to have been terrestrial types."[52] Bird saved the season for Brown by making this tremendous discovery, one that would indelibly etch his name in the honor rolls of paleontology.

Ironically, the Paluxy and other trackways played a role as late as 1980 in twentieth-century debates between evolutionists and creationists.[53] Some poorly preserved dinosaur tracks, carved or etched into limestone, were claimed by creationists to have been made by gigantic humans, who in their view must therefore have lived alongside the dinosaurs—"proving" that extinct dinosaurs are not as old as paleontologists claim. Although few modern creationists cling to this thoroughly debunked argument, a few still do, as a Google search on "Paluxy, human, dinosaur track" will attest.

As for Brown's activities during 1938, the annual report simply states:

> Late in the fall an opportunity was presented to make an aerial geological survey in Montana and Alberta....
> On this survey, covering all of Southern Alberta, two new localities for Cretaceous dinosaurs were examined, with good prospects on the headwaters of the Red Deer and Battle River. Two hundred aerial photographs were taken and valuable geological data secured.[54]

With field funds from AMNH again scarce in 1939, Brown signed on for more petroleum prospecting, this time with the North Continental Oil and Gas Corporation of Canada. The company picked up the tab for work conducted by Brown, Bird, and collector Harold Vokes in Alberta and Montana. The trip ran from June into October, with the first two months dominated by the collection of rock samples and the measuring of stratigraphic sections in southern Alberta. The first part of August found Bird in western

Alberta along the Upper Elbow River, primarily in Paleozoic strata, where he amassed a large collection of invertebrates and "a large fish skull from the Fernie Formation (Jurassic)."[55] The trio spent the latter part of August in the Middle Cambrian exposures of the Burgess Shale, discovered by Charles D. Walcott of the Smithsonian Institution in 1909. Brown reported that they had collected about 25,000 fossils of the remarkable soft-bodied invertebrates that exemplify the site. Vertebrates finally became the priority in September and October, near the Sweet Grass Hills in northern Montana and the Hell Creek region along the Big Dry, where Bird collected a *Triceratops* skull near Brown's old hunting grounds around the Twitchell ranch.

Fortunes turned flush again in 1940, when Sinclair offered up a $2,000 grant for the summer field season. This time the target was Texas and the approach was two-pronged, with Bird focusing on the sauropod tracks at Glen Rose and Bandera County west of San Antonio, while Brown, along with Schlaikjer, would bear down along the border of the Big Bend region.[56]

Bird arrived in Texas in early February. There, he was joined by a crew of twelve local men, placed at his disposal by the Works Progress Administration. After evaluating several potential outcrops of the Glen Rose Formation, Bird settled first on investigating sauropod tracks exposed in the southwest portion of Bandera County, at a locale known as the Davenport Ranch. Bird described the trackway thus:

> The animals had been wading in comparatively shallow water, and as the preservation of the tracks was excellent, it was hoped that one trail might supply us with a double stride favorable for a base mount, to be displayed in conjunction with the skeleton of *Brontosaurus* [*Apatosaurus*].
> But here we were partly disappointed ... by a superabundance of them. No less than seventeen different individuals had crossed an area 120 feet × 52 feet.... Many of these resulting trails occasionally traversed each other, and ... seven additional carnivores had also chosen to wander randomly over this common mud flat previous to the arrival of the sauropods....
> ... Charts and photographs were ... made, and a 3500 pound slab ... was taken for the University of Texas. An interesting number of small tracks made by young individuals hardly the size of modern elephants were noted. In addition, for the first time, the heavy drag impression of a sauropod's tail was uncovered for some sixteen feet.[57]

Today, this remarkable trackway is interpreted to represent a herd of twenty-three sauropods, all moving within a narrow corridor about fifty feet (15 m) wide. The overlapping of the tracks suggests that the larger sauropods were leading the way, with younger, smaller individuals following in line. The herd was moving at a modest walking pace, veering from right to left.[58] Unfortunately, most of the trackway, though carefully documented, was never collected, and much of it has suffered degradation from the elements.

Still seeking to place a trackway on exhibition, Bird returned to Glen Rose and assembled a new WPA crew of ten men to excavate the tracks he had discovered in 1938. Once a dike was constructed to divert the river, excavations along the banks revealed that "three other sauropod and several carnivore tracks were disclosed along a 300 foot exposure."[59] Although the "tempermental" river hampered the quarrying operations until late June, ultimately eight monumental track slabs, weighing more than 80,000 pounds and containing forty-nine footprints, were excavated from the riverbed. One slab measuring twenty-nine feet by eight feet and containing six fore- and six hind-foot impressions of a brachiosaurid sauropod, along with several theropod prints, was collected for the AMNH *Apatosaurus* exhibit. Although some have suggested that the theropods were stalking the sauropod, it's impossible to say when the theropods came along after the sauropod had passed by. The slab still resides in the renovated Hall of Saurischian Dinosaurs, having been reconstructed by Bird himself when he was called out of retirement by Brown's successor, Edwin H. Colbert, to help assemble a new version of the dinosaur halls in the mid-1950s. Other slabs went to several other institutions that were willing to pay the costs of shipping, including the University of Texas, the Smithsonian, Brooklyn College, Baylor University, and Southern Methodist University. In early September, Bird closed the quarry and joined Brown in his venture at Big Bend.

Two factors triggered Brown's exploration at Big Bend. First was another grant of unspecified size from Sinclair. Second was reconnaissance conducted the preceding year by Schlaikjer, a professor at Brooklyn College, supported by a grant from William O. Sweet.[60] Schlaikjer left New York at the end of July 1940 for Wyoming to pick up Brown's field vehicle and drive to Glen Rose, where Brown, after checking on Bird's progress, met him in early August. The two then drove west to Marathon, which served as their base of operations. From there they fanned out to the south toward the exposures of the Late Cretaceous Aguja Formation around the Chisos Mountains and along the Rio Grande, where the year before Schlaikjer had marked some specimens. Although the 200-foot-thick stack of fluvial sediments that composed the Aguja were rich in isolated dinosaur fossils, associated specimens were difficult to come by. The crew therefore extended their search into the

FIGURE 39. R. T. Bird and his WPA crew excavating the sauropod trackways from the bed of the Paluxy River near Glen Rose, Texas, during the AMNH expedition of 1940 (128434, American Museum of Natural History Library)

overlying, 1,600-foot-thick sequence of Late Cretaceous clays that form the Tornillo Formation (now called the Javelina Formation).

In his year-end report, Brown provided a summary of the discoveries: "The Big Bend area," he wrote, "adds a new chapter in our dinosaur work as it is the most southern point in the United States that dinosaurs have been found and the fauna is new to science. In all, eleven important specimens were excavated in this region." The specimens from the Aguja included a partial skull and associated limb bones of a ceratopsian related to *Brachyceratops;* a pelvis, sacrum, and associated dorsal vertebrae resembling *Pentaceratops;* and a complete skull of an ankylosaur similar to *Palaeoscincus*. Brown concluded that these forms represented "a faunal facies comparable in age to that of the Judith River and the Mesaverde Cretaceous of the northern states." A sectioned skull of *Edmontonia* collected during Brown's stay in the Big Bend region is still on display in the Hall of Ornithischian Dinosaurs. In addition, Brown noted the collection of

> two very large sauropod bones—a humerus and a cervical vertebra—Alamosaurus?—... from the Tornillo beds, the vertebra being the largest cervical that has been recorded.

FIGURE 40. Brown, Schlaikjer, and Bird (from right) with the largely reconstructed skull of *Deinosuchus* (left) collected during the AMNH expedition to the Big Bend region of Texas in 1940 contrasted with the skull of a modern crocodile (318634, American Museum of Natural History Library)

> The sauropod dinosaurs ... show that climatic conditions favorable for the life of sauropods persisted in this southern region millions of years after that group of dinosaurs had become extinct in the northern states.

The most intriguing fossil collected, however, was not a dinosaur at all, but rather "an incomplete [as in very] but enormous crocodile skull having teeth three inches long and an inch in diameter; probably D[e]inosuchus, which was as large as a medium sized dinosaur."[61]

Beyond these fossil reptiles, Brown and Bird also collected a suite of invertebrate specimens from Paleozoic and Mesozoic formations in the area, as well as several mammalian fossils from Tertiary (Tornillo) strata, thought to be Oligocene, about sixteen miles southwest of Porvenir. Titanotheres, *Dinictis, Mesohippus,* and oreodonts had previously been collected in this area by the University of Oklahoma, but Brown does not list the taxa he and Bird found.

In early November, the pair returned to Glen Rose to finish packing the trackway slabs, which were then hauled to the railhead at Walnut Springs. The crates from Brown and Bird's operations weighed in at 44,000 pounds, reportedly filling a full boxcar. Quite satisfied, Brown and his sidekick triumphantly set off for New York in late November.

Brown's sense of accomplishment during the year was surely heightened by a telegram he received from a characteristically jocular Lilian. In what can only be described as a rather ironic twist of fate, the telegram read:

CONGRATULATIONS to the BEST of the Five Model Husbands
from
The Prize-Winning Wife[62]

Barnum shared this Husband of the Year award, conferred by the Divorce Reform League, with such other somewhat questionable recipients as Franklin Delano Roosevelt and Lou Gehrig.[63] No documents have been found detailing Barnum's reaction or stipulating how the candidates were chosen. However, the Associated Press article announcing the awards quoted the director of the league as saying, "Too many husbands... want to be the center of gravity in married life with their wives revolving around them as so many unimportant and dependent satellites. This causes inferiority complexes in some wives which causes serious resentment and ends in rifts. In other wives, it causes disrespect for the husband's intelligence and leads to intellectual incompatibility."[64] Given Barnum's earlier concerns about losing Lilian to her writing career, it is at least somewhat ironic that winning the award more accurately reflected his overcoming of personal insecurities regarding their marriage, rather than the other way around.

The following year again brought great sorrow to Brown and his AMNH colleagues, in the form of "the untimely death of Walter Granger" at the age of sixty-eight. A charter member and eventual co-head (with Brown) of the Vertebrate Paleontology Department, Granger had served the museum for fifty years, including helping to spearhead the most famous expedition in AMNH history, the Central Asiatic Expeditions. His loss was deeply mourned and described as "the most important event" of 1941 for Brown and his department.[65]

Brown, meanwhile, rambled onward. In mid-November 1941, supported by a $600 allocation from the Frick Field Fund, he flew to Calgary to examine the type Kootenai Formation before heading south to Great Falls, Montana, to investigate a possible new dinosaur find nine miles east of the city in Cretaceous beds of the Kootenai. No dinosaur specimens had previously been reported from these alternating clays and sandstones. Brown confirmed that "the variegated phase in this Kootenai series . . . does appear to be similar to the Cloverly and the identification of this dinosaur will enable us to say definitely whether or not the beds are of the same age."[66] Excavation led Brown to suspect that it was a large dinosaur, about the size of *Ankylosaurus*, and possibly a new taxon.

From Great Falls, Brown headed south to Billings, then east to Ekalaka, where locals collecting in the Hell Creek exposures were giving AMNH a "new *Troodon* skull."[67] The skull turned out to be a *Pachycephalosaurus*, at the time the finest specimen known, which is still on display in the museum's dinosaur halls. Brown continued on to the ranch of William Winkley nine miles east of Powderville, where the specimen had been discovered the previous July, in order to collect additional elements that remained in the ground. In reopening the quarry and combing the scree slope below, more fragments of the skull and lower jaw were retrieved.

Brown was impressed by the locals' large collection of paleontological and historical objects, which had been assembled by the Carter County Geological Society under the leadership of W. H. Peck (one of the discoverers of the skull, along with T. G. Nielsen). Noting that the population of Ekalaka totaled less than 700, Brown proclaimed the collection to represent "one of the finest museum displays of its kind to be seen in the United States."[68] The collection still exists today and is being expanded thanks to the stewardship of the Carter County Museum in Ekalaka.

After investigating some unusual invertebrate tracks in the form of "eight-pointed stars" found in marine strata underlying the Eagle Sandstone south of Billings, Brown hopped a plane to continue his aerial surveys, this time flying over the Little Belt, Big Belt, and Snowy Mountains. In addition to locating promising outcrops west of Buffalo and the Judith River gap, he discovered "a meteor-like crater three miles south of Armington . . . 200 feet in diameter, 75 feet deep and filled with buffalo bones in the bottom."[69] The depression may have been the site of a bison jump, used by Native Americans to trap their prey. Brown returned to New York by plane on Christmas Eve.

While on this expedition, the nation had been rocked by the surprise attack on Pearl Harbor. As we have seen, one result of the looming war was that Brown arranged the sale of the type *Tyrannosaurus* specimen, collected in 1902, to the Carnegie Museum, in part to make sure that a record of the animal would survive if the AMNH was bombed. Perhaps a more immediate reason for the sale is revealed in the annual report: "The proceeds of this sale [$7,000] are to be set aside as an endowment for the Department of Paleontology, the interest to be used for expeditions and the purchase of specimens."[70]

Before he had left for Montana, Brown, now sole head of the department after Granger's death, sent Wayne Faunce, the acting director of the museum, his budget recommendations for 1942.[71] His salary for 1941 was listed at $5,625 (about $77,000 in today's currency), with a potential increase to $6,500. The staffing list included six preparators, among them Brown's field partners Albert Thomson, Carl Sorensen, and Roland Bird. But now, with war on the horizon, Brown's plans would have to change.

Upon his return to New York, Brown got word from the museum's administration that cuts were in the offing. He therefore helped rally a defense of all the scientific departments, especially his own. Quoting the previous director, Trubee Davidson, he wrote Faunce that the work of the scientific departments "'... is the keystone of the Museum's usefulness.' Yet in spite of this, during the past few years, there has been a trend to contract scientific departments to even a greater relative degree than the various non-scientific divisions of the Museum." Brown went on to argue that his department had reached peak employment in 1916, with subsequent deaths, retirements, and cuts reducing the size of the lab force to six. He further lamented the museum's inability to fund fieldwork at adequate levels, which forced curators to spend considerable time and energy seeking outside support. Most annoying was the museum's intention to cut one preparator and one artist—especially since the preparator on the low end of the seniority scale was his own protégé, Bird. "In spite of such calamitous pictures," he further noted, "it is proposed to raise the salaries of most of the survivors. Naturally, I would welcome additional salary allowance, having always been the lowest paid departmental head in the Museum. But I would not wish to have this at the price of weakening the department as a whole."[72]

Although Brown may well have been the lowest-paid departmental head, his salary was equal to that of Granger: departmental records show that from 1926 through 1928 both men received a salary of $5,000 per year, and in 1929 each was awarded a raise to $5,625 per year, which they received until 1941, the year Granger passed away. By modern standards, twelve years is an exceptionally long period to go without a raise, which may explain Brown's financial frustration, though he also enjoyed unusually free rein to augment his museum salary through consulting jobs for oil and mining companies. Adding to Brown's frustration was a coming change in museum policy:

> Regarding [mandatory] retirement at the age of sixty-eight to make way for younger career men—I would say that I, too, was a young career man, when I entered the service here. During the years... I have been offered opportunities for advancement outside, at a far greater salary. I rejected these offers upon the advice of former President Henry Fairfield Osborn, with the clear understanding that my position in the Museum was secure so long as I possessed the vigor and capacity to fill it. My record speaks for itself. My ability to accomplish results is undiminished.[73]

The reason for Brown's invocation of Henry Fairfield Osborn became clear three days later, when Faunce gave Brown permission to make his case directly to the first vice-president of the museum's trustees, A. Perry Osborn—HFO's son. Perry Osborn respectfully responded to his father's former protégé, extolling the value of Brown's department and its collections as "the glory of the Museum." Nonetheless, he felt obliged to counter Brown's "earnest plea":

> Making a budget for the Museum is a difficult problem, as I know, since for the past five years I have been the spearhead in raising money for the Museum.... During the war we must recognize that only essential activity can be maintained, because inevitably our contributions will be diverted to war efforts. What are we going to do? Many departments are on a far more skeletonized basis than V.P. Furthermore, V.P. Collections do not deteriorate.... It becomes a question of emphasis... and I think the present budget will reflect the considered policies of the Trustees. If you study it you will see there is a swing from Administration to Science of over $40,000 [almost $500,000 in today's currency].
>
> My understanding is that compulsory retirement, if adopted, does not in any way prevent an eminent scientist like yourself from continuing his work at the Museum. To sever connections with men like yourself... would be a calamity, but our Survey Committee and everyone who appeared before it, with the exception of your good self, were unanimously in favor of compulsory retirement age. Many urged 65 and some 68, but the Trustees will probably settle on 70.[74]

In the end, the annual report, under the section entitled "Staff Changes," contains the announcement that "Dr. Barnum Brown retired from active duty as of July 1, 1942, and was given the title of Curator Emeritus." He was sixty-nine years old. A two-page tribute replete with superlatives was included in his honor, listing his titles and summarizing his fieldwork, research, and contributions to the museum's exhibition programs: "In the history of the science of Vertebrate Paleontology there has never been a more energetic or more successful collector than Dr. Brown.... In fact, the present position that this Museum holds as a repository for dinosaurian fossils, as well as the absolutely unequalled display of these animals in our two Dinosaur Halls is owing primarily to the efforts of Dr. Brown."[75]

Brown was apparently mollified enough that he would maintain a relationship with AMNH for the rest of his life. In the months immediately following his retirement, however, he brushed off any slight he may have felt and redirected his considerable energies at a true and larger adversary, for as the annual report also states, "Late in the year, Dr. Brown went to Washington to serve the War Department in an advisory capacity. His appointment was for an indefinite period of time."[76] Not surprisingly, Brown still held several tricks up his sleeve—as well as under other parts of his wardrobe—as some of Hitler's minions would soon learn.

SIXTEEN

Brown as a Spy, Movie Consultant, and Showman at the World's Fair

(1942–1963)

WITH WORLD WAR II RAGING on through the spring of 1941 in Europe and Asia, the legendary explorer and then president of the museum, Roy Chapman Andrews, relayed an urgent request from the U.S. government to several paleontologists on the museum staff, including George Gaylord Simpson, Bobb Schaeffer, and Barnum Brown. The government was vitally interested in learning where these curators had done fieldwork, so as to reap the benefits of their knowledge about remote yet strategically critical areas. Brown was only too happy to oblige, citing his travels, with dates, through the United States, Canada, Cuba, Mexico, Patagonia, France, England, Turkey, Greece, the Aegean Islands, Ethiopia, Egypt, Somaliland, Arabia, India, and Burma:

> Many photographs were taken in all of these countries. My travels took me over the untravelled as well as the travelled sections of these countries.
> In Alberta, Canada, and the United States of Montana, Wyoming, South Dakota, Colorado, Utah, Arizona and New Mexico, I took and could supply many hundreds of aerial photographs.[1]

In late 1942, Brown's extensive knowledge of both domestic and international geology so impressed a Colonel Donovan that he was offered a job in Washington, D.C., as a member of the Office of Strategic Services (OSS), the forerunner of today's CIA.[2] Leaving Lilian in New York to "man the home front," he moved in with Frances, who was serving with the Red Cross

FIGURE 41. Brown in his now famous beaverskin coat next to Cretaceous deposits near Sweetwater, Montana, ca. 1916 (19508, American Museum of Natural History Library)

at their headquarters in the capital. According to Frances, Brown played a pivotal role in war planning at the end of 1942 and through most of 1943. His intelligence responsibilities derived from his intimate knowledge of the Aegean Sea and primarily involved planning one of the "four possible invasion routes up through the islands and on to the 'soft belly' of Europe."[3]

Frances welcomed the chance to spend time with her peripatetic father. This interlude represented, practically speaking, "the first and only time . . . Barnum and Frances had lived together, complete with a tiger cat that Barnum shortly christened 'Little Joe Big Stink' and taught to retrieve as a dog. It was a happy experience for both father and daughter."[4]

Yet it would be a mistake to infer that Barnum focused all his attention on Frances and her cat, or that his "Best Husband" award had altered his overall outlook on life. Despite the gravity of his OSS assignments, Brown still had time for other activities, as Frances recounted:

> In spite of all that gloom [surrounding the war], or perhaps because of it, there was much to do socially in the capital, a kind of frenzied round of receptions, dances, parties of all kinds, concerts, and so on. . . . Early on, [we] met a gorgeous and delightful blond beauty with a Teutonic accent who had no trouble at all captivating Barnum. Immediately, his "Brunnhilde" was all Barnum could talk about, and most of his spare time was spent

in plotting and keeping assignations. In the early meetings at dances and parties, [we] made it a threesome, but that did not last very long. Barnum was too besotted to want a chaperone, especially one that might be interfering. Not that [I] was bothered by Barnum's affairs with women, but [I] quickly learned that this particular lady was, in all probability, a very competent Nazi spy. With the secret war work that Barnum was doing, [I] was really fearful that his complete captivation by his Brunnhilde might well lead to his spilling information which would be disastrous for him and his country. No amount of reasoning seemed to penetrate, and so [I] spent a good many weeks dreading an international incident before the lady, somewhat suddenly, departed from the Washington scene and, presumably, went back to Germany not as successful as she had hoped to be.[5]

After completing his duties with the OSS, Brown remained in Washington working as a consulting geologist for the Board of Economic Warfare from 1943 to 1945.[6] This provided the opportunity to carry out yet another aerial survey, evaluating oil prospects in Alberta over an area encompassing 45,000 square miles. Perhaps during this stint, or perhaps during his earlier evaluation of potential oil fields for the Canadian North Continental Oil and Gas Company, Brown came to the attention of Britain's duke of Windsor, who asked him to investigate prospects for oil on his 4,000-acre ranch in Alberta.[7] In 1936, the duke, then King Edward VIII, had stunned the British people and parliament by abdicating the throne to marry Wallis Simpson, an American woman in the process of getting a divorce. Upon hearing of the duke's interest in an oil survey, Brown happily agreed. At one point, as Frances recounts, "Barnum and the Duke were on the floor of the latter's office studying a map of the whole terrain... [when] the duchess walked in. After a few minutes of pleasantries, she left the room; whereupon Barnum turned a wicked eye on the duke and remarked: 'Your highness, I, too, would have given up a kingdom for that lady!'"[8]

The project was at least partly successful; Brown recorded in his dossier that he had, under difficult circumstances related to garnering supplies, supervised the drilling of a test well for the duke and a New York banker, Elisha Walker. The well "yielded one barrel per hour of 56 gravity oil at a depth of approximately 4,250 feet."[9] The test well was then capped until such time as further drilling, an expensive operation at the best of times, could be carried out more economically. According to Frances, Brown's other major responsibilities for the Board of Economic Warfare involved "interpreting aerial photographs to detect camouflage in areas of Africa, India, Burma, and the Mediterranean Islands."[10]

After World War II ended, Brown spent a good part of 1946 at AMNH supervising preparation of yet-to-be-mounted specimens and finishing up a few scientific manuscripts.[11] But he was desperate to get back into the field, and an opportunity soon arose in the form of another oil consulting job, primarily under the sponsorship of Sohio Petroleum Company.[12]

From 1947 to 1952, Brown returned south of the border, this time to the jungles of Guatemala. Initially, his work was funded by a trio of American promoters who had secured "the Strapper-Osborne concession in the region around El Peten." Eventually, this concession was acquired by Sohio. Among Brown's major tasks was a 130-flight aerial survey that covered 44,000 square miles and documented the entire country of Guatemala. This overhead view was supplemented by a ground-based reconnaissance by jeep and horse to explore all the country's roads and trails, and by boat to cover all the river systems in the northern part of the country around El Peten, from mouth to source and back again. In 1951 and 1952, Brown also turned his attention to finding fossils in the region. These forays, funded in part by Sohio and in part by himself, were difficult and not especially successful. Primarily Pleistocene fossils were recovered, including a "boxcar" of fragments from mastodons and giant ground sloths and one fairly complete skeleton. One sloth pelvis exhibited three "knife cuts," which Brown suspected might have been made by humans. "If these cuts prove to have been made by man, as I and many associates contend," he observed, "it places man back thousands of years contemporaneous with the giant land sloths on the American continent."[13]

Both Frances and Lilian joined Brown on some of his collecting trips in Guatemala. At about this time, too, Lilian's writing career finally blossomed. She published three books about her travels to far-off lands with Dodd, Mead: *I Married a Dinosaur* (1950), about her and Barnum's adventures in colonial India and Burma; *Cleopatra Slept Here* (1951), about their year in Samos and Greece; and *Bring 'Em Back Petrified* (1956), about a six-month sojourn in the jungles near El Peten. Older and more knowledgeable about what fieldwork in foreign countries involved, Lilian charmed the locals in and around Santa Amelia on the banks of the Río Pasión. But one person proved to be a thorn in their sides, a female shaman of sorts who wielded great influence over the native population. Fearful that Barnum's collecting of fossil bones would disturb the spirits on whom she relied, she incessantly lobbied the locals not to help him; at one point, she even convinced a few of her disciples to destroy some of Barnum's fossils when he and Lilian were away from camp.[14]

Holiday visits by Frances were shorter but no less entertaining, as father and daughter toured Mayan ruins and plantations, largely by air. Frances remarked, "Barnum was delighted to be able to show his daughter all of Guatemala that could be covered in three-week periods." On one excursion, they got more excitement than they had bargained for, when they

> joined a sightseeing planeload headed for Copan in British Honduras whose Mayan ruins were spectacular. Unfortunately, a lively revolution was going on in that country . . . so the landing plane was met by armed guards who said that all the passengers would be jailed and the plane impounded. However, since the jail arrangements had to be made, the passengers would be allowed to disembark and go enjoy the Mayan ruins. [Afterward] the sightseers wandered with some trepidation back toward the plane—and a happy surprise. The guards informed them that they would be allowed to leave after all. It seemed that there were insufficient jail facilities for such a large group![15]

Similar political unrest soon overwhelmed the oil companies' plans for drilling in Guatemala, and Brown's work in the region came to an end in 1952.[16]

The next year, back in the United States, Brown got word that trouble was brewing at Dinosaur National Monument. Throughout the early history of the park, tensions had simmered between those who wanted to keep the monument undeveloped, as a natural preserve for both fossils and the modern biota along the Green and Yampa Rivers, and those who sought to build dams to facilitate more commercial activities. Alarmed, Brown set out from New York to survey the region by car and speak with the players himself. He spent several months in the region, and Lilian joined him for some of that time. In the end, Congress, in part because of Brown's arguments, vetoed "the dam proposals of the reclamation engineers . . . upholding the principle of park values. This congressional action saved the ancient Indian petroglyphs on some of the canyon walls and the magnificent canyons themselves."[17]

In 1955 Brown, now eighty-three years old, returned to his stomping grounds in Montana for his last formal field season, this time focusing his attention on the Cretaceous Claggett Shale exposed on the Van Haur farm north of Lewiston.[18] He succeeded in collecting an impressive plesiosaur skeleton from the marine unit (AMNH 2802 or 2502), but this exercise almost cost him his life. In late October, about a month after Brown returned to New York, Lilian wrote to George Shea, Brown's old friend in Billings with whom he had stayed while in Montana, to see if Shea could provide some insight regarding an ailment Brown was suffering. He was, she explained, "very lame in one leg... had severe aches and pains in his shoulders, back and throughout his body."[19] Although doctors had tried several drugs, including cortisone, streptomycin, and sulphur, the only one that seemed to have any beneficial effect was quinine, making them suspect a relapse of malaria. Yet blood tests came out negative for malaria. Horrified to hear of Barnum's illness, Shea telegraphed on October 29, alerting Lilian that "Barnum had some tick bites July 24. Other trouble came after September first."[20] Shea's revelation led to the proper diagnosis, as Frances later related: "Barnum fell victim to a tick producing Rocky Mountain spotted fever. Somehow he managed to get himself back to New York where once again Lilian nursed him back from death's door. That he slowly, fully recovered is a tribute to her skillful loving care and also to the tremendous stamina and will power of the octogenarian."[21]

Two years later, Brown began to plan his next audacious adventure. Although he had now more than satisfied his penchant for aerial surveys, he still had not fulfilled his dream of using a helicopter to collect, and that was next on his "bucket list"—in a rather literal way. "What he had in mind," Frances explains, "was to use a helicopter to search out and collect dinosaurs from the sheer sea cliffs of the Isle of Wight.... When he had prospected from the helicopter and determined the most promising site, he would land on top of the sea cliff. There he would strap himself, complete with tools, into a bosun chair and be lowered over the cliff to the spot where he wished to dig. To Barnum this plan seemed a perfectly reasonable one for a seasoned paleontologist who was approaching his ninetieth year."[22] To our knowledge, however, this grand scheme was never realized.

Much of the last twenty-five years of Brown's life was devoted to projects that would bring the many dinosaurs he had discovered to life for the public who dubbed him "Mr. Bones." One of those projects involved a movie that is deeply ingrained in the memories of baby-boomers. In December 1940, Brown gave a lecture about the project to the New York Academy of Sciences, entitled "The Methods of Walt Disney Productions." The published notes from the lecture reveal his role in Disney's animated classic, *Fantasia*:

> About three years ago, I had considerable correspondence with the Walt Disney Studios about dinosaurs, as to typical forms in the different geologi-

cal periods, their history and relationship[s], the various kinds of life that could be associated with dinosaurs... and the more primitive species that preceded the era of dinosaurs.

On the basis of these data, as well as models, restorations, etc., furnished through my efforts, a number of artists... proceeded to make scenes representing the dinosaurs and other typical associated reptiles. The plan was to run a continuous story from the beginning of life... up to the close of the dinosaur era....

When a great many drawings had been made, I visited the Disney Studios and found their walls covered with very credible and accurate types of prehistoric life including the associated floras....

In the meantime, while different sequences were being created... Walt Disney, Deems Taylor, and Leopold Stokowski selected eight outstanding musical compositions that, in their opinion, could be coordinated with the corresponding sections of the spectacle... after which the Philadelphia Symphony played the entire program through, making a recording of 450,000 feet of film. This was cut to the 18,000 used during the time that the complete spectacle of *Fantasia* was shown....

Thus, Walt Disney has taken the findings of the scientist to show the culmination of dinosaur life through change of environment and final extinction.[23]

The dinosaurs are contained within a twenty-minute-long segment of the film set to the music of Igor Stravinsky's ballet *The Rite of Spring*. The piece encompasses the birth of the cosmos, the origin of our galaxy, and the formation of the Earth, along with the origin and evolution of life. Within the evolutionary sequence featuring the Mesozoic Era, pterosaurs, plesiosaurs, mosasaurs, and dinosaurs—a mix of animals from various geologic periods—cavort across the seascapes and landscapes until a tyrannosaur (with the three-fingered hands of an allosaur) appears on scene to attack a defiant, but eventually conquered, stegosaur. The movie won an Honorary Academy Award in 1942 and today is fondly acclaimed as a classic.[24]

In the last year of his life, Brown renewed his productive relationship with Sinclair Oil, working as the primary consultant for the company's *Dinoland* pavilion at the 1964 New York World's Fair. It was an ambitious plan that involved the sculpting of ten life-size models ranging from the seventy-foot-long "*Brontosaurus*" and twenty-foot-tall *T. rex* to the diminutive, four-foot-long *Ornitholestes*. The animals were constructed in the Hudson, New York, studios of Louis Paul Jonas, beginning with

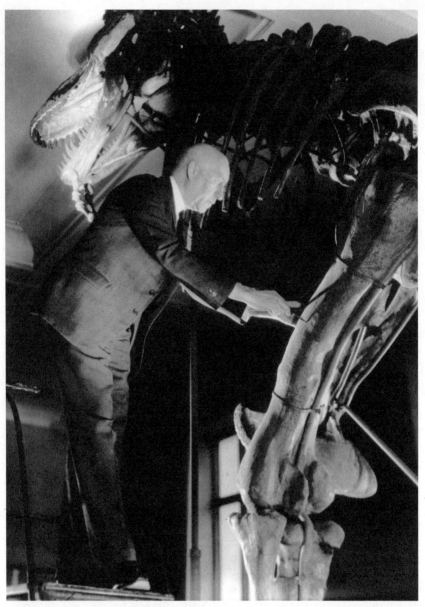

FIGURE 42. Brown measuring the femur of *Tyrannosaurus rex* in the Cretaceous Hall of the American Museum of Natural History, ca. 1938 (AMNH Vertebrate Paleontology Archive, 5:6 Portrait box)

two one-tenth-scale models, one of which was sliced into cross-sections to provide precise measurements for the final, full-scale replica, which was cast in fiberglass. Although the initial cost estimate of $250,000 eventually soared to $400,000 (or about $2.6 million in today's currency), the results were spectacular. When the models were eventually finished, they were loaded on a barge and floated down the Hudson River to New York, whose citizens commemorated the event by declaring the day "Dinosaur Day." *Dinoland* proved to be the most popular exhibit at the fair, drawing over ten million visitors, and Brown's work on the pavilion capped a steadfast and lifelong commitment to bring the wonders of dinosaurs directly to a fascinated public.

While working on the exhibition, Brown commuted between New York City and Hudson, and occasionally he would take the opportunity to continue on to Cambridge, Massachusetts, to visit Frances, then a dean at Radcliffe College. Brown had yet to lose his keen eye for the opposite sex, she revealed, remarking that "he cut his usual wide swath with the Radcliffe girls."[25]

In the end, however, his consulting on such projects as *Fantasia* and *Dinoland* pale in comparison to his dinosaur halls at AMNH. As Brown neared his ninetieth birthday, his successor at the museum, Edwin H. Colbert, wrote: "There are, in our Tyrannosaur Hall, thirty-six North American dinosaurs on display.... You collected twenty-seven, an unsurpassed achievement."[26] In all, Brown collected more than 1,200 crates of fossils for the museum between 1896 and 1942 on dozens of expeditions that cost more than $1.25 million in today's currency (see Appendix 3), and through the years more than 100 million people have passed through the AMNH fossil halls and stared in rapt disbelief at the spectacular evolutionary relics that he resurrected.

For months throughout late 1962 and early 1963, Lilian, along with Brown's colleagues at the museum, labored feverishly to plan an immense party. On February 12, for the ninetieth time, Brown would celebrate his birth in the forested hills and grassy valleys of eastern Kansas. But the upcoming anniversary was to be no normal fête. More than 150 members of the nation's scientific elite, including Ernst Mayr and George Gaylord Simpson, the most prominent evolutionary biologists of their time, planned to join him at a gala befitting a legend. Although Brown was not supposed to know in advance, he had picked up on the carelessly dropped clues and was eagerly awaiting the event.

FIGURE 43. Headstones of Marion Brown, Barnum Brown, and Lilian Brown in Oxford, New York, 2008 (Courtesy of Carl Mehling)

On the first of February, however, inevitability caught up with him as he sat down to dinner. Frances describes the event:

> Barnum suddenly laid down his fork at the supper table, leaned his head back, and told Lilian he was very tired. She managed to get him to bed, but during the night he slipped into a coma from which he never roused. He died in St. Luke's Hospital on February 5, one week short of his ninetieth birthday. The funeral service was held in All Angels Episcopal Church.... His body lay in state at the church before the service and afterward was sent to the Oxford, New York, cemetery to be laid to rest beside his first wife, Marion. Eight years later Lilian died, and her body was buried on the other side of Barnum, who undoubtedly would have had a good chuckle over being sandwiched between his two wives.[27]

The graves of Barnum, Marion, Lilian, and now Frances as well all lie in River View Cemetery near the edge of a small bluff overlooking the Chenango River. The arrangement of the graves is telling in two respects. Although Barnum lies between his two wives, the spacing is not regular, for Lilian, it seems, had her headstone placed several inches closer to Barnum's than Marion's is. Equally telling of the familial relationships is the fact that Frances chose not to be interred in the same row as her parents and stepmother. Instead, her headstone lies down the slope between Marion's parents, the grandparents who raised her. The graves speak volumes about how Lilian and Frances saw their lives in relation to Barnum.

During his last years between 1958 and 1962, Brown seems to have spent much of his time struggling to compose an autobiography.[28] Although he never completed the work, he drafted many sections, focusing especially on his years as a child and student in Kansas. After his death, Lilian, Barnum's former field assistant G. Edward Lewis, and a prospective editor at Dodd, Mead labored for several years, trying to finish the book. Yet their efforts ended in vain. Throughout this period, Lilian's feelings about the project waxed and waned. In February 1967, for example, Lilian wrote to Lewis with great excitement: "***GOOD NEWS: Have just 'phoned Dodd, Mead and talked with Editor, Allen Klot.... CHEERS. I told him I was working with you, and he was very pleased, saying to send them one or two chapters to see how things are going.... Also he liked [the idea of the book being] written by 'Barnum and Lilian Brown.'... Allen was very enthusiastic about the whole affair."[29]

As the months and years rolled on, however, Lilian felt weighted down by the responsibility, even with the help of Lewis, to whom she wrote: "Personally, I feel that I have done my share of glamorising the name of 'The great God Brown' in the three happy books I have written about him ... so WHY another????? fraught with the tragedy of his sudden death???"[30] Although Lilian survived Barnum by eight years, few details of her life after Barnum's death are recorded in the museum archives. But somewhere along the line, the idea of ghostwriting her husband's story was buried along with him.

Epilogue

WHAT ARE WE TO MAKE of Barnum Brown? First and foremost, he remains the greatest dinosaur collector the world has ever known. Through Brown's efforts, dinosaurs gained a strong foothold in the psyche of both the scientific community and the general public.

In writing this biography, a century after Brown's most famous discovery of *Tyrannosaurus* and almost a half century after his death, we still profoundly feel his presence. Specimens that Brown collected populate the cabinets and study tables of our offices. Books from his library, inscribed with his characteristic bold signature and passed down by our predecessors, now populate our bookshelves. Fifty-seven specimens that he personally collected form the foundation of the exhibition halls that we helped renovate in the 1990s (see Appendix 1). Many of the sites that Brown excavated and visited are common haunts for our own fieldwork and travels. Brown, a man known for discovery, science, and intrigue, still looms large.

Yet despite his daunting influence, drawing a comprehensive understanding of Brown's life is like tracking a ghost. Although barely visible, Brown's name can still be faintly read on the glass window of his old office door on the fifth floor of our museum. Beyond his field correspondence and technical writings, which allow us to chronicle his paleontological career, he left behind only a few misty glimpses and tantalizing hints of amorous adventures and clandestine governmental reconnaissance. Like one surviving in wild and remote regions, Brown cautiously covered his tracks.

Although his autobiographical notes document that he greatly admired his parents, Brown writes little about the tenor of their personal relationship. Was his father guilty of incest? Brown volunteers nothing, and we are left only court records. He also remains stoic about how this event affected him and his family. In relating the adventures associated with the ensuing odyssey to Yellowstone, he provides no clues. Was Brown's father fleeing the scene of the crime? Quite probably he was, given the circumstantial evidence, but we have no definitive statement of this. Had Brown's enraged mother temporarily exiled his father from their farm in Carbondale while the legal ramifications and public revulsion subsided? Again, that seems quite possible, but there is no documentation to confirm it. Although we have no record of Barnum's emotions regarding this scandal, it's hard to imagine that it had no impact on his later personal life.

We do know that as Barnum's career began to blossom under the guidance of Williston and Osborn, his personal life progressed in fits and starts, punctuated by a devastating tragedy. Perhaps the greatest misimpression lingering today is that Brown represented a kind of unsophisticated, social simpleton, focused solely on how many women he could seduce. To be sure, he loved women, and although we have yet to find evidence of any illegitimate offspring, as members of the paleontological community often casually claim, there is ample evidence to suggest that he played the field, both after the death of his first wife, Marion, and during his second marriage with Lilian.

Although information is sparse about Marion, Brown's correspondence as well as their daughter's later testament imply that Marion was, in many ways, the love of Brown's life. Described as a beautiful and vivacious blond, her journal documenting their field season in Montana radiates her love of nature and her keen powers of biological observation. She seemed up to any physical challenge, and Barnum loved having her along in the field, to the point of going toe to toe with the museum's administration to ensure her participation. Her unexpected and untimely death clearly devastated Barnum and created a chasm between him and Frances that took most of his life to bridge. He healed his heart by avidly pursuing his fieldwork, and despite his absence from her life, Frances eventually came to proclaim pride in her father and his accomplishments.

Even when viewed through the limited lens of archival documents, which form the core of this account, Brown clearly lived large. His adventurous exploits and unconventional relationships must have glowed incandescently against the more conservative social backdrop of his day.

Later in life, even his taste in women—personified by the partying, outgoing, and flirtatious Lilian—was extreme. Lilian's correspondence reveals that, at least when he was single, Brown fostered several simultaneous relationships with the feminine objects of his affection, and he was certainly not above stretching the facts to string along his quarry.

In the wake of the suit filed against Brown, apparently by a spurned lover, he, like his father, fled the scene in hopes that the dust would eventually settle. Lilian's determination to chase Barnum across the globe finally vanquished his fear of intimate vulnerability, triggered by the death of Marion. Tempestuous and confident in her own abilities, Lilian was also up to almost any challenge and certainly seemed capable of giving as good as she got. She persevered through Barnum's legal entanglements and apparent promiscuity during their courtship. Throughout their sojourn in Samos, even though she may well have had a fling of her own, she tried to reassure Barnum that he wouldn't lose her, as he desperately feared, if she spent time away from him while visiting her relatives in social sin-bins such as Hollywood. Eventually, they developed a rapport that even allowed them to joke about their unorthodox relationship with members of the field crew.

The passage in Frances's book about Brown's "Brunnhilde" seems to represent hard evidence, at least in that one instance, that Barnum did engage in extramarital relationships while married to Lilian. Frances writes forthrightly in characterizing her father: "A compliment from [Brown's mother] which he always treasured was her oft-repeated statement that Barnum was the best 'girl' she ever had. Modern-day readers of this memoir had better not put a 1980s interpretation on this remark of Clara's. If they did it would be about as far from the truth as anything could be. Barnum, in his prime, was not even monogamous!"[1] Yet even when Barnum did stray, the strong bond between Barnum and Lilian was not severed.

However, establishing the extent of Brown's alleged infidelity still requires the unearthing of further hard evidence . . . ironically, a task for which Brown was supremely proficient. At least figuratively, it may well be that Brown camouflaged and buried some skeletons as adroitly as he excavated others. Even if we never resolve the extent of Brown's infidelity, it's much easier to discern how Brown's childhood laid the foundation for his professional career. As revealed in his notes and correspondence, his superior abilities as an explorer and fossil collector developed under the guiding support of his parents on their homestead in eastern Kansas. From his father, he acquired a lifelong sense of responsibility, a determined work ethic, and streetwise business acumen. As Barnum recalled at the end of his life, his father "had a deep, abiding love for horses and other livestock, for the soil, and for his country. . . . Father's pioneering was purposeful: he was hard-working, with a good head for business; he sought and found promising opportunities worthy of the heavy investment of thought, time, and labor that he poured into them.... [In all] Father was a remarkably good business man and manager. He knew how to turn adversity into advantage."[2]

Indeed, it was his father who took Barnum on his first expedition during their marathon trek to Montana in search of a possible new domicile, and both of his parents supported their son's interest in collecting invertebrate fossils in the coal-laced strata of their Carbondale farm. His mother instilled in him her deep love for the flora and fauna that abounded in the local woods and fields, even as she nurtured in him acute abilities for observing natural phenomena. Seminal in Brown's success were the skills he learned from his parents in managing crews of workers on the farm, whether organizing the tasks to be accomplished or assuring that the workers were well fed and compassionately cared for. Finally, despite the resulting hardships in cost and lost labor, Brown's parents sacrificed the skills that Brown had learned on the farm to a greater good by sending him to school and college so that he could pursue his long-held dream of becoming a geologist and paleontologist. These childhood experiences clearly allowed Brown to excel as a fossil collector.

Osborn hired the young Barnum Brown primarily for his prowess in the field. Osborn guessed correctly, and Brown went on to become one of the most prolific and successful fossil collectors ever. Yet Osborn also lobbied hard for his young protégé to attend Columbia University and obtain a Ph.D. Although Barnum never graduated, he went on to write some of the more lucid and important papers of his generation. As a paleontological research scientist Brown is given far less credit than he deserves.

It is true that he flunked out of Columbia and never finished a formal dissertation, which, by his own admission to Osborn, were bitter pills to swallow. Toward the end of his career, both Edwin H. Colbert, who succeeded Brown at AMNH, and Wann Langston, the University of Texas paleontologist who met Brown while a student, described Brown as a loner and somewhat arrogant. It may well be that some of this "arrogance" flowed not only from his acknowledged success as a collector but also from the sense of inferiority that haunted him in the wake of his academic shortcomings.

Many modern paleontologists complain about Brown's limited folio of field notes and the paucity of publications he produced, but those criticisms are not entirely fair. His notes, to be sure, are perfunctory, usually tersely to the point, just museum business. Often, our best insights into Brown and his activities derive from the writings of colleagues and wives. It is amusing to read the scattered reports of Brown and contrast them with descriptions of the same places and events by the more effusive Marion and, especially,

Lilian. Whereas Brown would mention a few fossils collected and compile a meticulous account of expenses, Lilian would pen pages detailing sights, smells, impressions, and experiences.

Nonetheless, many details of Brown's expeditions are preserved in the field correspondence that underpins the prose of this study; although he didn't keep many meticulous daily journals, he did take time to record his actions with considerable fidelity for his supervisors. Beyond that, his bibliography contains about a hundred publications, the majority of which pertain to his scientific research, supplemented by a suite of more popular articles requested by the museum to inform the public about his exploration and adventures.[3] The scope of those publications, along with the sheer determination and field innovations that made them possible, is immense.

Brown named myriad species of fossil animals. But good paleontology isn't just about naming new kinds of extinct animals. As now, paleontological research is at its best when seminal problems are addressed from new perspectives or with novel analytical techniques, and Brown brought both an innate curiosity and a refreshing eye to many issues. His observations on the boundary between the Cretaceous and Tertiary formations of the western interior, on the camping style of Paleolithic humans in the Southwest, on construction principles of pueblo dwellers at Chaco Canyon, on microfossils too small to be seen by the naked eye, on how Pleistocene mammals accumulated in sink holes, and on how the faunas of the Siwalik Hills fit together chronologically all provided fresh scientific insights. To us, one work in particular stands out: a monograph with Erich Schlaikjer about the diminutive horned dinosaur *Protoceratops,* then thought to have laid the eggs that became the Central Asiatic Expeditions' most famous discovery.[4] Interestingly, after the initial descriptions by Granger and Gregory, Osborn charged Brown and his younger colleague with writing the definitive paper on the first and most ubiquitous dinosaur found at Mongolia's Flaming Cliffs by the Central Asiatic Expeditions—field trips in which Brown played no part. Unlike most other detailed descriptions of dinosaurs at the time, Brown's study went far beyond a bone-by-bone anatomical description. In many ways, it framed the template of how such papers are written today by paying careful attention to variations and changes in form during growth, as well as to how different anatomical characteristics may indicate gender and behavior.

Yet above all, Brown's true genius and his professional heart were focused primarily on fieldwork. Part of any scientist's job is to understand how technological advances can be applied to the work at hand, and at this Brown was particularly adept. Although he did not "invent" the technique of plastering vertebrate fossils in jackets, as he seemed to believe, he nonetheless introduced this technique to the AMNH. In addition, he pioneered the use of other technological innovations for collecting fossils that are still used today. The heavy equipment that the Union Pacific Coal Company put at his disposal in the Mesaverde project resonates with modern use of backhoes and bulldozers to facilitate the excavation and field transport of large specimens. Aerial surveys and support remain key ingredients of many fossil-collecting expeditions, either to locate new outcrops, transport crew members and equipment to inaccessible areas, or lift heavy jackets out of regions without roads. Thus, his professional accomplishments were even more sweeping than is commonly acknowledged, because for almost any obstacle in the field, Brown could devise an elegant, yet pragmatic, solution: Need to get a fossil out of the ground and safely back to New York? Treat it like a broken bone and encase it in plaster. No place to camp on the Red Deer River? Why not employ a flat boat, along with a motorboat, to propel the entire campsite directly to the fossil localities? Need to make detailed records of localities and locals, but don't like to (or don't want to) take detailed field notes? Shoot an exceptional number of photographs. No roads to adequately prospect vast expanses of potentially rich outcrops? Do an aerial survey. No labor to collect weighty dinosaur tracks? Work a deal for a WPA crew. And so it goes. But he was also capable of doing things the old-fashioned way, and from his dispatches it is clear that no one worked harder than Brown.

Beyond that, to be successful in dinosaur paleontology today, one must balance scientific work with a public persona. Done well, fossil collecting involves the discovery of both appealing exhibition specimens and specimens for research, which, if strategically described and publicized through the media, can be used to fuel public interest and funding for continued work. Following Osborn's lead, Brown was one of the first to fully exploit this strategy, initially through his newspaper dispatches and later, after Osborn's death, through his use of the nascent electronic medium of radio in broadcasting his results. Especially notable from our perspective was Brown's fascination with the modernization of dinosaur displays at the World's Fair and other venues that brought dinosaur science to the public in an easily understandable fashion.

Yet in the end, especially for paleontologists, it was Barnum Brown who opened the world's eyes to the wonders of the imperial dinosaurian carnivores that roamed the world in the Late Cretaceous. Although more recent press reports abound with discoveries of larger and more allegedly intimidating predators, all those dwell under the long shadow of Barnum Brown and *Tyrannosaurus*.

Although Brown was a loner of sorts, especially in the field, his close colleagues regarded him with utmost respect. Before he died, Brown's longtime supervisor, W. D. Matthew, had offered this characteristically sober evaluation, which was printed in the annual report of 1942: "Brown's energy, initiative, and persistence, with a discriminating judgment as to what prospects would best repay the work of collecting, expensive and time-consuming in these dinosaur skeletons, have made him the most successful in this line. In the preparation, study and exhibition of the collections of fossil reptiles he has been no less successful."[5]

An equally characteristic yet more playful testament came from his friendly competitor Roy Chapman Andrews. Written as the foreword to Lilian's first book, it provides an apt postscript for Brown's legendary career:

> Barnum has been my friend for well nigh forty years.... I have known him to disappear from the museum, just fading out like the Vanishing American, and none of the staff knew where he had gone.... But invariably his whereabouts was disclosed by a veritable avalanche of fossils descending in carload lots upon the museum.... He has discovered many of the most important and most spectacular specimens in the whole history of paleontology. When he ceases to look for bones on this earth, the celestial fossil fields may well prepare for a thorough inspection by his all-seeing eyes. He'll arrive in the Other World with a pick, shellac, and plaster, or else he won't go.[6]

Barnum Brown lived fast and worked hard. He was a wilderness gourmand, a natty dresser, a drinker, gambler, and smoker. Yet despite these wild streaks, he almost always seemed in control. Lilian's assessment that "it was a different time and Brown was a different kind of guy" seems most appropriate. He was sex, dinosaurs, and science all wrapped around an enigmatic private life, and we have just begun brushing the loose sediment off the surface to reveal what lies buried beneath.

APPENDIX ONE

List of Major Specimens Collected by Barnum Brown on Display in the AMNH Fossil Halls

(57 total)

Scientific Name	Specimen Number	Elements	Age	Formation	Collection Date	Locality
ORIENTATION CENTER (1 TOTAL)						
cf. Barosaurus	AMNH 7535	skull and neck of juvenile	Late Jurassic	Morrison Fm.	1934	Howe Quarry, WY
HALL OF VERTEBRATE ORIGINS (7 TOTAL)						
Buettneria perfecta	AMNH 6759	skull	Late Triassic	Chinle Fm.	1929	Outside of Cameron, AZ
Champsosaurus laramiensis	AMNH 982	skull	Late Cretaceous	Hell Creek Fm.	1902	near Jordan, MT
Gavialis browni	AMNH 6279	skull	Miocene	Middle Siwaliks	1922	near Nathot, India
Teleorhinus robustus	AMNH 5850	skull and mandibles	Early Cretaceous	Cloverly Fm.	1904	Beauvais Creek, MT
Protosuchus richardsoni	AMNH 3024	skeleton (cast)	Late Triassic	Moenave Fm.	1931	near Cameron, AZ

Machaeroprosopus gregorii	AMNH 3060	skull and mandibles	Late Triassic	?Chinle Fm.	1936	near Cameron, AZ
Geochelone atlas	AMNH 6332	free-standing skeleton	Miocene	Upper Siwaliks	1922	near Chandigarh, India

HALL OF SAURISCHIAN DINOSAURS (6 TOTAL)

Albertosaurus libratus	AMNH 5458	skeletal plaque mount	Late Cretaceous	Dinosaur Park Fm.	1914	Red Deer River, Canada
Tyrannosaurus rex	AMNH 5027	free-standing skeleton with skull	Late Cretaceous	Hell Creek Fm.	1908	north of Jordan, MT
Deinonychus antirrhopus	AMNH 3015	free-hanging skeleton	Early Cretaceous	Cloverly Fm.	1931	Beauvais Creek, MT
Struthiomimus altus	AMNH 5339	skeletal plaque	Late Cretaceous	Dinosaur Park Fm.	1914	Red Deer River, Canada
Struthiomimus altus	AMNH 5421	skeletal plaque	Late Cretaceous	Dinosaur Park Fm.	1913	Red Deer River, Canada
Psilopterus australis	AMNH 9157	skull and mandibles	Miocene	Santa Cruz Fm.	1899	Argentina

HALL OF ORNITHISCHIAN DINOSAURS (28 TOTAL)

Sauropelta edwardsi	AMNH 3032	caudal series	Early Cretaceous	Cloverly Fm.	1932	Beauvais Creek, MT
Sauropelta edwardsi	AMNH 3036	skeleton with scutes	Early Cretaceous	Cloverly Fm.	1932	Beauvais Creek, MT
Ankylosaurus magniventris	AMNH 5214	skull and tail club	Late Cretaceous	Horseshoe Cyn. Fm.	1910–11	Red Deer River, Alberta
Edmontonia rugosidens	AMNH 3076	sectioned skull	Late Cretaceous	Aguja Fm.	1940	Brewster Co., TX
Euoplocephalus tutus	AMNH 5404	skull	Late Cretaceous	Dinosaur Park Fm.	1913	Red Deer River, Alberta
Euoplocephalus tutus	AMNH 5337	pelvis	Late Cretaceous	Dinosaur Park Fm.	1914	Red Deer River, Alberta
Tenontosaurus tilletti	AMNH 3034	free-standing skeleton	Early Cretaceous	Cloverly Fm.	1932	Mott Creek, MT
Saurolophus osborni	AMNH 5220	skeletal plaque	Late Cretaceous	Horseshoe Cyn. Fm.	1911	Red Deer River, Alberta

(continued)

Major specimens collected by Barnum Brown on display in AMNH fossil halls *(continued)*

Scientific Name	Specimen Number	Elements	Age	Formation	Collection Date	Locality
hadrosaur	AMNH 5350	mandible	Late Cretaceous	Dinosaur Park Fm.	1914	Red Deer River, Alberta
Prosaurolophus maximus	AMNH 5386	skull	Late Cretaceous	Dinosaur Park Fm.	1915	Red Deer River, Alberta
Kritosaurus navajovius	AMNH 5799	skull	Late Cretaceous	Ojo Alamo Fm.	1904	AZ
Lambeosaurus lambei	AMNH 5353	skull	Late Cretaceous	Dinosaur Park Fm.	1914	Red Deer River, Alberta
Lambeosaurus lambei	AMNH 5373	skull	Late Cretaceous	Dinosaur Park Fm.	1915	Red Deer River, Alberta
Corythosaurus casuarius	AMNH 5348	skull crest	Late Cretaceous	Dinosaur Park Fm.	1914	Red Deer River, Alberta
Corythosaurus casuarius	AMNH 5240	skeletal plaque with skin	Late Cretaceous	Dinosaur Park Fm.	1912	Red Deer River, Alberta
Corythosaurus casuarius	AMNH 5360	skin patch	Late Cretaceous	Dinosaur Park Fm.	1914	Red Deer River, Alberta
Corythosaurus casuarius	AMNH 5338	skeletal plaque	Late Cretaceous	Dinosaur Park Fm.	1914	Red Deer River, Alberta
Anatotitan copei	AMNH 5886	free-standing skeleton	Late Cretaceous	Hell Creek Fm.	1902	Crooked Creek, MT
Corythosaurus/Lambeosaurus	AMNH 5340	juvenile skeletal plaques	Late Cretaceous	Dinosaur Park Fm.	1914	Red Deer River, Alberta
Centrosaurus apertus	AMNH 5239	skull	Late Cretaceous	Dinosaur Park Fm.	1912	Red Deer River, Alberta
Centrosaurus apertus	AMNH 5351	skeletal plaque	Late Cretaceous	Dinosaur Park Fm.	1914	Red Deer River, Alberta
Monoclonius cutleri (= *?Centrosaurus*)	AMNH 5427	partial skeleton with skin	Late Cretaceous	Dinosaur Park Fm.	1913	Red Deer River, Alberta
Chasmosaurus kaiseni	AMNH 5401	skull and mandibles	Late Cretaceous	Dinosaur Park Fm.	1913	Red Deer River, Alberta

Chasmosaurus belli	AMNH 5402	skull and mandibles	Late Cretaceous	Dinosaur Park Fm.	1913	Red Deer River, Alberta
Styracosaurus albertensis	AMNH 5372	skeletal plaque	Late Cretaceous	Dinosaur Park Fm.	1915	Red Deer River, Alberta
Triceratops horridus	AMNH 5033	free-standing partial skeleton	Late Cretaceous	Hell Creek Fm.	1909	Sand Creek, MT
Triceratops serratus	AMNH 970	skull	Late Cretaceous	Hell Creek Fm.	1902	Hell Creek, MT
Hadrosaur	AMNH 3650	trackway of "Mystery Dinosaur"	Early Cretaceous	Mesaverde Fm.	1937	States Mine, CO

HALL OF MAMMALS AND THEIR EXTINCT RELATIVES (3 TOTAL)

Propalaehoplophorus minor	AMNH 9197	skull and carapace	Miocene	Santa Cruz Fm.	1899	Monte Leon, Argentina
Hapalops ruetimeyeri	AMNH 9250	free-standing skeleton	Miocene	Santa Cruz Fm.	1899	Rio Gallegos, Argentina
Megalocnus rodens	AMNH 16876	free-standing skeleton	Pleistocene	?	1911	Cienfuegos, Cuba

HALL OF ADVANCED MAMMALS (12 TOTAL)

Hystrix primigenia	AMNH 20551	skull	Miocene	Mytilini Fm.	1923	Samos, Greece
Ictitherium viverrinum	AMNH 20695	skull and lower jaws	Miocene	Mytilini Fm.	1924	Samos, Greece
Ramoceros osborni	AMNH 9476	free-standing skeleton	Miocene	Pawnee Ck. Fm.	1901	Cedar Creek, CO
Prostrepsiceros houttumschindeleri	AMNH 20575	skull cap with horns	Miocene	Mytilini Fm.	1924	Samos, Greece
Hypohippus osborni	AMNH 9407	free-standing skeleton	Miocene	Loup Fork Beds	1901	Pawnee Buttes, CO
Diadiaphorus majusculus	AMNH 9291	skull and lower jaws	Miocene	Santa Cruz Fm.	1899	Santa Cruz, Argentina

(continued)

Major specimens collected by Barnum Brown on display in AMNH fossil halls *(continued)*

Scientific Name	Specimen Number	Elements	Age	Formation	Collection Date	Locality
Diadiaphorus majusculus	AMNH 9196	hind foot	Miocene	Santa Cruz Fm.	1899	Santa Cruz, Argentina
Thoatherium minisculum	AMNH 9167	hind foot	Miocene	Santa Cruz Fm.	1899	Argentina
Nesodon imbricatus	AMNH 9234	skull and lower jaws	Miocene	Santa Cruz Fm.	1899	Halliday Estancia, Argentina
Mammuthus sp.	AMNH 1922	sectioned molar	Pliocene	Siwaliks	1922	Chandigarh, India
Mammuthus sp.	AMNH 19821	sectioned molar	Pliocene	Upper Siwaliks	1922	Chandigarh, India
Bison bison antiquus	Anthro 20.2/5865	lower jaw, partial hind limb, point	Pleistocene	Folsom Fm.	1927–28	Folsom, NM

APPENDIX TWO

Memoirs of Barnum Brown

Discovery, Excavation, and Preparation of the Type Specimen *Tyrannosaurus rex* (AMNH No. 973) Discovered 1902, Completely Excavated 1905

Sold to Carnegie Museum 1941 at cost [$7,000] after we had made casts of the limb bones. The transaction was accomplished because the American Museum was afraid that German airships might bomb this [the American] Museum and destroy the second *Tyrannosaurus rex* skeleton now mounted here [AMNH 5027] and that at least one specimen might be preserved during the first [actually second] World War.

In 1902 while I was Assistant Curator of Fossil Reptiles in the American Museum of Natural History, Dr. Richard Swan Lull of Yale University, Philip Brooks (volunteer) and I were sent to eastern Montana to search for dinosaur bones in the Cretaceous Hell Creek beds of the Lance formation.

Our outfitting point and base of supplies was Miles City, Montana, a point on the railroad one hundred and thirty miles from the "badlands" near Jordan, Montana, our destination.

After five long days by wagon across undulating prairie, past numerous flocks of sheep and fewer herds of cattle, we arrived at the little log post-office [of] Jordan. Here also was a saloon, the latter owned and run by Nigger Bob who was not a negro, but he was sunburned and very dark. His saloon, which sold only beer, was a "Mecca" for all cow-hands in the surrounding country for Bob kept the beer in the cool cellar under the saloon and it was a welcome change from the tepid water of the creeks. Bob's sister operated the restaurant furnishing excellent food and Sundays conducted Sunday School for the ranchers' children when the cattlemen came in to trade.

We made camp near the old Max Sieber house. Sieber was a buffalo hunter who said when people came in to settle within ten miles of him that it was getting crowded and he would have to move on.

Near the house there is a high sandstone hill, Mount Sheba, with a stream [Hell

Creek] running at its base. It was here we camped, and before the cook called us for dinner I had found bones that had fallen down the hill into the stream and traced them up the hillside to their point of origin.

A few miles beyond this point at the head of a small stream tributary to the Missouri River, the prairie abruptly changes to a panorama of wonderful "badlands"; a variety of variegated sculpted cliffs and domes intersected by deep canyons with scattered pine trees and pockets of junipers; while on the hillsides in the broader valleys, lines of cottonwoods mark the stream courses. That fall when the cottonwood leaves had turned yellow it was such a striking scene my saddle horse stopped to look.

The dullness of the denuded earth is relieved by bright-colored clay in bands traceable on the same level for miles. Hard globular sandstones of all sizes are scattered among the layers of sand, and groups of them clutter the slopes of the hills.

Returning to the bones on Sheba Mountain we attacked the hillside with plow and scraper but soon the sand was so hard that the plow was no longer effective so we sent to Miles City for dynamite to drill and blow off sections of the hillside above the bone layer; one can dynamite safely within eight inches of the bones.

It took two seasons to complete this excavation, hot dirty work with the thermometer ranging up to 110° fahrenheit at noon without any shade, when Bob's cool beer at Jordan was a welcome interlude. Drinking three bottles would make a man see a mirage of beer bottles on all the distant peaks.

Some of the blocks containing bones were of enormous size. The one containing the pelvis weighed 4150 pounds after we had chipped off all surplus rock, which was as hard as granite, and when boxed it made a load requiring 4 horses to transport it to Miles City.

It was hot tedious work and when completed we left a scar on Mount Sheba thirty feet long, thirty feet wide and twenty five feet deep. And worth all of our effort for this dinosaur proved to be the type specimen of *Tyrannosaurus rex*.

Well do I remember the preparation of this specimen when it reached the Museum. Pete Kaisen, my assistant, and I were down in [the] yard chipping off the hard granite-like rock surrounding and in the cavities of the pelvis, practically breaking it into pieces but saving all bone so that we could put it together again when we had removed all the matrix.

Adam Herman, then head of the laboratory, came down where we were at work. "You are ruining the specimen," exclaimed Mr. Herman. "I'll have Professor Osborn stop this work." "Please call him down," I replied.

When Professor Osborn came down I explained how it was necessary to break the block to pieces, saving all pieces of bone so that the pelvic bones could be reassembled without matrix and he could study, figure and describe it. He readily understood agreeing to our method of preparation. Thus all parts of the specimen were freed of matrix.

The preparation of this specimen was even more difficult than its excavation but eventually it was accomplished and it was the second specimen I found near the John Willis ranch on the Big Dry Creek in Montana, No. [5027], now mounted in the American Museum, [that] enabled Professor Osborn to fully characterize this reptile.

Tyrannosaurus rex is a giant reptile distantly related to lizards, crocodiles and birds. It hind legs are formed like those of birds and the bones are pneumatic. It was a powerful creature, doubtless swift of movement when occasion demanded speed and capable of destroying any of the contemporary creatures, a king of the period and monarch of its race.

Tyrannosaurus rex stands 18½ feet high and is 45 feet in length, with enormous hind feet 3 feet in length and 3 feet wide, which [together] with its powerful neck and sharp dagger-like teeth enabled it to kill and devour any of its contemporaries. He is now the dominant figure in the Cretaceous Hall to awe and inspire young boys when they grow up to search for the same or other fossils, for almost every expedition turns up some remains of new or more complete prehistoric creature[s].

APPENDIX THREE

Summary of Fossil Collections by Barnum Brown and His AMNH Crews

Year	Locality	Number of Crates	Weight (in pounds)	Cost (in dollars)	Adjusted Cost (in today's dollars)
1896	NM, WY	11	?	1,672	38,000
1897	WY	60 or 80	?	1,192	27,700
1898	CO	27	3,451	?	
1899	Patagonia	14	9,000	823 (£169)	18,700
1900	SD, WY	31	40,000	801	18,200
1901	AZ, CO	?	?	?	
1902	MT	21	12,500	1,345	30,500
1903	SD, MT, AR	29	?	1,448	31,500
1904	SD, MT, NM	26	?	?	
1905	MT	21	?	1,557	33,800
1906	MT	33	?	1,610	34,300
1907	no fieldwork				
1908	MT	15	?	1,187	24,700
1909	MT	21	?	1,067	22,700
1910	MT, Alberta	42	6,500	1,506	30,700
1911	Cuba,	?	?	900	18,400
	Alberta	22	7,300	924	18,900
1912	Alberta	20	?	4,671	93,400
1913	Alberta	80	30,000	3,562	72,700
1914	Alberta	83	?	3,363	67,200
1915	Alberta	65	?	2,870	57,400
1916	MT	28+	?	?	

Year	Locality	Number of Crates	Weight (in pounds)	Cost (in dollars)	Adjusted Cost (in today's dollars)
1917	no fieldwork (WW I)				
1918	Cuba	16	?	?	
1919	OK, TX	?	?	?	
1920	Abyssinia	10	?	?	
1921–23	India	83+ or 42	?	11,890	143,300
1923	Burma	3	?	4,930	58,000
1923–24	Samos, Greece	60	?	?	
1925	NE	?	?	?	
1926	no fieldwork				
1927	MT to NM	2	?	?	
1928	NM	37	?	?	
1929	CO, NM, AZ	?	?	1,324	15,600
1930	AZ	1	?	2,739	33,000
1931	AZ, UT, MT	7	?	4,000	53,300
1932	WY, MT	24	?	2,282	33,600
1933	WY, MT, SD	1	?	1,322	20,700
1934	WY	145	?	7,438	112,700
1935	TX, OK	?	?	271	4,000
1936	AZ, IN	?	?	300	4,300
1937	WY, CO	61	25,000	7,733	109,000
1938	MT, TX	3+	?	400	5,700
1939	MT, Alberta	?	?	?	
1940	TX	150	44,000	2,000	29,000
1941	MT	4	?	496	6,800
1942	retired				
TOTALS		1,215+		$76,800+ £169	1,267,800

NOTES: Conversions are based on R. Sahr, Consumer Price Index (CPI) conversion factors 1800 to estimated 2016 to convert to dollars of 2006, Political Science Department, Oregon State University, 2006. Generally, the figures provided in this table are cited in the Annual Reports (1:1 Administration/Annual Reports), except for the following:

1897: Two figures for the total number of crates, 60 and 80, are given in the annual reports. The 80 is stated for all four expeditions that season. The 60 may well apply specifically to the quarrying at Como Bluff, which is also described as filling two carloads (2:5 B2 F3). In our total, 60 is used.

1900: About 1899, one pound sterling was worth $4.87 (http://eh.net/atp/answers/0789.php).

1904: The 26 crates cited in 2:3 B2 F14 do not seem to include the fossils from Conrad Fissure, Arkansas.

1916: Crate number is from 2:3 B4 F5, 8/28/1916, but several specimens were collected afterward in that season, according to a letter written on 10/3/1916.

1920: No vertebrate fossils were found, but invertebrate fossils and living or modern organisms were collected (2:3 B4 F10, 3/9/1921).

1921–23: The case number for the Indian expedition is confusing. A letter dated 8/5/1922 (2:5 B2 F5) documents that Brown reported storing 83 cases partway through the expedition. However, his final letter about the collection (dated 2/3/1923) states that he shipped 42 cases. Perhaps he repacked the specimens stored earlier. In our total, 42 is used.

1923: In a letter of 8/1/1923 (2:5 B2 F6), Brown states that 3 cases were shipped from Burma.

1923–24: The 60 cases from Samos are documented in 2:5 B2 F7, 10/3/1924.

1930: The one case is cited in Brown's report on the Arizona expedition (2:4 B6 F1).

1931: The cost of the field season is mentioned in 2:5 B2 F7, Memo of 6/11/1932.

1932: The number of cases is cited in 2:4 B6 F10, 10/27/32.

1933: The cases and number of specimens are cited in Brown's report located in 2:4 B7 F1.

1938: Number of cases cited in 2:4 B7 F9, 9/21/1938.

1940: Number of cases is extrapolated from Bird (1985: 195), who indicates the *Deinosuchus* case was numbered 150. The cost of the expedition is cited in 2:4 B8 F2, 7/11/1941.

1941: The cost for the expedition is cited in 2:4 B8 F3, Financial Statement.

NOTES

Citations such as 2:6 B4 F2 represent the filing code for documents or photos in the American Museum of Natural History (AMNH) Vertebrate Paleontology Archives or the AMNH Central Archives and designate the shelf location (2:6), box number (B4), and folder number (F2), sheet number (S2), or envelope (E) where the document or photo is stored.

PROLOGUE

1. Preston 1984: 101–105.
2. F. Brown 1987: 31.
3. F. Brown 1987: 13–14.
4. 2:6 B4 F2, "Patagonia Land's End": 1.
5. F. Brown 1987: 14.
6. F. Brown 1987: 14.
7. 2:6 B4 F2, "Patagonia Land's End": 2.
8. F. Brown 1987: 14–15.

1. CHILD OF THE FRONTIER

1. 2:5 B1 F5, "My Most Unforgettable Character." This document is the source for all quotes in chapter 1 unless otherwise specified.
2. Snell and Metzler 1972: 25.
3. F. Brown 1987: 1.
4. H. G. O'Connor, Geologic Map of Osage County, Kansas ([1955] 2007), with geologic formation boundaries adjusted by D. R. Collins to fit 1:24,000 topographic base of the U.S. Geological Survey: Kansas Geological Survey, Map M-54, scale 1:50,000; at www.kgs.ku.edu/General/Geology/County/nop/osage.html.
5. Complaint of Incest, filed by Melissa Brown Taylor in Osage County,

State of Kansas vs. William Brown, April 25, 1889. Graciously provided by Scott Williams.

6. The Brown Family Record in the AMNH archives (2:5 B1 F, Brown Family History) documents that Melissa did marry a man named Taylor, but no date or full name is provided.

7. Throughout this book, idiosyncratic spelling and usage have been preserved in direct quotations. Clear typographic errors, in contrast, have been silently corrected.

8. Document for Criminal Action, filed in Osage County, Legal Case No. 11, State of Kansas vs. William Brown, April 25, 1889. Graciously provided by Scott Williams.

9. Document for Forfeiture of Recognizance, filed in District Court, Osage County, Document No. 541, State of Kansas vs. William Brown, May 20, 1889. Graciously provided by Scott Williams.

10. Conversions of past dollar amounts to sums in current U.S. dollars in this book are based on R. Sahr, Consumer price index (CPI) conversion factors 1800 to estimated 2016 to convert to dollars of 2006, Political Science Department, Oregon State University, 2006.

2. STUDENT . . . OF SORTS

1. Kohl et al. 2004: 28.
2. 2:5 B1 F4, "Dossier of Barnum Brown."
3. Kohl et al. 2004: 20.
4. Colbert 1984: 66–70.
5. Colbert 1984: 70–73.
6. Jaffe 2000.
7. Colbert 1984: 73–75.
8. Colbert 1984: 70, 72.
9. Kohl et al. 2004: 20.
10. Kohl et al. 2004: 21.
11. Kohl et al. 2004: 19.
12. 2:5 B1 F3, "Journal of 1894 KU Expedition," B. Brown.
13. 2:5 B1 F3, "Journal of 1894 KU Expedition," B. Brown.
14. 2:5 B1 F3, 12/21/62.
15. 2:5 B1 F3, 12/21/62.
16. 2:5 B1 F3, "Journal of 1894 KU Expedition," B. Brown.
17. Barbour 1892.
18. Barbour 1895. Today, *Daemonelix* are recognized as fossil burrows of an extinct genus of fossil beaver named *Palaeocastor;* starting in 1904 during an expedition by the Carnegie, more skeletons of this beaver began to be discovered

inside these corkscrew-shaped burrows. Fuchs therefore basically prevailed in the debate. See http://eobasileus.blogspot.com/2008/02/odd-prehistoric-rodents-part-iii-devils.html.
19. 2:5 B1 F3; 12/21/62.
20. 2:5 B1 F3, "Journal of 1894 KU Expedition," B. Brown.
21. Kohl et al. 2004: 24.
22. 2:5 B1 F4, B. Brown Notes.
23. 2:5 B1 F2, B. Brown Notes.
24. 2:5 B1 F2, B. Brown Notes.
25. 2:5 B1 F4, B. Brown Notes.
26. Kohl et al. 2004: 23–24. See also Schuchert and LeVene 1940: 213–214.
27. Colbert 1984: 88.
28. Kohl et al. 2004: 24.
29. Kohl et al. 2004: 24.
30. Kohl et al. 2004: 37.
31. Kohl et al. 2004: 38.
32. Kohl et al. 2004: 38.
33. Kohl et al. 2004: 43.
34. Kohl et al. 2004: 44.
35. Kohl et al. 2004: 82–84.
36. Kohl et al. 2004: 82–83.
37. Kohl et al. 2004: 83–84.
38. Kohl et al. 2004: 86.
39. Kohl et al. 2004: 122.
40. 2:5 B1 F2, 9/20/1895.
41. 2:5 B1 F3, 9/26/1895.
42. 2:5 B1 F1, Williston letter to Wortman, 2/18/96.

3. APPRENTICE EXTRAORDINAIRE

1. Kohl et al. 2004: 29; Rainger 1991: 17, 68.
2. Colbert 1984: 149.
3. Rainger 1991: 75.
4. Colbert 1984: 149.
5. 2:5 B1 F1, Williston letter to Wortman, 2/18/1896.
6. 2:5 B1 F3, 9/4/1918.
7. Colbert 1984: 145–149.
8. Rainger 1991: 24.
9. Rainger 1991: 25.
10. Rainger 1991: 73–74.
11. Rainger 1991: 78.

12. Regal 2002: xviii.
13. Rainger 1991: 28.
14. Rainger 1991: 30.
15. Rainger 1991: 28.
16. Colbert 1984: 148.
17. Rainger 1991: 28; Colbert 1984: 147.
18. Rainger 1991: 29.
19. Rainger 1991: 29.
20. Rainger 1991: 62.
21. Rainger 1991: 31.
22. Rainger 1991: 31, 61.
23. Rainger 1991: 68.
24. Rainger 1991: 69.
25. Rainger 1991: 68, 89.
26. 2:5 B1 F3, 4/2/1896.
27. 2:5 B3 F11, Expedition into the San Juan Basin, 1896.
28. 2:5 B3 F11, Expedition into the San Juan Basin, 1896.
29. 2:5 B1 F4, B. Brown notes on 1896 expedition.
30. 2:5 B3 F11, Expedition into the San Juan Basin, 1896.
31. 2:5 B3 F11, Expedition into the Big Horn and Wind River Basins, 1896.
32. 2:5 B3 F11, Expedition into the Big Horn and Wind River Basins, 1896.
33. 2:5 B3 F11, Expedition into the Big Horn and Wind River Basins, 1896.
34. 2:5 B1 F1 and 2, B. Brown notes, 1896.
35. 2:5 B1 F4, B. Brown notes on 1896 expedition.
36. 2:5 B2 F3, 11/15/1896.
37. Rainger 1991: 94–97 (quotes pp. 94, 96).
38. 2:5 B2 F3, 11/15/1896.
39. 2:5 B2 F3, 3/24/1897.
40. 2:5 B2 F3, 3/30/1897.
41. 2:5 B2 F3, 3/30/1897.
42. 2:5 B2 F3, 3/31/1897.
43. 2:5 B1 F1, Barnum Brown: Additional Information.
44. 2:5 B2 F3, 4/20/1897.
45. 2:5 B2 F3, 5/1/1897.
46. 2:5 B2 F3, 5/6/1897.
47. 2:5 B2 F3, 8/8/1897.
48. 2:5 B3 F11, Expedition into the Jurassic of Wyoming.
49. 2:5 B2 F3, 6/14/1897.
50. 2:5 B3 F11, Expedition into the Jurassic of Wyoming.
51. 2:5 B3 F11, Expedition into the Jurassic of Wyoming.
52. 2:5 B2 F3, 8/15/1897.

53. 2:5 B3 F11, Expedition into the Jurassic of Wyoming.
54. 2:5 B1 F1, Barnum Brown: Additional Information.
55. Bird 1985: 213.
56. 2:5 B1 F1, letter apparently from Osborn to Dean R. S. Woodward, 4/30/1898.
57. 2:5 B2 F3, 8/15/1897.

4. TO LAND'S END: PATAGONIA

1. Colbert 1992: 4–10.
2. Colbert 1992: 5.
3. Colbert 1992: 22–30.
4. Colbert 1992: 9.
5. Colbert 1992: 42.
6. Colbert 1992: 44–47.
7. Rainger 1991: 63, 88.
8. 2:5 B3 F11, Report of Museum Expedition of 1898 into Kansas, Nebraska, and Colorado.
9. Rainger 1991: 91.
10. Regal 2002: 31.
11. Regal 2002: xv.
12. Colbert 1992: 182–183.
13. Rainger 1991: 76.
14. 2:5 B3 F11, Report of Museum Expedition of 1898 into Kansas, Nebraska, and Colorado.
15. 2:5 B2 F3, 8/18/1898.
16. 2:5 B2 F3, 9/19/1898.
17. 2:5 B2 F3, 9/19/1898.
18. 2:5 B3 F11, Osborn Report, 1898.
19. Rainger 1991: 92–93.
20. Rainger 1991: 93. There was and is still great interest in the strange, Tertiary mammalian faunas of South America that evolved in isolation when South America was separated from other continents, as well as in the distribution of marsupials across the continents of the Southern Hemisphere, from South America to Australia. As mammalian paleontologists, both Osborn and Matthew were intently interested in acquiring collections of these taxa, both for their studies and for exhibition.
21. Rainger 1991: 75.
22. Rainger 1991: 81.
23. www.peabody.yale.edu/archives/ypmbios/hatcher.html; Rea 2001: 22.
24. 6:2 B1 F8, B. Brown notebook. Regarding dates for the voyage from New

York to Punta Arenas: In the draft for a popular article that apparently was not published, "Patagonia: Land's End," the departure and arrival dates are listed as 12/8/1898 and 1/10/1899. In *Let's Call Him Barnum,* Frances Brown lists them as 12/7/1898 and 1/5/1899. We've chosen to use the dates listed in the detailed log of the voyage contained in BB's notebook, which we take to have been recorded during the voyage.

25. 2:6 B4 F2, "Patagonia, Land's End": 3.
26. 2:6 B4 F2, "On the Straits of Magellan."
27. 2:6 B4 F2, "Patagonia, Land's End": 3.
28. Darwin 1909: chap. 5
29. 2:6 B4 F2, "Patagonia, Land's End": 3.
30. 2:5 B3 F11, Brown Report, Expedition to Patagonia: 1–2; 2:6 B2 F2, map of route.
31. 2:6 B4 F2, "Patagonia, Land's End": 4.
32. 2:6 B4 F2, "Patagonia, Land's End": 4.
33. 2:5 B3 F11, Brown Report, Expedition to Patagonia: 2.
34. 2:6 B4 F2, "Patagonia, Land's End": 5.
35. 2:5 B3 F11, Brown Report, Expedition to Patagonia: 2–3.
36. 2:6 B4 F2, "Patagonia, Land's End": 5.
37. 2:5 B3 F11, Brown Report, Expedition to Patagonia: 3.
38. www.peabody.yale.edu/archives/ypmbios/hatcher.html.
39. 2:6 B4 F2, "Patagonia, Land's End": 2–3.
40. www.peabody.yale.edu/archives/ypmbios/hatcher.html.
41. 2:5 B2 F3, 8/15/1899.
42. 2:5 B2 F3, Brown Patagonian contract.
43. 2:5 B2 F3, 8/15/1899.
44. Rainger 1991: 77–79.
45. 2:5 B2 F3, 8/15/1899.
46. Brown 1903.
47. 2:5 B2 F3, 11/25/1899.
48. 2:5 B3 F11, Brown Report, Expedition to Patagonia: 3.
49. 2:6 B4 F2, "Patagonia, Land's End": 7.
50. 2:5 B2 F3, 11/25/1899.
51. 2:5 B2 F3, 12/29/1899.
52. 2:6 B4 F2, "Patagonia, Land's End": 8.
53. 2:6 B4 F2, "Patagonia, Land's End": 6.
54. 2:6 B4 F2, "Patagonia, Land's End": 10.
55. 2:6 B4 F2, "Patagonia, Land's End": 10–11.
56. 2:6 B4 F2, "Patagonia, Land's End": 8–9.
57. 2:5 B2 F3, 2/15/1900.
58. 2:6 B4 F2, "Patagonia, Land's End": 13.

59. 2:6 B4 F2, "Patagonia, Land's End": 13–14.
60. 2:6 B4 F2, Genevieve: 1.
61. It is unclear precisely when Brown's voyage around the Horn with Saltpeare occurred. Both synthetic accounts of the expedition, "Patagonia, Land's End" and *Let's Call Him Barnum* (F. Brown 1986), say that it was at the end of the expedition, after all the fieldwork in the Santa Cruz was completed. However, a page of notes entitled "Shipwreck near Spaniard Harbor" (2:6 B4 F2) suggests that it was at the start, since Brown is said to have told Hatcher that they would meet up at Punta Arenas in a couple of weeks, presumably to start their trip to the Andes. Since this latter account is only partial and disconnected from the rest of the story, we've chosen to rely on the two more comprehensive synthetic accounts, especially given that the entries in Brown's field notebook about the trip begin only on Februrary 28, 1900 (6:2 B1 F8).
62. 2:5 B4 F2, Genevieve.
63. The identity of this third crew member is unclear. In the document entitled "Shipwreck near Spaniard Harbor" (2:6 B4 F2), the one-armed skipper, Smith, declines to go, stating that he cannot navigate any farther than the harbor; the one-armed skipper's first mate, however, agrees to go. In "Patagonia, Land's End" (14), though, the one-armed skipper who had stolen Saltpeare's cutter is said to have been "not unduly disturbed" by being caught and "said he would go with us" around the Horn. A third account appears in a document entitled "Patagonia Saltpeare & Wreck," a transcript of a conversation recorded using a dictation device called Genevieve (2:6 B4 F2), which Brown seems to have wanted to buy. In this account, the third crew member agrees to go. We've chosen to go with the account in Genevieve, but the reality is open to question.
64. 2:6 B4 F2, "Patagonia, Land's End": 15.
65. 2:6 B4 F2, "Patagonia, Land's End": 15; 2:6 B4 F2, Genevieve: 1–2.
66. 2:6 B4 F2, "Patagonia, Land's End": 16.
67. 2:6 B4 F2, "Patagonia, Land's End": 17.
68. 2:6 B4 F2, "Patagonia, Land's End": 19.
69. F. Brown 1987: 22.

5. TO THE DEPTHS OF HELL CREEK

1. 2:6 B4 F2, "Patagonia, Land's End": 19.
2. 1:1 Dept. Vertebrate Paleontology, Ann. Report 1900.
3. 2:3 B1 F10, 7/7/1900.
4. Rainger 1991: 21.
5. Rainger 1991: 97; www.carnegiemnh.org/vp/history.html; Colbert 1984: 164.
6. www.carnegiemnh.org/vp/history.html.
7. www.fieldmuseum.org/research_collections/geology/history.htm.

8. 2:3 B1 F10, 6/29/1900.
9. 2:3 B1 F10, 8/9/1900.
10. 1:1 Dept. Vertebrate Paleontology, Ann. Report 1900.
11. 2:3 B1 F10, 9/1/1900.
12. 2:3 B1 F10, 10/4/1900.
13. 2:3 B1 F10, 10/17/1900.
14. Berger 2004; 1:1 Admin./Ann. Report, 1900, Field Notes (for collecting in Westin County, Wyoming).
15. 2:3 B1 F10, 12/1/1900.
16. 2:3 B1 F10, 12/10/1900.
17. 2:3 B1 F10, 12/26/1900.
18. 2:3 B1 F10, 12/10/1900.
19. 2:3 B1 F10, 1/25/1901.
20. 1:1 Dept. Vertebrate Paleontology, Ann. Report 1900.
21. 2:3 B2 F3, 5/20/1901.
22. 2:3 B2 F3, 7/13/1901.
23. 2:3 B2 F3, 7/22/1901.
24. 2:3 B2 F3, 8/4/1901.
25. 2:3 B2 F3, 8/31/1901.
26. Hornaday 1925: 5–6.
27. Hornaday 1925: 47–51.
28. Lepley and Lepley 1992: 84.
29. Hornaday 1925: 79–80.
30. Colbert 1984: 82–83.
31. Colbert 1984: 84–85.
32. 2:3 B2 F6, 5/29/1902.
33. http://lib.montana.edu/collect/spcoll/montana1900/montana1900.html.
34. 2:3 B2 F6, 6/17/1902.
35. 2:3 B2 F6, 6/19/1902.
36. 2:3 B2 F6, 6/19/1902.
37. 2:3 B2 F6, 7/7/1902.
38. Jordan 2003: 139–140.
39. 2:3 B2 F6, 7/12/1902.
40. 2:3 B2 F6, 7/25/1902.
41. 2:3 B2 F6, 7/19/1902.
42. 2:3 B2 F6, 8/2/1902.
43. 2:3 B2 F6, 8/12/1902.
44. 2:3 B2 F6, 8/12/1902.
45. 2:3 B2 F6, 9/3/1902.
46. Berger 2004.
47. 2:3 B2 F6, 10/13/1902.

48. 2:3 B2 F13, 5/8/1903.
49. 1:1 Dept. Vertebrate Paleontology, Ann. Report 1903.
50. 2:3 B2 F13, 6/7/1903.
51. 2:3 B2 F13, 6/11/1903.
52. 2:3 B2 F13, 5/24/1903.
53. 2:3 B2 F13, 5/31/1903.
54. 2:3 B2 F13, 7/3/1903.
55. Colbert 1984: 234.
56. 2:3 B2 F13, 7/29/1903.
57. 2:3 B2 F13, 7/23/1903.
58. 1:1 Dept. Vertebrate Paleontology, Ann. Report 1903: 120–121.
59. Brown acknowledges Osborn's letter in a response written 9/23/1903 (2:3 B2 F13). The letter from Osborn has not been found.
60. 1:1 Dept. Vertebrate Paleontology, Ann. Report 1903: 121–123.

6. LOVE

1. 2:5 B1 F2, Scattered Notes: 3.
2. F. Brown 1987: 23–24.
3. F. Brown 1987: 23–24.
4. 2:6 B7 F1, M. R. Brown, 1904, Log Book of the Bug Hunters: 1.
5. Rainger 1991: 71.
6. 2:3 B2 F14, 6/10/04.
7. 1:1 Dept. Vertebrate Paleontology, Ann. Report 1904.
8. 2:6 B7 F1, M. R. Brown, 1904, Log Book of the Bug Hunters: 4.
9. 2:6 B7 F1, M. R. Brown, 1904, Log Book of the Bug Hunters: 6.
10. 2:6 B7 F1, M. R. Brown, 1904, Log Book of the Bug Hunters: 6.
11. 2:6 B7 F1, M. R. Brown, 1904, Log Book of the Bug Hunters: 14.
12. 2:6 B7 F1, M. R. Brown, 1904, Log Book of the Bug Hunters: 16–17.
13. 2:6 B7 F1, M. R. Brown, 1904, Log Book of the Bug Hunters: 20.
14. 2:6 B7 F1, B. Brown notes for Log Book of the Bug Hunters: 5 (ten pages of supplementary notes at end of Marion's account).
15. 2:6 B7 F1, M. R. Brown, 1904, Log Book of the Bug Hunters: 23; 1:1 Dept. Vertebrate Paleontology, Ann. Report 1904.
16. 2:6 B7 F1, M. R. Brown, 1904, Log Book of the Bug Hunters: 24–25.
17. 2:6 B7 F1, M. R. Brown, 1904, Log Book of the Bug Hunters: 26.
18. 2:6 B7 F1, B. Brown notes for Log Book of the Bug Hunters: 1.
19. 2:6 B7 F1, M. R. Brown, 1904, Log Book of the Bug Hunters: 27, 28.
20. 2:6 B7 F1, M. R. Brown, 1904, Log Book of the Bug Hunters: 30.
21. 1:1 Dept. Vertebrate Paleontology, Ann. Report 1904.
22. 2:6 B7 F1, M. R. Brown, 1904, Log Book of the Bug Hunters: 31.

23. 2:6 B7 F1, M. R. Brown, 1904, Log Book of the Bug Hunters: 33.
24. 2:6 B7 F1, M. R. Brown, 1904, Log Book of the Bug Hunters: 35–36.
25. 2:6 B7 F1, M. R. Brown, 1904, Log Book of the Bug Hunters: 41, 44; 2:6 B7 F1, B. Brown notes for Log Book of the Bug Hunters: 5
26. Brown 1910.
27. 1:1 Dept. Vertebrate Paleontology, Ann. Report 1904.
28. 2:6 B7 F1, B. Brown notes for Log Book of the Bug Hunters: 8.
29. F. Brown 1987: 25.
30. 2:6 B7 F1, B. Brown notes for Log Book of the Bug Hunters: 9.
31. 2:6 B7 F1, B. Brown notes for Log Book of the Bug Hunters: 10.
32. 2:6 B7 F1, B. Brown notes for Log Book of the Bug Hunters: 10.
33. Osborn 1905: 477; 1906: 281.
34. Osborn 1905: 477.
35. Osborn 1905.
36. 2:3 B2 F16, 6/5/1905.
37. 1:1 Dept. Vertebrate Paleontology, Ann. Report 1905.
38. 2:3 B2 F16, 6/24/1905.
39. 2:3 B2 F16, 6/24/1905.
40. 2:3 B2 F16, 7/6/1905.
41. 2:3 B2 F16, 7/2/1905.
42. 2:3 B2 F16, 7/2/1905.
43. 1:1 Dept. Vertebrate Paleontology, Ann. Report 1905.
44. 2:3 B2 F16, 7/2/1905.
45. 2:3 B2 F16, 7/15/1905.
46. 2:3 B2 F16, 7/25/1905.
47. 2:3 B2 F16, 8/8/1905.
48. 2:3 B2 F16, 8/22/1905.
49. 2:3 B2 F16, 9/22/1905.
50. 2:3 B2 F16, 9/22/1905.
51. 2:3 B2 F15, 10/14/1905.
52. Osborn 1906: 281.
53. 2:3 B2 F16, 9/22/1905.

7. LOSS

1. 2:3 B3 F1, 6/28/1906.
2. 1:1 Dept. Vertebrate Paleontology, Ann. Report 1906.
3. 2:3 B3 F1, 7/17/1906.
4. 2:3 B3 F1, 8/23/1906.
5. 1:1 Dept. Vertebrate Paleontology, Ann. Report 1906.

6. Osborn 1906: 281.
7. Osborn 1906: 290–291.
8. Osborn 1906: 281.
9. Osborn 1906: 296.
10. 2:3 B3 F1, 10/10/1906.
11. 1:1 Dept. Vertebrate Paleontology, Ann. Report 1907.
12. 1:1 Dept. Vertebrate Paleontology, Ann. Report 1907; Brown 1907.
13. Alvarez et al. 1980.
14. Dingus and Rowe 1997.
15. 6:2 B3 E4, B. Brown 1908 field notebook, 6/21/1908.
16. 6:2 B3 E4, B. Brown 1908 field notebook, 7/1/1908.
17. 6:2 B3 E4, B. Brown 1908 field notebook, 7/3/1908.
18. Norell et al. 1995: 190.
19. 6:2 B3 E4, B. Brown 1908 field notebook, 7/4/1908.
20. 2:3 B3 F4, 7/8/1908.
21. 2:3 B3 F4, 7/8/1908.
22. 2:3 B3 F4, 7/15/1908.
23. 2:3 B3 F4, 7/30/1908.
24. 6:2 B3 E4, B. Brown 1908 field notebook, 7/21/1908.
25. 2:3 B3 F4, 8/1/1908.
26. 2:3 B3 F4, 8/10/1908.
27. 2:3 B3 F4, 8/10/1908.
28. 6:2 B3 E4, B. Brown 1908 field notebook, 10/8/1908, 10/10/08.
29. 1:1 Dept. Vertebrate Paleontology, Ann. Report 1908.
30. Osborn 1912.
31. Rainger 1991: 70.
32. Berger 2004; Osborn 1916: 761–762.
33. Berger 2004.
34. Brown 1915: 322.
35. AMNH Central Archives, 1209, 1941–1949: 7/7/1941.
36. 2:5 B2 F4, Memoirs of Barnum Brown: Discovery, excavation, and preparation of the type *Tyrannosaurus rex* AMNH No. 973. (See also Appendix 2.)
37. 1:1 Dept. Vertebrate Paleontology, Ann. Report 1909.
38. 1:1 Dept. Vertebrate Paleontology, Ann. Report 1909.
39. 2:3 B3 F5, 8/15/1909.
40. 1:1 Dept. Vertebrate Paleontology, Ann. Report 1909; 2:3 B3 F5, 9/27/1909.
41. F. Brown 1987: 25–26.
42. AMNH Central Archives, Osborn, B3 F26, 5/11/1910.
43. F. Brown 1987: 24–25.

8. THE CANADIAN DINOSAUR BONE RUSH

1. F. Brown 1987: 27.
2. Colbert 1984: 178–179.
3. Colbert 1984: 181–182.
4. Osborn 1902; Lambe 1902.
5. Eberth 1997a,b; Eberth 2005.
6. 1:1 Dept. Vertebrate Paleontology, Ann. Report 1910.
7. 2:3 B3 F6, 7/24/1910.
8. In what follows, unless otherwise specified, quotations and details about the expedition are from the Annual Report of 1910 (1:1 Dept. Vertebrate Paleontology).
9. 2:6 B4 F1, P. Kaisen Diary, 1910–1916: 5.
10. 2:3 B3 F6, 8/1/1910.
11. 2:3 B3 F6, 8/6/1910.
12. 2:3 B3 F6, 7/31/1910.
13. 2:3 B3 F6, 8/6/1910.
14. Quoted in Colbert 1984: 188.
15. Personal communication, D. Tanke.
16. 1:1 Dept. Vertebrate Paleontology, Ann. Report 1910.
17. 2:6 B4 F1, P. Kaisen Diary, 1910–1916: 9.
18. 2:3 B3 F6, 9/12/1910.
19. 2:3 B3 F6, 9/22/1910.
20. 1:1 Dept. Vertebrate Paleontology, Ann. Report 1910.
21. 2:3 B3 F6, 10/25/1910.
22. 1:1 Dept. Vertebrate Paleontology, Ann. Report 1911.
23. 1:1 Dept. Vertebrate Paleontology, Ann. Report 1911.
24. 1:1 Dept. Vertebrate Paleontology, Ann. Report 1911; Brown 1913: 227.
25. 2:6 B4 F1, P. Kaisen Diary, 1910–1916: 21.
26. 1:1 Dept. Vertebrate Paleontology, Ann. Report 1911.
27. 2:6 B4 F1, P. Kaisen Diary, 1910–1916: 24.
28. 2:3 B3 F8, 9/3/1911.
29. 1:1 Dept. Vertebrate Paleontology, Ann. Report 1911.
30. 2:5 B1 F4, B. Brown notes.
31. 1:1 Dept. Vertebrate Paleontology, Ann. Report 1911.
32. 1:1 Dept. Vertebrate Paleontology, Ann. Report 1911.
33. 2:6 B4 F1, P. Kaisen Diary, 1910–1916: 28.
34. 2:3 B3 F10, 7/28/1912.
35. Colbert 1984: 190.
36. Dingus and Norell 2008: 24.
37. 2:3 B3 F10, 7/9–12/1912; Spalding 2001.

38. 2:6 B4 F1, P. Kaisen Diary, 1910–1916: 30; http://research.amnh.org/paleontology/reports/1912.html.
39. 2:6 B4 F1, P. Kaisen Diary, 1910–1916: 32.
40. 2:3 B3 F10, 9/22/1912.
41. 2:6 B4 F1, P. Kaisen Diary, 1910–1916: 34.
42. 1:1 Dept. Vertebrate Paleontology, Ann. Report 1912.
43. 2:3 B3 F10, 7/28/1912; Darren Tanke, personal communication.
44. 1:1 Dept. Vertebrate Paleontology, Ann. Report 1912.
45. 2:3 B3 F10, 8/11/1912.
46. 2:3 B3 F10, 9/22/1912.
47. 2:6 B4 F1, P. Kaisen Diary, 1910–1916: 37.
48. 2:3 B3 F10, 8/11/1912.
49. 1:1 Dept. Vertebrate Paleontology, Ann. Report 1912.
50. 2:6 B4 F1, P. Kaisen Diary, 1910–1916: 41.
51. 2:3 B3 F10, 9/5/1912.
52. 2:3 B3 F10, 9/17/1912.
53. 1:1 Dept. Vertebrate Paleontology, Ann. Report 1912.
54. Chiappe and Dingus 2001.
55. 1:1 Dept. Vertebrate Paleontology, Ann. Report 1912.
56. 2:3 B3 F10, 10/23/1912.
57. 2:3 B4 F2, 6/22/1913.
58. 1:1 Dept. Vertebrate Paleontology, Ann. Report 1913.
59. 2:3 B3 F2, 7/6/1913.
60. 2:6 B4 F1, P. Kaisen Diary, 1910–1916: 48.
61. 2:3 B3 F2, 7/6/1913.
62. 2:3 B4 F2, 7/14/1913.
63. 2:3 B4 F2, 7/14/1913.
64. 2:3 B4 F2, 7/21/1913.
65. 2:3 B4 F2, 8/3/1913.
66. 2:3 B4 F2, 7/23/1913.
67. 2:3 B4 F2, 8/3/1913.
68. 2:3 B4 F2, 7/23/1913.
69. 2:6 B4 F1, P. Kaisen Diary, 1910–1916: 49.
70. 1:1 Dept. Vertebrate Paleontology, Ann. Report 1913.
71. 1:1 Dept. Vertebrate Paleontology, Ann. Report 1913.
72. 2:3 B4 F2, 9/2/1913.
73. 2:3 B4 F2, 9/23/1913.
74. 1:1 Dept. Vertebrate Paleontology, Ann. Report 1914.
75. 2:6 B4 F1, P. Kaisen Diary, 1910–1916: 57.
76. 2:3 B4 F3, 7/3/1914.
77. 2:3 B4 F3, 7/3/1914.

78. 2:5 B1 F4, B. Brown, Red Deer River Work 1914.
79. 2:3 B4 F3, 7/3/1914.
80. 2:3 B4 F3, 8/5/1914.
81. 2:3 B4 F3, 7/3/1914.
82. 2:3 B4 F3, 8/10/1914.
83. 2:3 B4 F3, 9/7/1914.
84. 2:3 B4 F3, 9/13/1914; Darren Tanke, personal communication.
85. 1:1 Dept. Vertebrate Paleontology, Ann. Report 1914.
86. 1:1 Dept. Vertebrate Paleontology, Ann. Report 1915.
87. 1:1 Dept. Vertebrate Paleontology, Ann. Report 1915.
88. 2:3 B4 F4, 8/26/1915.
89. 2:3 B4 F4, 7/25/1915.
90. 2:3 B4 F4, 10/25/1915.
91. 1:1 Dept. Vertebrate Paleontology, Ann. Report 1915; 2:3 B4 F4, 10/25/1915.
92. Brown 1919: 427.
93. Rainger 1991: 95.
94. Currie 2005: 5–7; Brown, Russell, and Ryan 2009; Sampson, Ryan, and Tanke 1997; Darren Tanke, personal communication.
95. Tanke 2005; Darren Tanke, personal communication.

9. CUBA, ABYSSINIA, AND OTHER INTRIGUES

1. 2:3 B4 F5, 7/27/1916.
2. 2:3 B4 F5, 7/27/1916.
3. 2:3 B4 F5, 7/27/1916.
4. 2:3 B4 F5, 8/15/1916 and 8/18/1916.
5. Eberth 1997b: 380.
6. 2:3 B4 F5, 7/27/1916.
7. 1:1 Dept. Vertebrate Paleontology, Ann. Report 1917.
8. F. Brown 1987: 27.
9. 2:5 B1 F1, 2:5 B1 F4, B. Brown dossier and field trip records.
10. 1:1 Dept. Vertebrate Paleontology, Ann. Report 1918.
11. 2:3 B4 F7, 4/30/1918.
12. 2:3 B4 F7, 6/1/1918.
13. 2:5 B2 F4, 4/20/1918.
14. 1:1 Dept. Vertebrate Paleontology, Ann. Report 1918.
15. 2:3 B4 F7, 6/14/1918.
16. 2:5 B1 F4, 10/17–22/1918.
17. 2:5 B1 F4, 10/17–22/1918.
18. 2:5 B4 F4, 12/3/1918.
19. 2:3 B4 F9, 6/18/1919.

20. 2:3 B4 F9, 5/24/1919.
21. 2:5 B2 F4, 12/28/1919.
22. 2:5 B2 F4, Matthew to Brown, 1/6/1920.
23. 2:5 B2 F4, Matthew to Osborn, 1/6/1920.
24. Carter 2003; www.exxonmobil.com/Europe-English/about_who_history_europe.aspx.
25. www.digitalhistory.uh.edu/historyonline/oil.cfm.
26. 5:3 B1 F2, Capt. H. F. Moon's Diary, 1920 Dudley Expedition: 72.
27. Brown 1925b: 607.
28. F. Brown 1987: 31–32.
29. Brown 1925b: 607.
30. 5:3 B1 F2, Capt. H. F. Moon's Diary, 1920 Dudley Expedition: 72.
31. 5:3 B1 F2, Capt. H. F. Moon's Diary, 1920 Dudley Expedition: 74.
32. 5:3 B1 F2, Capt. H. F. Moon's Diary, 1920 Dudley Expedition: 78.
33. 5:3 B1 F2, Capt. H. F. Moon's Diary, 1920 Dudley Expedition: 80.
34. 5:3 B1 F2, Capt. H. F. Moon's Diary, 1920 Dudley Expedition: 81.
35. Brown 1925b: 609.
36. www.onwar.com/aced/data/mike/madmullah1899.htm; www.chakoten.dk/mad_mullah.html.
37. 5:3 B1 F2, Capt. H. F. Moon's Diary, 1920 Dudley Expedition: 82.
38. 5:3 B1 F2, Capt. H. F. Moon's Diary, 1920 Dudley Expedition: 85.
39. 5:3 B1 F2, Capt. H. F. Moon's Diary, 1920 Dudley Expedition: 89.
40. 5:3 B1 F2, Capt. H. F. Moon's Diary, 1920 Dudley Expedition: 92.
41. Brown 1925b: 609–612.
42. 5:3 B1 F2, B. Brown's Journal of Camel Caravans, Abyssinia, 1921–1922: 32–33.
43. 5:3 B1 F2, B. Brown's Journal of Camel Caravans, Abyssinia, 1921–1922: 38.
44. 5:3 B1 F2, B. Brown's Journal of Camel Caravans, Abyssinia, 1921–1922: 46.
45. 5:3 B1 F2, B. Brown's Journal of Camel Caravans, Abyssinia, 1921–1922: 47.
46. 2:5 B1 F4, Dossier of Barnum Brown, Issa Somalis.
47. 5:3 B1 F2, B. Brown's Journal of Camel Caravans, Abyssinia, 1921–1922: 51.
48. Josephson 1952.
49. F. Brown 1987: 35.
50. 2:6 B7 F3, 12/31/1920.
51. 2:6 B7 F3, 1/8/1921.
52. 2:6 B7 F3, 1/18/1921.
53. 2:6 B7 F3, 2/18/1921.
54. 2:3 B4 F11, 5/18/1921.
55. 2:3 B4 F11, 5/20/1921.
56. F. Brown 1987: 33.
57. 2:3 B4 F11, 5/23/1921.

58. 2:3 B4 F11, 6/6/1921.
59. 2:5 B2 F4, 6/3/1921.
60. 2:5 B2 F4, 9/1/1921.
61. Barton 1941: 310.
62. 2:6 B7 F3, 6/30/1921.

10. JEWELS FROM THE ORIENT: RAJ INDIA

1. 2:5 B2 F5, 5/14/1922.
2. Colbert 1935: iii.
3. www.wku.edu/~smithch/chronob/PILG1875.htm.
4. Eldredge and Gould 1972.
5. Kennedy and Ciochon 1999.
6. Hartwig 2002: 369.
7. Hartwig 2002: 369.
8. Regal 2002: xi.
9. Lewin 1987: 53–55.
10. Lewin 1987: 56.
11. 2:6 B7 F4, 12/25/1921.
12. 2:6 B7 F4, 12/25/1921.
13. 2:6 B7 F4, 1/1/1922.
14. 2:6 B7 F4, 1/4/1922.
15. Lindsay et al. 1980.
16. Barry 1995; Pilbeam et al. 1996; Barry et al. 2002; Behrensmeyer and Barry 2005; L. J. Flynn and J. C. Barry, personal communication.
17. 2:5 B2 F5, 1/10/1922.
18. 2:5 B2 F5, 1/25/1922.
19. 2:5 B2 F5, 1/27/1922.
20. 2:5 B2 F5, 1/27/1922.
21. 2:5 B2 F5, 2/9/1922.
22. 2:5 B2 F5, 2/23/1922; L. J. Flynn and J. C. Barry, personal communication; Tassy 1983: 337.
23. 2:6 B7 F4, 2/14/1922.
24. 2:5 B2 F5, 2/23/1922.
25. 2:6 B7 F4, 2/25/1922.
26. 2:5 B2 F5, 3/7/1922.
27. 2:6 B7 F4, 3/31/1922.
28. 2:6 B7 F4, 4/1–3/1922.
29. 2:5 B2 F5, 4/23/1922.
30. 2:5 B2 F5, 4/23/1922.

31. 2:6 B7 F4, 4/24/1922.
32. 2:5 B2 F5, 5/14/1922.
33. 2:5 B2 F5, 6/1/1922.
34. 2:5 B2 F5, 6/15/1922.
35. Regal 2002: 154–162.
36. 2:5 B2 F5, 6/21/1922.
37. AMNH Central Archives, Osborn, B3 F26, 6/17/1922.
38. 2:6 B7 F4, 6/30/1922.
39. 2:5 B2 F5, 7/13/1922.
40. 2:5 B2 F5, 7/14/1922 and 8/1/1922.
41. 2:6 B7 F4, 7/22/1922.
42. 2:6 B7 F4, 7/22/1922.
43. 2:5 B F5, 8/5/1922.
44. 2:5 B2 F5, 8/8/1922.
45. 2:5 B2 F5, 8/11/1922.
46. 2:6 B7 F4, 8/14/1922.
47. 2:6 B7 F4, 8/14/1922.
48. 2:5 B2 F5, 10/10/1922.
49. 2:6 B7 F4, 10/17/1922.
50. 2:6 B7 F4, 10/17/1922.
51. 2:5 B2 F4, 10/19/1922.
52. 2:5 B2 F5, 10/20/1922.
53. 2:5 B2 F5, 11/3/1922.
54. 2:6 B7 F4, 11/21/1922.
55. 2:5 B2 F5, 12/1/1922.
56. 2:5 B2 F5, 12/7/1922.
57. 2:5 B2 F5, 12/8/1922.
58. 2:5 B2 F5, 12/15/1922.
59. Rainger 1991: 102.
60. 2:5 B2 F5, 12/28/1922.
61. Matthew 1929; Colbert 1992: 202–205.
62. Colbert 1935: iii, 70–320; Pilgrim 1937; Tassy 1983; J.C. Barry and L.J. Flynn, personal communication. Brown's Siwalik holotypes comprise *Rhizomys punjabiensis* = *Rhizomyides punjabiensis* AMNH 19762 (Colbert 1935: 70); *Sivacanthion complicatus* AMNH 19626 (Colbert 1935: 73); *Martes lydekkeri* AMNH 19407 (Colbert 1935: 94); *Amphiorycteropus browni* (Colbert 1933) AMNH 29997 (Colbert 1935: 128); *Gaindatherium browni* AMNH 19409 (Colbert 1935: 183); *Pecarichoerus orientalis* AMNH 29955 (Colbert 1935: 214); *Cervus punjabiensis* AMNH 19911 (Colbert 1935: 320).

Tassy (1983) lists at least four proboscidean holotypes based on AMNH specimens: *Zygolophodon metachinjiensis* (AMNH 19414); *Protanancus chinjiensis*

(AMNH 19421); *Gomphotherium browni* (AMNH 19417); *Paratetralophodon hasnotensis* (AMNH 19838).

And Pilgrim (1937) lists the following bovids as having AMNH holotypes. Boselaphinae: *Selenoportax vexillarius* (AMNH 19748); *Helicoportax praecox* (AMNH 19476); *Strepsiportax gluten* (AMNH 19746); *Strepsiportax chinjiensis* (AMNH 19450); *Tragoportax salmontanus* (AMNH 19467); *Tragocerus browni* (AMNH 19662); *Sivaceros gradiens* (AMNH 19448). Gazellinae: *Gazella lydekkeri* (AMNH 19663); *Antilope subtorta* (AMNH 19989).

63. 2:6 B7 F4, 2/2/1923.
64. 2:5 B2 F6, 1/15/1923.
65. 2:6 B7 F4, 2/2/23.
66. 2:6 B7 F4, 1/2/1923.
67. 2:6 B7 F4, 2/2/1923.
68. F. Brown 1987: 37.
69. 2:6 B7 F4, 2/2/1923.
70. 2:6 B7 F4, 2/2/1923.

11. PERILS AND PEARLS UP THE IRRAWADDY: BURMA

1. Ciochon and Gunnell 2002: 157.
2. Ciochon 1985.
3. 2:5 B2 F6, 2/3/1923.
4. 2:6 B7 F8, 2/23/1923.
5. 2:6 B7 F8, 2/23/1923.
6. 2:6 B7 F8, 2/23/1923.
7. Brown 1925a: 295–296.
8. 2:5 B2 F6, 3/5/1923.
9. 2:5 B2 F6, 3/28/1923 and 3/31/1923.
10. L. Brown 1950: 179.
11. L. Brown 1950: 215–218.
12. L. Brown 1950: 228.
13. 2:5 B2 F6, 4/28/1923.
14. Ciochon 1985.
15. Colbert 1937; Pilgrim 1927.
16. 2:5 B2 F6, 6/1/1923.
17. F. Brown 1987: 36–37.
18. L. Brown 1950: 252–258.
19. 2:5 B2 F6, 6/1/1923.
20. 2:6 B7 F8, 7/4/1923.
21. 2:5 B2 F6, 6/1/1923.
22. 2:5 B2 F6, 6/1/1923.

23. 2:5 B2 F6, 6/16/1923.
24. 2:6 B7 F8, 7/4/1923.
25. 2:5 B2 F6, 7/31/1923.
26. 2:5 B2 F6, 8/1/1923.
27. Ciochon 1985: 34.
28. 2:5 B2 F6, 8/1/1923.

12. SAMOS: ISLE OF INTRIGUE

1. Solounias and Mayor 2004; Solounias and Ring 2007.
2. Solounias and Ring 2007.
3. Matthew 1915; Colbert 1992: 177.
4. Solounias and Ring 2007; Solounias et al. 1999.
5. 2:5 B2 F6, 8/26/1923.
6. 2:5 B2 F6, 8/26/1923.
7. 2:6 B8 F2, 8/23/1923.
8. 2:5 B2 F6, 8/26/1923.
9. 2:5 B2 F6, 8/26/1923.
10. 2:6 B8 F2, 9/2/1923.
11. 2:5 B2 F6, 9/10/1923.
12. 2:5 B2 F6, Request for Permission to Excavate Fossils [in Greece].
13. 2:5 B2 F6, 9/16/1923.
14. 2:5 B2 F6, 9/19/1923.
15. 2:5 B2 F6, 9/28/1923.
16. Rainger 1991: 70–71.
17. Rainger 1991: 217–226, 230–231.
18. Rainger 1991: 239.
19. 2:5 B2 F6, 9/28/1923.
20. 2:6 B8 F2, 10/10/1923.
21. Brown 1927: 30.
22. L. Brown 1951: 49.
23. 2:5 B2 F6, 10/17/1923.
24. 2:5 B2 F6, 10/25/1923.
25. 2:5 B2 F6, 11/8/1923.
26. 2:5 B2 F6, 11/10/1923.
27. 2:5 B2 F6, 11/24/1923.
28. 2:5 B2 F6, 12/4/1923.
29. Bernor et al. 1996; Kostopoulos et al. 2003.
30. Solounias and Ring 2007.
31. 2:5 B2 F6, 12/4/1923.
32. 2:5 B2 F6, 12/8/1923.

33. 2:6 B8 F2, 12/25/1923.
34. 2:6 B8 F2, 1/21/1924.
35. 2:5 B2 F7, 1/24/1924.
36. 2:5 B2 F7, 1/24/1924.
37. 2:6 B8 F2, 2/13/1924.
38. 2:6 B8 F2, 2/16/1924.
39. 2:5 B2 F7, 2/14/1924.
40. 2:5 B2 F7, 2/14/1924.
41. 2:5 B2 F7, 2/29/1924.
42. 2:6 B8 F2, 3/5/1924.
43. 2:5 B2 F7, 3/12/1924.
44. 2:5 B2 F7, 3/15/1924.
45. 2:6 B8 F2, 3/23/1924.
46. 2:5 B2 F7, 4/4/1924.
47. 2:6 B8 F2, 4/4/1924.
48. 2:5 B2 F7, 4/17/1924.
49. 2:5 B2 F7, 5/15/1924.
50. 2:6 B8 F2, 5/30/1924.
51. 2:6 B8 F2, 6/17/1924.
52. 2:5 B2 F7, 7/1/1924.
53. 2:5 B2 F7, 7/1/1924.
54. 2:6 B8 F2, 8/7/1924.
55. 2:5 B2 F7, 8/13/1924.
56. 2:5 B2 F7, 9/10/1924.
57. 2:5 B2 F7, 8/29/1924.
58. 2:5 B2 F7, 9/1/1924.
59. F. Brown 1987: 48.
60. 2:5 B2 F7, 9/10/1924.
61. 2:6 B8 F2, 10/2/1924.
62. 2:6 B8 F2, 10/23/1924.
63. 2:6 B8 F2, 11/9/1924.
64. 2:6 B8 F2, 11/12/1924.
65. 2:5 B2 F7, 10/3/1924.
66. 2:5 B2 F7, 11/15/1924.
67. Solounias and Ring 2007.
68. 2:5 B2 F7, 11/15/1924.
69. 2:6 B8 F2, 11/9/1924.
70. 2:6 B8 F2, 11/9/1924.
71. L. Brown 1950: 261–268.
72. L. Brown 1950: 263.
73. L. Brown 1950: 267.

13. ANCIENT AMERICANS HUNTING BISON? BIRDS AS DINOSAURS?

1. 2:5 B1 F5, B. Brown, List of field projects.
2. 2:5 B2 F7, 5/23/1925.
3. 1:1 Dept. Vertebrate Paleontology, Ann. Report 1925.
4. Wolf and Mellett 1985.
5. Regal 2002: 146.
6. Osborn 1922a: 2.
7. Regal 2002: 148.
8. Osborn 1922b: 246.
9. Regal 2002: 148.
10. Osborn 1922b: 245–246.
11. Gregory and Hellman 1923a: 14.
12. Gregory and Hellman 1923b: 518.
13. 2:3 B5 F8, 6/4/1925.
14. 2:3 B5 F8, 6/14/1925.
15. 2:3 B5 F8, 6/15/1925.
16. 2:3 B5 F8, 6/23/1925.
17. 2:3 B5 F8, 7/3/1925.
18. 2:3 B5 F8, Accounting documents.
19. 2:3 B5 F8, 8/25/1925.
20. Wolf and Mellett 1985.
21. 2:3 B5 F8, 8/27/1925.
22. 2:3 B5 F8, 8/31/1925.
23. Regal 2002: 149.
24. Gregory 1927.
25. 1:1 Dept. Vertebrate Paleontology, Ann. Report 1927.
26. Wolf and Mellett 1985; Matthew and Cook 1909: 390.
27. Colbert 1992: 216.
28. 2:5 B2 F7, 9/21/1927.
29. Ostrom 1970: 12–13 (fig. 3); Darton 1904.
30. Meltzer 2006: 24–50.
31. Cook 1925.
32. Meltzer 2006: 30.
33. Meltzer 2006: 34.
34. Meltzer 2006: 36.
35. Cook 1927; Figgins 1927.
36. Brown to Hrdlička, 3/16/28, AH/NAA, in Meltzer 2006: 39.
37. Meltzer 2006: 39.
38. 1:1 Dept. Vertebrate Paleontology, Ann. Report 1927.

39. 1:1 Dept. Vertebrate Paleontology, Ann. Report 1928.
40. 1:1 Dept. Vertebrate Paleontology, Ann. Report 1928. On the dating of the site, see Meltzer 2006: 151.
41. 1:1 Dept. Vertebrate Paleontology, Ann. Report 1928.
42. 1:1 Admin./Ann. Report 1929; 2:4 B6 F5, Walter Herring Endowment Expedition, 1929.
43. 1:1 Dept. Vertebrate Paleontology, Ann. Report 1929.
44. 2:4 B6 F5, 7/9/1929.
45. Meltzer 2006.
46. 2:4 B6 F5, 7/19/1929.
47. 2:4 B6 F8, Mother "Neo's" Bequest, 10/13/1930.
48. 2:4 B6 F8, Mother "Neo's" Bequest, 10/13/1930.
49. 2:4 B6 F8, Colgate-Arizona Expedition Report, 1930.
50. 2:4 B6 F8, Colgate-Arizona Expedition Report, 1930.
51. 1:1 Dept. Vertebrate Paleontology, Ann. Report 1930. *Protosuchus* is now known to have been collected from the Early Jurassic Moenave or Kayenta Formations.
52. 2:4 B6 F10, 1931 Reports.
53. 2:4 B6 F10, 1931 Reports; 2:4 B6 F10, 6/9/1931.
54. 2:4 B6 F10, 7/6/1931.
55. 2:4 B6 F10, 1931 Reports.
56. 2:4 B6 F10, 7/6/31.
57. 2:4 B6 F10, 7/31/1931.
58. 2:4 B6 F10, 1931 Report, Expedition to Montana, New Mexico, and Arizona, 1931.
59. Ostrom 1970: 11–50; Moberly 1960.
60. Peck and Craig 1962; Ostrom 1970: 142–144.
61. Maxwell 1997; Chen and Lubin 1997; Burton et al. 2006.
62. 2:4 B6 F10, 7/31/1931.
63. 2:4 B6 F10, 8/4/1931.
64. 2:4 B6 F10, 1931 Report, Expedition to Montana, New Mexico, and Arizona, 1931.
65. 2:4 B6 F10, 1931 Report, Expedition to Montana, New Mexico, and Arizona, 1931.
66. Ostrom 1969.
67. Huxley 1870.
68. J. S. McIntosh, personal communication, 11/8/2006.
69. Conversations between Norell and the late Bobb Schaeffer, who received his Ph.D. at the musem (through Columbia) in the 1930s.
70. Ostrom 1969, 1970.

14. DIGGING—AND FLYING—FOR DINOSAURS

1. 2:4 B6 F12, 8/2/1932.
2. 2:5 B2 F7, 1932 memorandum.
3. 2:4 B6 F12, Financial Statement for 1932 Expedition.
4. 2:4 B6 F12, 8/2/1932.
5. 2:4 B6 F12, 8/6/1932.
6. 2:4 B6 F12, 8/7/1932.
7. 2:4 B6 F12, 8/8/1932 and 8/11/1932.
8. 2:4 B6 F12, 9/10/1932.
9. Regal 2002: 179.
10. AMNH Central Archives, Osborn, B3 F26.
11. Regal 2002: 180.
12. 2:4 B6 F10, 9/28/1932.
13. 2:4 B6 F10, 9/28/1932.
14. 2:4 B6 F10, 9/28/1932.
15. 2:4 B6 F10, 10/27/1932.
16. 2:4 B6 F10, 10/27/1932.
17. Brown 1933.
18. 2:4 B6 F10, 9/28/1932.
19. 2:4 B7 F1, Financial Statement: Expedition to Lower Cretaceous of Montana, 1933.
20. 2:4 B7 F1, 8/7/1933.
21. "Hidden Motors Give Life to Prehistoric Monsters," *Popular Science,* June 1933; available at http://blog.modernmechanix.com/category/robots/page/2.
22. 2:4 B7 F1, 8/17/1933.
23. 2:4 B7 F1, 8/17/1933.
24. 2:4 B6 F12, Howe vs. Brown Case No. 4458; 2:4 B7 F1, Report of Dinosaur Expedition 1933.
25. 2:4 B7 F1, 9/5/1933.
26. 2:4 B7 F1, 9/18/1933.
27. 2:4 B7 F1, 9/18/1933.
28. 1:1 Dept. Vertebrate Paleontology, Ann. Report 1933.
29. Makovicky and Sues 1998.
30. 2:4 B7 F1, Report of Dinosaur Expedition 1933.
31. F. Brown 1987: 59.
32. 2:4 B7 F3, Financial Statement 1934 American Museum-Sinclair Expedition, 7/31/1936.
33. 2:4 B7 F3, 5/26/1934.
34. 2:4 B7 F3, 6/4/1934.
35. 2:4 B7 F3, 6/11/1934.
36. 2:4 B7 F3, 5/31/1934.

37. 2:4 B7 F3, 6/4/1934.
38. 2:4 B7 F3, 6/7/1934.
39. Bird 1985: 17.
40. Bird 1985: 20.
41. Bird 1985: 34.
42. Bird 1985: 35.
43. Bird 1985: 54; E. S. Gaffney, personal communication.
44. Bird 1985: 41–45.
45. Bird 1985: 47.
46. 2:4 B7 F3, 6/10/1934.
47. 2:4 B7 F3, 6/10/1934.
48. F. Brown 1987: 59.
49. 2:4 B7 F3, 6/15/1934.
50. 2:4 B7 F3, 6/15/1934.
51. 2:4 B7 F3, 6/20/1934.
52. 2:4 B7 F3, 7/3/1934 and 7/12/1934.
53. 2:4 B7 F3, 7/26/1934.
54. Bird 1985: 62.
55. L. Brown 1936: 139.
56. L. Brown 1936: 141.
57. Brown 1935a: 4.
58. 2:4 B6 F12, Howe vs. Brown—Case No. 4458.
59. Brown 1935b: 95.
60. Brown 1935b: 95.
61. Brown 1935b: 96.
62. Brown 1935b: 96.
63. 2:4 B7F3, Air Log.
64. Brown 1935b: 96.
65. Brown 1935b: 97.
66. Brown 1935b: 98.
67. Brown 1935b: 98.
68. Brown associates this feature, which he describes as "a long, narrow streak, beginning nowhere and ending nowhere," with the Anasazi because it borders "Chaco Canyon, near Pueblo Bonito . . . and was evidently a ceremonial way, or race course" (Brown 1935b: 115).
69. 2:4 B7 F3, Report 1/25/35 and Air Log.
70. 2:4 B7 F3, 9/20/1934.
71. 2:4 B7 F3, 10/6/1934.
72. L. Brown 1936: 142.
73. 2:4 B7 F3, 10/11/1934.
74. Bird 1985: 54.

75. 2:4 B7 F3, 11/12/1934.
76. 2:4 B7 F3, Report 1/25/35.
77. 2:4 B7 F3, 11/20/1934.
78. 2:4 B7 F3, 12/17/1934.
79. 2:4 B6 F12, Howe vs. Brown—Case No. 4458.
80. 2:4 B7 F3, 7/31/1936.
81. Memo to Richard Tedford, chairman of Vertebrate Paleontology, 1992.

15. TOWARD THE GOLDEN YEARS

1. Regal 2002: 187–188.
2. 2:4 B7 F4, Report 10/29–11/18/1935.
3. 1:1 Dept. Vertebrate Paleontology, Ann. Report 1936.
4. Bird 1985: 95.
5. 1:1 Dept. Vertebrate Paleontology, Ann. Report 1937.
6. 2:4 B7 F7, 12/17/1936.
7. 1:1 Dept. Vertebrate Paleontology, Ann. Report 1937, "Expedition to Hell Creek and Judith River 6–10, 1937."
8. 2:4 B7 F7, 5/22/1937.
9. 2:4 B7 F7, 6/2/1937; 2:4 B7 F7, 6/3/1937; 2:4 B7 F7, Permit, 7/3/1937.
10. 2:4 B7 F7, 7/10/1937, Brown to Adamson.
11. 2:4 B7 F7, 7/10/1937, Brown to Granger.
12. 2:4 B7 F7, "The Mystery Dinosaur: An Account of the American Museum-Sinclair Expedition of 1937."
13. 2:4 B7 F7, 7/16/1937.
14. 2:4 B7 F7, 7/23/1937.
15. 2:4 B7 F7, 8/7/1937.
16. For another account, which is both entertaining and descriptive, see Bird 1985: 96–109.
17. 2:4 B7 F7, "The Mystery Dinosaur: An Account of the American Museum-Sinclair Expedition of 1937."
18. 2:4 B7 F7, "The Mystery Dinosaur: An Account of the American Museum-Sinclair Expedition of 1937."
19. 2:4 B7 F7, "The Mystery Dinosaur: An Account of the American Museum-Sinclair Expedition of 1937."
20. 2:4 B7 F7, "The Mystery Dinosaur: An Account of the American Museum-Sinclair Expedition of 1937."
21. 2:4 B7 F7, 11/1/1937.
22. 2:4 B7 F7, 8/2/1937.
23. 2:4 B7 F7, 8/18/1937 and 8/23/1937.
24. 2:4 B7 F7, 8/25/1937.

25. 2:4 B7 F7, 9/4/1937.
26. 2:4 B7 F7, 8/23/1937.
27. 2:4 B7 F7, 8/23/1937.
28. 2:4 B7 F7, 9/4/1937.
29. 2:4 B7 F7, 9/7/1937.
30. 2:4 B7 F7, 9/10/1937.
31. 2:4 B7 F7, 9/23/1937 and 10/21/1937.
32. 2:4 B7 F7, 10/2/1937.
33. 2:4 B7 F7, Union Pacific freighting bill.
34. 2:4 B7 F7, press release and *New Horizons* summary, 11/8/37.
35. 2:4 B7 F7, Expenditures American Museum-Sinclair 1937 Expedition.
36. 2:4 B7 F7, 4/26/1938.
37. 2:4 B7 F9, 4/5/1938.
38. 2:4 B7 F9, 4/22/1938.
39. 2:4 B7 F9, 8/9/1938 and 8/10/1938.
40. 2:4 B7 F9, 8/21/1938.
41. 2:4 B7 F9, 8/4/1938 and 8/26/1938.
42. 2:4 B7 F9, 9/9/1938.
43. 2:4 B7 F9, 9/12/1938.
44. 2:4 B7 F9, 9/16/1938.
45. 2:4 B7 F9, 9/19/1938.
46. 2:4 B7 F9, 9/21/1938.
47. 2:4 B7 F9, 10/2/1938.
48. 2:4 B7 F9, 10/29/1938.
49. 2:4 B7 F9, 11/20/1938.
50. 2:4 B7 F9, Thanksgiving, 1938; the story is also recounted in Bird 1985: 146–189.
51. 2:4 B7 F9, 11/30/1938.
52. 2:4 B7 F9, 11/29/1938.
53. Morris 1980.
54. 2:4 B7 F9, Dept. Vertebrate Paleontology, Ann. Report 1938.
55. 1:1 Dept. Vertebrate Paleontology, Ann. Report 1939.
56. 1:1 Dept. Vertebrate Paleontology, Ann. Report 1940.
57. 1:1 Dept. Vertebrate Paleontology, Ann. Report 1940, "Report for 1940 by Roland T. Bird."
58. Lockley 1991.
59. 1:1 Dept. Vertebrate Paleontology, Ann. Report 1940.
60. 1:1 Dept. Vertebrate Paleontology, Ann. Report 1940.
61. 1:1 Dept. Vertebrate Paleontology, Ann. Report 1940.
62. 2:5 B1 F6, 3/1/1940.
63. Stucker 1951: 467–468.

64. *Los Angeles Times*, 2/17/1940.
65. 1:1 Dept. Vertebrate Paleontology, Ann. Report 1941.
66. 1:1 Dept. Vertebrate Paleontology, Ann. Report 1941, "Field Activities of Barnum Brown."
67. 1:1 Dept. Vertebrate Paleontology, Ann. Report 1941, "Field Activities of Barnum Brown."
68. 1:1 Dept. Vertebrate Paleontology, Ann. Report 1941, "Field Activities of Barnum Brown."
69. 1:1 Dept. Vertebrate Paleontology, Ann. Report 1941, "Field Activities of Barnum Brown."
70. 1:1 Dept. Vertebrate Paleontology, Ann. Report 1941.
71. 2:4 B3 F3, 12/18/1941.
72. 2:4 B3 F1, 1/12/1942.
73. 2:4 B3 F1, 1/12/1942.
74. 2:4 B8 F3, 1/19/1942.
75. 1:1 Dept. Vertebrate Paleontology, Ann. Report 1942.
76. 1:1 Dept. Vertebrate Paleontology, Ann. Report 1942.

16. BROWN AS A SPY, MOVIE CONSULTANT, AND SHOWMAN AT THE WORLD'S FAIR

1. 2/5 B1 F1, 4/28/1941.
2. 1:1 Dept. Vertebrate Paleontology, Ann. Report 1942; F. Brown 1987: 61.
3. F. Brown 1987: 61.
4. F. Brown 1987: 61.
5. F. Brown 1987: 62.
6. F. Brown 1987: 62.
7. F. Brown 1987: 60; 2:5 B1 F1, Dossier of B. Brown.
8. F. Brown 1987: 60.
9. 2:5 B1 F1, Dossier of B. Brown.
10. F. Brown 1987: 62.
11. F. Brown 1987: 64.
12. 2:5 B1 F1, Dossier of B. Brown.
13. 2:5 B1 F1, Dossier of B. Brown.
14. L. Brown 1956: 56–65.
15. F. Brown 1987: 66.
16. F. Brown 1987: 74.
17. F. Brown 1987: 77.
18. 2:5 B1 F1, B. Brown Fieldwork summary; F. Brown 1987: 77.
19. 2:6 B8 F6, 10/25/1955.

20. 2:6 B8 F6, Shea telegram, 10/29/1955.
21. F. Brown 1987: 77.
22. F. Brown 1987: 78.
23. Brown 1941: 100–101.
24. www.imdb.com/title/tt0032455/awards; www.youtube.com/results?search_query=fantasia+dinosaurs&search_type=&aq=0&oq=fantasia+dinos. =
25. F. Brown 1987: 79.
26. 2:5 B1 F1, 1/31/1963.
27. F. Brown 1987: 79.
28. 2:5 B1 F1, 4, 5, B. Brown Field Activities, 1894–1960.
29. 2/6 B8 F7, 2/27/1967.
30. 2/6 B8 F7, 2/6/1964.

EPILOGUE

1. F. Brown 1987: 6.
2. 2:1 B1 F5, Autobiographical notes, "My Most Unforgettable Character."
3. Lewis 1964.
4. Brown and Schlaikjer 1940.
5. 1:1 Dept. Vertebrate Paleontology, Ann. Report 1942.
6. L. Brown 1950: x–xi.

BIBLIOGRAPHY

Alvarez, L., et al. 1980. Extraterrestrial cause for the Cretaceous-Tertiary extinction. *Science* 208: 1095–1108.

Barbour, E. H. 1892. Notes on a new order of gigantic fossils. *University Studies* 1 (4): 301–313.

———. 1895. Is *Daemonelix* a burrow? *American Naturalist* 29: 517–527.

Barry, J. C. 1995. Faunal turnover and diversity in the terrestrial Neogene of Pakistan. In E. S. Vrba, G. H. Denton, T. C. Partridge, and L. H. Burckle, eds., *Paleoclimate and Evolution with Emphasis on Human Origins*, 115–134. New Haven: Yale University Press.

Barry, J. C., et al. 2002. Faunal and environmental change in the Late Miocene Siwaliks of northern Pakistan. *Paleobiology* 28, mem. 3: 1–71.

Barton, D. R. 1941. Father of the dinosaurs. *Natural History* 47: 308–312.

Behrensmeyer, A. K., and J. C. Barry. 2005. Biostratigraphic surveys in the Siwaliks of Pakistan: A method for standardized surface sampling of the vertebrate fossil record. *Palaeontologia Electronica* 8 (1); at http://palaeo-electronica.org/paleo/2005_1/behrens15/issue1_05.htm.

Berger, B. 2004. "*Tyrannosaurus rex*." Published on the AMNH Vertebrate Paleontology website, now decommissioned with copy filed under author's name in Osborn Library, AMNH.

Bernor, R. L., N. Solounias, C. C. Swisher III, and J. Van Couvering. 1996. The correlation of three classical "Pikermian" faunas—Maragha, Samos and Pikermi—with the European MN unit system. In R. L. Bernor, R. L. V. Fahlbusch, and H. W. Mittman, eds., *The Evolution of Western Eurasian Neogene Mammal Faunas*, 137–156. New York: Columbia University Press.

Bird, R. T. 1985. *Bones for Barnum Brown: Adventures of a Dinosaur Hunter*. Fort Worth: Texas Christian University Press.

Brown, Barnum. 1903. A new species of fossil edentate from the Santa Cruz Formation of Patagonia. *Bulletin of the American Museum of Natural History* 19: 453–457.

———. 1907. The Hell Creek beds of the Upper Cretaceous of Montana. *Bulletin of the American Museum of Natural History* 23: 823–845.
———. 1910. The Cretaceous Ojo Alamo beds of New Mexico, with description of the new dinosaur genus *Kritosaurus*. *Bulletin of the American Museum of Natural History* 28: 267–274.
———. 1913. Some Cuban fossils. *American Museum Journal* 13: 221–228.
———. 1915. *Tyrannosaurus*, a Cretaceous carnivorous dinosaur. *Scientific American* 113: 322–323.
———. 1919. Hunting big game of other days: A boating expedition in search of fossils in Alberta, Canada. *National Geographic* 35: 407–429.
———. 1925a. Byways and highways in Burma. *Natural History* 25: 295–308.
———. 1925b. Through the land of Sheba. *Natural History* 25: 602–617.
———. 1927. Samos, romantic isle of the Aegean. *Natural History* 27: 19–32.
———. 1933. Stratigraphy and fauna of the Fuson-Cloverly Formation in Montana, Wyoming, and South Dakota. *Bulletin of the Geological Society of America* 44: 74.
———. 1935a. Sinclair dinosaur expedition, 1934. *Natural History* 36: 3–15.
———. 1935b. Flying for dinosaurs. *Natural History* 36: 95–116.
———. 1941. The methods of Walt Disney Productions. *Transactions of the New York Academy of Sciences,* ser. 2, 3 (4): 100–105.
Brown, B., and E. M. Schlaikjer. 1940. The structure and relationships of *Protoceratops*. *Annals of the New York Academy of Sciences* 40: 133–266.
Brown, C. M., A. P. Russell, and M. J. Ryan. 2009. Pattern and transition of surficial bone texture of the centrosaurine frill and their ontogenetic and taxonomic implications. *Journal of Vertebrate Paleontology* 29: 132–141.
Brown, F. R. 1987. *Let's Call Him Barnum.* New York: Vantage Press.
Brown, L. 1936. On safari in America. *Natural History* 37: 139–148.
———. 1950. *I Married a Dinosaur.* New York: Dodd, Mead.
———. 1951. *Cleopatra Slept Here.* New York: Dodd, Mead.
———. 1956. *Bring 'Em Back Petrified.* New York: Dodd, Mead.
Burton, D., et al. 2006. New radiometric ages from the Cedar Mountain Formation, Utah, and the Cloverly Formation, Wyoming: Implications for contained dinosaur faunas. *Geological Society of America Abstracts with Programs* 38 (7): 52.
Carter, P. D. 2003. Petrotyranny—the present day. *Island Tides* 15 (5).
Chen, Z.-Q, and S. Lubin. 1997. A fission track study of the terrigenous sedimentary sequences of the Morrison and Cloverly Formations in northeastern Bighorn Basin, Wyoming. *Mountain Geologist* 34: 51–62.
Chiappe, L. M., and L. Dingus. 2001. *Walking on Eggs.* New York: Scribner.
Ciochon, R. L. 1985. Fossil ancestors of Burma. *Natural History* 94: 26–36.
Ciochon, R. L., and G. F. Gunnell. 2002. Eocene primates from Myanmar: His-

torical perspectives on the origin of Anthropoidea. *Evolutionary Anthropology* 11: 156–168.

Colbert, E. H. 1933. The presence of tubulidentates in the Middle Siwalik beds of northern India. *American Museum Novitates*, no. 604: 1–10.

———. 1935. Siwalik mammals in the American Museum of Natural History. *Transactions of the American Philosophical Society*, n.s., 26: i–401.

———. 1937. A new primate from the Upper Eocene Pondaung Formation of Burma. *American Museum Novitates*, no. 951: 1–18.

———. 1984. *The Great Dinosaur Hunters and Their Discoveries.* New York: Dover.

———. 1992. *William Diller Matthew: Paleontologist. The Splendid Drama Observed.* New York: Columbia University Press.

Cook, H. J. 1925. Definite evidence of human artifacts in the American Pleistocene. *Science* 62: 459–460.

———. 1927. New geological and paleontological evidence bearing on the antiquity of mankind in America. *Naural History* 27: 240–247.

Currie, P. J. 2005. History of research. In P. J. Currie and E. B. Koppelhus, eds., *Dinosaur Provincial Park: A Spectacular Ancient Ecosystem Revealed*, 3–33. Bloomington: Indiana University Press.

Darton, N. H. 1904. Comparison of the stratigraphy of the Black Hills, Bighorn Mountains, and Rocky Mountain Front Range. *Bulletin of the Geological Society of America* 15: 379–448.

Darwin, C. R. 1871. *The Descent of Man, and Selection in Relation to Sex.* 2 vols., 1st ed. London: John Murray.

———. 1909 [1839]. *The Voyage of the Beagle.* The Harvard Classics, vol. 29. New York: Collier.

Dingus, L., and M. Norell. 2008. *The Dinosaur Hunters.* London: Carlton.

Dingus, L., and T. Rowe. 1997. *The Mistaken Extinction: Dinosaur Evolution and the Origin of Birds.* New York: W. H. Freeman.

Eberth, D. A. 1997a. Edmonton Group. In P. J. Currie and K. Padian, eds., *Encyclopedia of Dinosaurs*, 199–204. San Diego: Academic Press.

———. 1997b. Judith River wedge. In P. J. Currie and K. Padian, eds., *Encyclopedia of Dinosaurs*, 379–385. San Diego: Academic Press.

———. 2005. The geology. In P. J. Currie and E. B. Koppelhus, eds., *Dinosaur Provincial Park: A Spectacular Ancient Ecosystem Revealed*, 54–82. Bloomington: Indiana University Press.

Eldredge, N., and S. J. Gould. 1972. Punctuated equilibria: An alternative to phyletic gradualism. In T. J. M. Schopf, ed., *Models in Paleobiology*, 82–115. San Francisco: Freeman Cooper.

Figgins, J. D. 1927. The antiquity of man in America. *Natural History* 27: 229–239.

Gregory, W. K. 1927. *Hesperopithecus* apparently not an ape nor a man. *Science*, n.s., 66: 579–581.

Gregory, W. K., and M. Hellman. 1923a. Notes on the type of *Hesperopithecus haroldcookii* Osborn. *American Museum Novitates*, no. 53: 1–16.

———. 1923b. Further notes on the molars of *Hesperopithecus* and of *Pithecanthropus*. *Bulletin of the American Museum of Natural History* 48: 509–530.

Hartwig, W. C., ed. 2002. *The Primate Fossil Record*. Cambridge Studies in Biological and Evolutionary Anthropology, no. 33. Cambridge: Cambridge University Press.

Hatcher, J. B. 1903. *Reports of the Princeton University Expeditions to Patagonia, 1896–1899*. Vol. 1: *Narrative of the Expeditions. Geography of Southern Patagonia*. Princeton: Princeton University Press.

Hornaday, W. T. 1925. *A Wild Animal Round-Up*. New York: Scribner.

Huxley, T. H. 1870. Further evidence of the affinity between the dinosaurian reptiles and birds. *Quarterly Journal of the Geological Society of London* 26: 12–31.

Jaffe, M. 2000. *The Gilded Dinosaur: The Fossil War between E. D. Cope and O. C. Marsh and the Rise of American Science*. New York: Crown.

Jordan, A. J. 2003. *Jordan*. Missoula, Mont.: Mountain Press.

Josephson, E. M. 1952. *Rockefeller "Internationalist": The Man Who Misrules the World*. New York: Chedney Press.

Kennedy, K. A. R., and R. L. Ciochon. 1999. A canine tooth from the Siwáliks: First recorded discovery of a fossil ape? *Journal of Human Evolution* 14: 231–253.

Kohl, M. F., L. D. Martin, and P. Brinkman. 2004. *A Triceratops Hunt in Pioneer Wyoming*. Glendo, Wyo.: High Plains Press.

Kostopoulos, D. S., S. Sen, and G. D. Koufus. 2003. Magnetostratigraphy and revised chronology of the late Miocene mammal localities of Samos, Greece. *International Journal of Earth Sciences* 92: 779–794.

Lambe, L. M. 1902. On Vertebrata of the Mid-Cretaceous of the North West Territory, part 2: New genera and species from the Belly River Series (Mid-Cretaceous). *Geological Survey of Canada, Contributions to Paleontology* 3: 23–81.

Lepley, J. G., and S. Lepley. 1992. *The Vanishing West: Hornaday's Buffalo, the Last of the Wild Herds*. Fort Benton, Mont.: River and Plains Society.

Lewin, R. 1987. *Bones of Contention: Controversies in the Search for Human Origins*. New York: Touchstone.

Lewis, G. E. 1964. Memorial to Barnum Brown (1873–1963). *Bulletin of the Geological Society of America* 75: 19–25.

Lindsay, E. H, N. M. Johnson, and N. D. Opdyke. 1980. Correlation of Siwalik faunas. In L. L. Jacobs, ed., *Aspects of Vertebrate History: Essays in Honor of Edwin H. Colbert*, 309–319. Flagstaff: Museum of Northern Arizona Press.

Lockley, M. 1991. *Tracking Dinosaurs*. Cambridge: Cambridge University Press.

Makovicky, P., and H.-D. Sues. 1998. Anatomy and phylogenetic relationships of the theropod dinosaur *Microvenator celer* from the Lower Cretaceous of Montana. *American Museum Novitates*, no. 3240: 1–27.

Matthew, W. D. 1915. Climate and evolution. *Annals of the New York Academy of Sciences* 24: 171–318.

———. 1929. Critical observations upon Siwalik mammals (exclusive of Proboscidea). *Bulletin of the American Museum of Natural History* 56: 437–560.

Matthew, W. D., and Harold J. Cook. 1909. A Pliocene fauna from western Nebraska. *Bulletin of the American Museum of Natural History* 26: 361–414.

Maxwell, W. D. 1997. Cloverly Formation. In P. J. Currie and K. Padian, eds., *Encyclopedia of Dinosaurs,* 128. San Diego: Academic Press.

Meltzer, J. J. 2006. *Folsom.* Berkeley: University of California Press.

Moberly, R., Jr. 1960. Morrison, Cloverly, and Sykes Mountain formations, northern Bighorn Basin, Wyoming and Montana. *Bulletin of the Geological Society of America* 71: 1137–1176.

Morris, J. 1980. *Tracking Those Incredible Dinosaurs.* San Diego: CLP Publishers.

Norell, M. A., E. S. Gaffney, and L. Dingus. 1995. *Discovering Dinosaurs in the American Museum of Natural History.* New York: Knopf.

Osborn, H. F. 1902. On Vertebrata of the Mid-Cretaceous of the North West Territory, part 1: Distinctive characters of the Mid-Cretaceous fauna. *Geological Survey of Canada, Contributions to Paleontology* 3: 5–21.

———. 1905. *Tyrannosaurus* and other Cretaceous carnivorous dinosaurs. *Bulletin of the American Museum of Natural History* 21: 475–479.

———. 1906. *Tyrannosaurus,* Upper Cretaceous carnivorous dinosaur (second communication). *Bulletin of the American Museum of Natural History* 22: 281–296.

———. 1912. Crania of *Tyrannosaurus* and *Allosaurus. Memoirs, American Museum of Natural History,* n.s., 1: 1–30.

———. 1913. *Tyrannosaurus* restoration and model of skeleton. *Bulletin of the American Museum of Natural History* 32: 91–92.

———. 1916. Skeletal adaptations of *Ornitholestes, Struthiomimus, Tyrannosaurus. Bulletin of the American Museum of Natural History* 35: 733–771.

———. 1922a. *Hesperopithecus,* the first anthropoid primate found in America. *American Museum Novitates,* no. 37: 1–5.

———. 1922b. *Hesperopithecus,* the first anthropoid primate found in America. *Proceedings of the National Academy of Sciences* 8: 245–246.

Ostrom, J. H. 1969. Osteology of *Deinonychus antirrhopus,* an unusual theropod from the Lower Cretaceous of Montana. *Bulletin of the Peabody Museum of Natural History,* no. 30: 1–165.

———. 1970. Stratigraphy and paleontology of the Cloverly Formation (Lower Cretaceous) of the Bighorn Basin area, Wyoming and Montana. *Bulletin of the Peabody Museum of Natural History,* no. 35: 1–234.

Peck, R. E., and W. W. Craig. 1962. Lower Cretaceous nonmarine ostracods and charophytes of Wyoming and adjacent areas. Guidebook, Wyoming Geological Association, Seventeenth Annual Field Conference, 33–43.

Pilbeam, D., M. Morgan, J. C. Barry, and L. Flynn. 1996. European MN units and the Siwalik faunal sequence in Pakistan. In R. L. Bernor, V. Fahlbusch, and H.-W. Mittman, eds., *The Evolution of Western Eurasian Neogene Mammal Faunas*, 96–105. New York: Columbia University Press.

Pilgrim, G. E. 1927. A *Sivapithecus* palate and other primate fossils from India. *Memoirs of the Geological Survey of India (Palaeontologica Indica)* 14: 1–26.

———. 1937. Siwalik antelopes and oxen in the American Museum of Natural History. *Bulletin of the American Museum of Natural History* 72: 729–874.

Preston, D. J. 1984. Barnum Brown's bones. *Natural History* 10: 101–105.

Rainger, R. 1991. *An Agenda for Antiquity: Henry Fairfield Osborn and Vertebrate Paleontology at the American Museum of Natural History, 1890–1935*. Tuscaloosa: University of Alabama Press.

Rea, T. 2001. *Bone Wars: The Excavation and Celebrity of Andrew Carnegie's Dinosaur*. Pittsburgh: University of Pittsburgh Press.

Regal, B. 2002. *Henry Fairfield Osborn: Race and the Search for the Origins of Man*. Burlington, Vt.: Ashgate.

Sampson, S. D., M. J. Ryan, and D. H. Tanke. 1997. Craniofacial ontogeny in centrosaurine dinosaurs (Ornithischia: Ceratopsidae): Taxonomic and behavioral implications. *Zoological Journal of the Linnean Society* 121: 293–337.

Schuchert, C., and C. M. LeVene. 1940. *O. C. Marsh: Pioneer in Paleontology*. New Haven: Yale University Press.

Snell, M. E., and R. Metzler. 1972. *Carbondale, Kansas, Centennial*. Carbondale: Centennial Association, Inc.

Solounias, N., and A. Mayor. 2004. Ancient references to the fossils from the land of Pythagoras. *Earth Sciences History* 23: 283–296.

Solounias, N., and U. Ring. 2007. Samos Island, part II: Ancient history of the Samos fossils and the record of earthquakes. In G. Lister, M. Forster, and U. Ring, eds., *Inside the Aegean Metamorphic Core Complexes, Journal of the Virtual Explorer*, electronic ed., vol. 27, paper 6.

Solounias, N., M. Plavcan, J. Quade, and L. Witmer. 1999. The Pikermian Biome and the savanna myth. In J. Agusti, P. Andrews, and L. Rook, eds., *Evolution of the Neogene Terrestrial Ecosystems in Europe*, 427–444. Cambridge: Cambridge University Press.

Spalding, D. A. E. 2001. Bones of contention: Charles H. Sternberg's lost dinosaurs. In D. H. Tanke and K. Carpenter, eds., *Mesozoic Vertebrate Life*, 481–503. Bloomington: Indiana University Press.

Stucker, G. 1951. Harvester of the past. *Nature*, November, 467–470.

Tanke, D. H. 2005. Identifying lost quarries. In P.J. Currie and E. B. Koppelhus, eds., *Dinosaur Provincial Park: A Spectacular Ancient Ecosystem Revealed*, 34–53. Bloomington: Indiana University Press.

Tassy, P. 1983. Les Elephantoidea miocènes du plateau du Portwar, groupe de Siwailik, Pakistan [The Miocene Elephantoidea of the Potwar Plateau, Siwalik Group, Pakistan]. *Annales de Paléontologie* 69 (2): 99–136; 69 (3): 235–297; 69 (4): 317–354.

Wolf, J., and S. Mellett. 1985. The role of "Nebraska man" in the creation-evolution debate. *Creation/Evolution* (National Center for Science Education) 16: 31–43.

ACKNOWLEDGMENTS

The development and production of this book has encompassed nearly a decade, and throughout that period, a great many individuals have provided assistance and encouragement. Without suggesting that they necessarily agree with all the results presented herein, we wish to thank them all unabashedly.

From the start, we have benefited from the sage advice and persistent efforts of our partner, Samuel Fleishman of Literary Artists Representatives. Equally key to the realization of this biography has been our patient and probing editor, Blake Edgar, along with his colleagues Jacqueline Volin and Lisa Tauber, at University of California Press, as well as Anne Canright.

A host of other individuals have provided information and assistance at key points in the project. Of the former, we are especially indebted to Scott Williams, who selflessly drew our attention to legal documents and images he gathered regarding the legal case involving Brown's father and the coal mining operations around Brown's hometown of Carbondale, Kansas. In addition, Jack McIntosh generously provided an account of an encounter with Brown and his specimen of "Daptosaurus."

While composing this account, the scope and accuracy of the manuscript was greatly enhanced through the suggestions of many reviewers. The cogent critiques of Gene Gaffney, Kevin Padian, and Tim Rowe are reflected throughout, as are the thoughts of two anonymous reviewers. The efforts of Larry Flynn and John Barry were key in improving the quality of the chapter on Brown's expedition to the Indian subcontinent. Darren Tanke graciously offered his knowledge and expertise to refine the chapter and update the taxonomic names for specimens collected during Brown's expeditions to the Red Deer River. Similarly, we acknowledge the efforts of Nikos Solounias to help us more accurately portray the context and results of Brown's enterprise on the Greek island of Samos. Finally, numerous colleagues aided in reviewing other portions of the work or hosting us in the field, including Dick Tedford, John Flynn, Bill Clemens, Don Brinkman, Joseph Hatcher, and Brian Regal.

The core of this story is preserved in documents housed in the archives of the American Museum of Natural History. Within the Division of Paleontology, Chris Norris, Susan Bell, Ivy Rutzky, and Ruth O'Leary provided critical support for accessing the essential field notes, correspondence, and imagery contained in the Vertebrate Paleontology Archives, and scanning of relevant photos was expertly handled by Frank Ippolito. Equally critical to our efforts was Barbara Mathe, who freely lent her vast experience in our efforts to glean information and imagery from the archives and photo collections of the American Museum of Natural History Library, along with Kelli Anderson and Gregory August Raml, who helped find and scan the images. The help of Helen Bergamo, Julie Kabelac, and Bridget Burke at Wells College proved invaluable in our efforts to find photos of Marion and Frances Brown. David Meltzer was kind enough to lend his support in securing a photo of Brown at Folsom. Matt Lamanna and Bernadette Callery of the Carnegie Museum of Natural History aided ably in the search for Hatcher's images from his Patagonian expeditions. Toni Bressler and Letha Johnson helped to secure permission for Brown images from the University of Kansas.

Throughout, essential assistance with imagery and fieldwork, along with ceaseless torrents of lit-up levity, radiated from a cohort of clownish colleagues, including Carl Mehling, Mick Ellison, and Dina Langis. In addition, Karin Fittante aided greatly in research involving Brown's prickly legal entanglements.

Our trip to Carbondale to seek out Brown's roots was facilitated by Wyona Hiatt and Marla Gleason, who also graciously offered their knowledge of Kansas history. Fred Lanfear, Charlotte Stafford, and Vicky House of the Oxford Historical Society also filled us in on local events and characters, and they were kind enough to show us around the town and guide us to the Brown family plot.

Finally, a great debt of gratitude is owed to Elizabeth Chapman, Vivian Pan, and Inga Norell, who patiently tolerated the unrelenting intrusions of Brown's immortal influence as we struggled to come to terms with his legendary deeds and perplexing personality.

INDEX

Abbott, Charles, 233
Abel, Othenio, 229
Abyssinia, Brown's activity in, 159–63, 164*map*, 165*fig.*, 166–71, 236
Acker, Karl, 214–15
Adamson, Hans, 256, 266, 268, 274, 275
aerial reconaissance, 248, 251, 257–60, 258*fig.*, 264, 272, 281, 285, 287, 288, 301
Africa, human origins in, 169, 176
Aguja Formation, Texas, 277, 278
Alberta, Canada, Brown's activity in, 125, 128–36, 130*map*, 136–52, 138*fig.*, 144*fig.*, 149*fig.*, 273, 275–76, 280, 287
Albertosaurus, 134, 135, 136, 138, 145, 146, 148, 149, 272
Albian Age, 242
alligator, fossilized, 136, 155, 192, 206
Allosaurus, 177, 237, 240, 291
Alvarez hypothesis, 114
Ameghino, Florentino, 62
American Anthropological Association, 234–35
American Association for the Advancement of Science, 23, 232
American Museum of Natural History (AMNH), 31, 39–40; Andrews as head of, 271, 285; anthropology department at, 237; Central Asiatic Expeditions (CAE) sponsored by, 43, 177, 189, 193, 194, 207, 213, 215, 218, 244, 246, 271, 280, 300; Cope's fossil collection at, 24, 60, 83, 123; geology department at, 227, 229, 232; Granger's work at, 43–44, 133, 177, 241, 280, 282; Gregory's work at, 212–13, 220; horse project at, 60–61, 83; Howe Quarry collection at, 262–63; Matthew's work at, 54–55, 59–62, 121, 122, 132, 194, 212, 213, 227, 229, 232; Osborn's work at, 40, 42, 47–48, 60–61, 62–63, 70, 80, 105–6, 112–13, 121–23, 129, 132–33, 177, 212, 215, 216, 227–32, 264, 283, 310–11; paleontological collections at, 24, 60, 79, 83, 122–23, 129, 136, 151, 174, 194, 205, 215, 216, 227, 243, 244–45, 262–63; paleontological exhibits at, 42, 47–48, 78, 83, 86, 93, 100, 103, 111, 113, 123–24, 136, 143, 151, 194, 223, 236, 237, 240, 243, 244, 249, 263, 265, 269, 277, 278, 292*fig.*, 293, 311; private benefactors of, 181, 238, 246, 251, 265, 266, 272, 273, 276, 280; Simpson's work at, 232; Thomson's work at, 229; vertebrate paleontology department at, 40, 42, 59, 80, 132–33, 227, 236–37, 265, 280; Walter Herring Endowment Expedition co-sponsored by, 237; Wortman's work at, 38, 42, 63, 70, 80, 121
American Museum of Natural History, Brown's employment at: and departmental funding, 236–37, 246, 249, 251, 282; and exhibitions, 113, 205, 225, 240, 284, 288, 292*fig.*, 293, 311; and fossil collections, 205, 225, 239–40, 243, 244–45, 253, 262–63, 284, 310–11; Granger's supervision of, 241, 242, 247, 249, 266, 268, 271–72; Matthew's supervision of (*see under* Matthew); Osborn's supervision of (*see under* Osborn); and radio broadcasts, 266, 268, 271, 272; and retirement, 283–84;

American Museum of Natural History, Brown's employment at *(continued)* and salary, 70, 71–72, 80, 136, 143, 156, 282; and scientific papers, 71, 113, 114, 122, 288; and staff expansion, 265, 272; and staff promotions, 132–33, 136, 227; and Stucker's purge of Brown's records, 33–34

American Museum of Natural History, Brown's fieldwork for: in Abyssinia, 159, 169, 236; in Alberta, Canada, 125, 128–36, 130*map*, 132*fig.*, 136–52, 138*fig.*, 144*fig.*, 149*fig.*, 273, 275–76, 280; in Arizona, 83, 103, 237, 238–39, 265; in Arkansas, 96, 104; in Burma, 198–207, 201*map*, 215, 217, 218, 224; in Colorado, 62, 83, 83*fig.*, 232, 236, 237, 268–69; in Cuba, 136, 155–57, 156*map*, 157*fig.*; in Florida, 136; in India, 180–95, 181*fig.*, 183*fig.*, 192*fig.*, 215, 217, 218, 224; in Indiana, 99; in Kansas, 264; in Mexico, 136; in Mississippi, 136; in Montana, 86–94, 92–93*figs.*, 100–102, 106–9, 108*fig.*, 111–12, 113, 114, 117–21, 120*fig.*, 122*fig.*, 124–25, 129, 131–32, 153–54, 232–33, 239–44, 247, 248–49, 250, 251, 280–81, 289, 309–11; in Nebraska, 229–31; in Nevada, 237, 238; in New Mexico, 43–45, 103, 232, 233, 234–36, 235*fig.*, 242–43; in Oklahoma, 264; in Patagonia, xii–xiv, 62–78; in Samos, Greece, 208–25, 212*map*, 214*fig.*; in South Dakota, 81–83, 94, 99–100; in Texas, 136, 264–65, 276–80; in Utah, 232, 237, 238, 265; in Wyoming, 45–46, 49–56, 54–55*figs.*, 70, 79, 95, 122, 124, 129, 232, 240, 246–47, 249, 250, 251–54, 254*fig.*, 255*fig.*, 256–57, 258*fig.*, 260–63, 265–68, 271–72

Amhiorycteropus, 194
ammonite shells, 102
Amphipithecus, 204, 207
Anasazi, 103, 239
Anatosaurus, 113
Anatotitan, 93
Anchiceratops, 151
Andrews, Roy Chapman, 43, 177, 189, 193, 194, 213, 215, 271, 285, 302
Anglo-American Oil Company, 159, 160, 169

Ankylosaurus, 114, 135, 137, 138, 141, 144, 146, 149, 150, 154, 177, 273, 278, 281
antelope, fossilized, 83, 185, 189, 193, 194, 223, 242
anthracothere, 203, 206, 207
Anthrocohyus, 203
anthropoid fossils, 175, 176, 189, 204, 207, 208, 220, 227–32
anthrothere, 207
Apatosaurus, 53, 237, 250, 276, 277
Aptian Age, 242
Archaeopteryx, 157
Argentina, 62
Argyrocetus, 69
Arizona, 83, 103, 104, 237, 238, 239, 240, 265
Arkansas, 95, 98*fig.*, 104
arrowheads. *See* projectile points
artiodactyl, 207
Astrapotherium, 73–74
Atatürk, Mustafa Kemal, 172
Athens Museum, 211, 217, 219, 221, 222
Australopitchecus afarensis, 169
autobiographical notes, Brown's, 1–5, 7–12, 16–17, 45–46, 56, 97, 154, 227, 295

Bailey, E. H. S., 30
Balfour, Francis, 42
Baluchistan, 182, 183, 191
Barbour, Erwin Hinckly, 26, 27–28
Barbour, Thomas, 155
Barnum, P. T., 5–6
Barosaurus, 239–40, 261, 263
Battle Formation, Alberta, Canada, 129
Baylor University, 277
Beagle, voyage of the, 62, 65
Bearpaw Formation: in Alberta, Canada, 129; in Montana, 102, 125, 154
Beauvais Creek, Montana, 95, 233, 241, 242, 245
beaver, fossilized, 104, 316n18
Belly River Group, Alberta, Canada, 128, 129, 134, 138, 143, 151, 154, 267
Belodon, 83
Big Bend, Texas, 276, 277–78
Big Dry Creek, Montana, 86, 106, 113, 116*fig.*, 117, 120*fig.*, 122*fig.*, 131, 276, 311

Bighorn Basin, Wyoming, 45–46, 46–47*figs.*, 232, 246, 250
Billings, Montana, 100, 101, 106, 248, 252, 258, 260, 281, 290
biography of Brown, by daughter Frances, 5–6, 78; and BB's Abyssinian expedition, 171; and BB's affair with German woman, 286–87, 298; and BB's bout with malaria in Burma, 204; and BB's death, 294; and BB's government service, 286–87; and BB's marriage to Marion Brown, 97, 98, 125, 126, 297; and BB's Patagonia expedition, 320n24, 321n61; and BB's visit to Radcliffe College, 293; and Lilian's experience in India, 195; and relations between BB and Lilian, 170
Bird, Junius, 252
Bird, Roland T., 57, 252–54, 261, 265, 266, 268–69, 270*fig.*, 273–77, 278*fig.*, 279, 279*fig.*, 280
birds: dinosaurian origins of, 243–45; fossilized, 95, 104, 136, 155, 157, 215, 242
bison, xii, 15–16, 31–32, 84–86, 233–36, 238, 242, 281
Black Hawk (Native American chief), 2
Black Hills, South Dakota, 79, 100
Bleeding Kansas, 3
Board of Economic Warfare, U.S., 287
Bone Cabin Quarry, Wyoming, 80, 88
Brachiosaurus, 80, 236
Brachyceratops, 154, 278
Bramatherium, 193
Bridger Basin, Wyoming, 41
Bridgerian Age, 43
British Museum of Natural History, London, 122–23, 131, 194, 219
brontosaur, 53, 55, 250, 276, 291
brontothere, 207
Bronx Zoo, 84, 86, 123, 148
Brooklyn College, 277
Broom, Robert, 95
Brown, Alice Elizabeth (BB's sister), 2, 6, 6*fig.*, 8
Brown, Barnum: birth of, 5–6; childhood of, 1, 6*fig.*, 6–18; as Columbia University student, 48–49, 57–58, 59, 72, 212, 299; culinary skills of, 225; death of, 33, 294; and Divorce Reform League award, 280; early education of, 10; forename of, 5–6; gravesite of, 294*fig.*, 294–95; high school education of, 19; honorary doctorate received by, 253; illnesses suffered by, 92, 204–6, 219, 290; as Kansas farm boy, 7–11, 17–18, 298; and marriage to Marion Brown, 55*fig.*, 98*fig.*, 98–105, 111–12, 114, 125–27; and marriage to Lilian McLaughlin (*see under* Brown, Lilian); romantic affairs of, 25, 33, 34, 141, 158–59, 179–80, 286–87, 297, 298; as target of lawsuits, 158–59, 160, 171–72, 173, 180, 187, 190, 262, 298; as University of Kansas student, 19–21, 20*fig.*, 29, 30–31, 42–43, 50, 52, 205; and wagon trip with his father, 12, 14–18, 299
Brown, Barnum, professional activities of. *See* American Museum of Natural History, Brown's employment at; Disney Studios, Brown as consultant to; government, U.S., Brown's work for; mining industry consultant, Brown's employment as; oil industry consultant, Brown's employment as
Brown, Barnum, writings of: autobiographical notes, 1–5, 7–12, 16–17, 45–46, 56, 97, 154, 227, 295; field notes, 9, 25–29, 32–35, 117–18, 299; lecture on film work, 290–91; memoirs, 124, 167, 309–11; popular articles, 68–69, 73–78, 123, 150–51, 161, 163, 199, 215, 217, 320n24, 321nn61,63; scientific papers, xi, xii, 71, 72, 113, 122, 243, 249, 299, 300
Brown, Charles W. (BB's father-in-law), 171, 188
Brown, Clara (*née* Silver; BB's mother), 2, 4, 5, 6*fig.*, 7, 8, 17, 96, 297, 298, 299
Brown, Frances (BB's daughter), 33, 135; BB's AMNH records received by, 34; birth of, 114; as boarding school student, 210; gravesite of, 294*fig.*, 294–95; and mother's death, 125–27; as Radcliffe College dean, 293; raised by mother's parents, 125, 126, 128, 171, 201, 210, 295; as Red Cross worker, 285–86; and relations with stepmother Lilian, 210; and residence in

Brown, Frances *(continued)*
  Washington D.C. with BB, 285–86; and trip to Guatemala with BB, 288, 289; and visits to AMNH as teenager, 201–2, 247; as Wells College student, 115*fig. See also* biography of Brown, by daughter Frances
Brown, Frank (BB's brother), 5–6, 6*fig.*, 8
Brown, John (abolitionist leader), 3
Brown, Lilian (*née* McLaughlin; BB's second wife): BB's initial acquaintance with, 169–73, 174; and BB's liaisons with other women, 179–80, 297, 298; as caregiver during BB's illnesses, 204–6, 219–20, 290; death of, 294; and death of BB, 294; and Divorce Reform League award, 280; gravesite of, 294*fig.*, 294–95; and marriage vows with BB in India, 187–88, 190, 195, 197; possible extramarital affair of, 223, 298; and relations with BB's daughter Frances, 210; and travel in Burma with BB, 198–99, 202–6; and travel in Egypt with aunt, 169–70; and travel in France, 225; and travel in Greece with BB, 209–10, 213–14, 216–26; and travel in Guatemala with BB, 288–89; and travel in India with aunt and BB, 177, 179–80, 181–82, 184, 185, 187–89, 190–91, 192, 192*fig.*, 195, 196*fig.*, 197; and travel in Turkey with BB, 172, 173; and trips to California, 251, 261; and trip to Chicago with BB, 249; and trip to Indiana to meet BB, 265; and trip to Montana with BB, 250; and trip to Utah with BB, 289; as writer, 170, 210, 220, 224, 256, 261, 280, 288, 295, 300, 302; and Wyoming fieldwork with BB, 254, 256, 257, 260–61, 266, 267, 271
Brown, Marion (*née* Brown; BB's first wife): BB's life with, 55*fig.*, 98*fig.*, 98–105, 111–12, 114, 125, 126, 297; BB's response to death of, 125–27, 128, 133, 210, 297, 298; gravesite of, 294*fig.*, 294–95
Brown, William (BB's father), 1–2, 4, 5, 6*fig.*, 9, 10, 12–18, 79, 113, 296–97, 298–99
Bryan, William Jennings, 104–5, 186–87, 228
Buddhism, 205–6

Buettneria, 237
Bugti Hills, in Baluchistan, 182, 183, 186, 189, 190, 191, 192, 193
Bumpus, Hermon C., 111
Burgess Shale, Canada, 276
Burma: Brown's bout with malaria in, 204–6; Brown's fossil-collecting activity in, 198–207, 201*map,* 215, 217, 218, 224; Brown's oil-industry work in, 198–99, 200*fig.*; geological formation of, 199–200
Burma, fossils discovered in: mammals, 200, 201, 203–4, 205, 206, 207, 215; primates, 204, 206, 207, 215

caiman, 155
Calcutta, India, 177, 179, 195, 198
Calgary Oil Company, 148
California, 2, 251, 261
Camarasaurus, 80, 95
Cambrian Period, 276
camels, fossilized, 60, 62, 78, 83, 180, 182, 189, 191–92, 193, 223, 242
Cameron, Arizona, 237, 238, 239, 240, 260, 265, 273
Camptosaurus, 95, 232, 241, 242, 247, 251, 273
Carbondale, Kansas, 4, 5, 7–8, 11–12, 19, 46, 79, 83, 96, 104, 135, 136, 139, 146, 297
Carbon Hill, Kansas, 2, 3, 4
Carlin, W.E., 23
Carnegie, Andrew, 79–80, 181, 240
Carnegie Museum of Natural History, Pittsburgh, 63, 69, 79–80, 106, 124, 240, 281, 309
Carruth, William H., 30
Carter County Geological Society, 281
Carter County Museum, Ekalaka, Montana, 281
Cashen Ranch, Montana, 95, 241, 242, 243, 246, 249, 260, 273
Cautley, Proby T., 175, 194
caves, Brown's exploration of, 104, 136, 156, 237, 238
CBS radio broadcasts, 266, 268, 271, 272
Cenozoic Era, 114, 174, 175
Central Asiatic Expeditions (CAE), sponsored by AMNH, 43, 177, 189, 193, 194, 207, 213, 215, 218, 244, 246, 271, 280, 300

Ceratopsia, xi, 32, 112, 134, 135, 141, 143, 146, 147, 149, 150, 151, 154, 260, 266, 278, 300
Ceratosaurus, 82
Cervus punjabiensis, 194
Chaffee, Robert, 266, 267, 272, 273
chalicothere, 30
Chalicotherium, 193
champsosaur, 148, 149, 236
Champsosaurus, 93, 124
charophyte, 242
Chasmosaurus, 151
Chester, Mike, 250
Cheyenne Indians, 16
Chicago, 2, 3, 249–50
Chicxulub crater, Yucatán, 115
chimpanzee, 212–13
China, 189, 194, 199, 211, 212, 215, 218, 224
Christman, Erwin S., 42, 123
Ciochon, Russel, 207
Civil War, U.S., 3–4, 22
clam, fossilized, 133
Claosaurus, 81, 82, 87
Cloverly Formation, in Montana and Wyoming, 95, 232–33, 241, 242, 243–45, 246, 248, 249, 251, 260, 266, 273, 281
coal deposits, 2, 10–11, 110, 114, 131
coal mines, 11, 60, 233
Colbert, Edwin H., 61, 139, 174, 194, 204, 265, 277, 299
Colgate, Sidney, 238
Colorado: Brown's fossil-collecting activity in, 62, 83, 83*fig.*, 232, 236, 237, 268–69; Brown's travels with wife Marion in, 102; Marsh's fossil-collecting activity in, 134
Colorado, fossils discovered in: dinosaurs, 22–23, 134, 236; mammals, 62, 83, 236; plants, 269
Colorado Museum of Natural History, Denver, 233, 234, 235, 236
Colton, Don B., 240
Columbia University: Brown as student at, 48–49, 57–58, 59, 72, 212, 299; Gregory as student at, 212; Matthew as student at, 59–60; Osborn's work at, 42, 59
Como Bluff, Wyoming, 23, 47, 49, 50–56, 54–55*figs.*, 70, 151, 205, 240, 247, 260
Congress, U.S., 240, 289

Conrad Fissure, Arkansas, 98*fig.*, 104, 114
Cook, Harold, 227, 229, 230, 232, 233–34
Cope, Edward Drinker, 21, 22–24, 41, 60, 61, 80, 83, 86, 123, 139
copper deposits, 155, 156
Coptic Christians, in Abyssinia, 160, 161, 162, 166
Coryphodon, 43, 45, 46–47*figs.*
Corythosaurus, 143, 144*fig.*, 148, 149, 151
costs and financing of Brown's expeditions, 238, 239, 240, 246, 249, 265, 266, 273, 280; to Alberta, Canada, 141, 146, 149; to Burma, 198, 206; to Indian subcontinent, 181, 187, 189, 193; to Montana, 91, 94, 110, 121; to New Mexico, 103; to Patagonia, 70; to South Dakota, 80, 83; summaries of, 312–14; to Texas, 276, 277; to Wyoming, 34, 52, 251, 262, 273
Crazy Horse (Native American chief), 16
creationism, 275
creodont, 211, 272
Cretaceous Hall (AMNH), 292*fig.*, 311
Cretaceous Period, 80, 81, 83, 84, 110, 114–16, 131, 241, 242
Cretaceous Period fossils: in AMNH collection, 129; scientific papers on, 23, 32
Cretaceous Period fossils, discovery of: in Alberta, Canada, 125, 129, 134, 138; in central Asia, 193; in Colorado, 134; in Montana, 86, 100, 124, 129, 241, 242, 249, 281, 286*fig.*, 289, 309; in New Mexico, 103; in Patagonia, 143; in South Dakota, 29, 79, 94; in Texas, 277–78; in Wyoming, 31–32, 79, 95, 124, 129, 266
Cretaceous-Tertiary boundary, extinction event at, 110, 114–16, 153, 300
crocodile, fossilized, xii, 100, 101, 106, 112, 136, 148, 194, 236, 237, 240, 279, 279*fig.*
Crow Indians, 16, 100, 101*fig.*, 106, 232, 240, 250
crustacean, fossilized, 125, 155
Cuba: Brown's fossil-collecting activity in, 136, 155–57, 156*map*, 157*fig*; Brown's work as industry consultant in, 155, 156–57
Cuba, fossils discovered in: birds, 136, 155, 157; dinosaurs, 157; fish, 157; mammals, 136, 155; plants, 155

Currie, Ethel D., 236
Currie, Philip, 134, 136
Custer, George, 16, 30, 86
Cutler, William E., 130–31

Daemonelix structures, 27–28, 316–17n18
"Daptosaurus," 243, 244, 245
Darrow, Clarence, 187
Darton, Nelson Horatio, 232–33
Darwin, Charles, 40, 41, 42, 62, 65, 175, 176
Darwinian theory of evolution, 61, 175, 176, 186, 228, 243
Deccan Traps, India, 115
Deinonychus, 243–45, 251
Deinosuchus, 279, 279*fig.*
de la Torre, Carlos, 136
Denver, Colorado, Brown's visit to, 102
Denver Formation, Colorado, 123, 134
Denver Museum of Nature and Science, 233, 234, 235, 236
Diadiaphorus, 78
Dimetrodon, 250, 265
Dinictis, 279
Dinoland pavilion at New York World's Fair, 291, 293
dinosaur bones, discovery of: in Alberta, Canada, 125, 128–36, 130*map*, 136–52, 138*fig.*, 144*fig.*, 149*fig.*; in Arizona, 237, 239, 240, 265; in Central Asia, 193; in Colorado, 22–23, 134, 236; in India, 177; in Montana, xi, 86–94, 92–93*figs.*, 102, 107–10, 108*fig.*, 111–13, 117–21, 120*fig.*, 122*fig.*, 124–25, 129, 153–54, 232, 241–42, 243, 247, 248–49, 250, 251, 266, 273, 276, 280–81, 289, 309–11; in New Mexico, 103; in Patagonia, 143; in South Africa, 95; in South Dakota, 79, 81–82, 94; in Tanzania, 131; in Texas, 277–79; in Utah, 237, 240, 241; in Wyoming, 23, 31–32, 34–36, 47, 49, 50, 52, 53, 55, 56, 79, 80, 95, 106, 122, 124, 129, 246–47, 253, 261, 266–67, 271, 272
dinosaur bones, scientific papers on: by Cope, 22; by Hatcher, 32; by Lull, 88; by Osborn, 94, 105–6, 112–13, 121–22, 212; by Ostrom, 245
dinosaur eggs, discovery of, 43–44, 223, 300

Dinosaur National Monument, Utah, 240–41, 259, 260, 265, 266, 271, 273, 289
Dinosaur Park Formation, Alberta, Canada, 129, 151
Dinosaur Provincial Park, Alberta, Canada, 151, 152
dinosaurs: birds evolved from, 243–45; depicted in Disney film, 290–91, 293; evolution of, 138; extinction of, xii, 110, 114–17, 138; public interest in, 48, 110, 113, 123, 256, 293
dinosaurs, exhibitions of: at American Museum of Natural History, 47–48, 86, 93, 103, 111, 123–24, 151, 263, 269, 292*fig.*, 293, 311; summary of, 304–8; at World's Fair in Chicago, 250; at World's Fair in New York, 291, 293, 301
dinosaur tracks, discovery of, 237, 266, 267, 268–69, 270*fig.*, 274–75, 276–77, 278*fig.*
Diplodocus, 53, 54–55*figs.*, 55, 56, 57, 80, 237, 251
Disney Studios, Brown as consultant to, 290–91, 293
Dissopsalis, 211
Ditmars, Raymond L., 123
Divorce Reform League award, 280
Donkali tribe, in Abyssinia, 67
Dorodon, 136
Douglas, Stephen A., 2, 3
Douglass, Earl, 240
Dryopithecus, 212, 220
Dubois, Eugene, 175, 229
duckbill (dinosaur), 43, 81, 93, 102, 103, 109, 110, 123, 151, 154, 236, 249, 266, 267, 271, 272
Dudley Expedition to Abyssinia, 159–63, 164*map*, 165*fig.*, 166–69
Duke of Windsor, 287
Dyche, Lewis Lindsay, 30
Dynamosaurus imperiosus, 106, 112, 122

Edaphosaurus, 265
edentate, 72, 78, 216
Edgemont, South Dakota, 81, 94, 99
Edmonton Formation, Alberta, Canada, 128, 129, 134, 135, 137, 138, 142, 150, 151

Edmontonia, 150, 278
Edward VIII (King of England), 287
Egypt, 170–71
Ekalaka, Montana, 260, 281
Eldredge, Niles, 175
elephants, fossilized, 99, 180, 182, 183*fig.*, 191, 194, 215
Elotherium, 29, 30
Empire Oil Company, 158
entelodont, 30
Eocene Epoch, 43, 45, 174, 193, 198, 203, 204, 207, 233, 236, 272
Eotriceratops xerinsularis, 134
Ethiopia, 159–63, 164*map*, 165*fig.*, 166–69
eugenics, 41
Euphorion, 208
evolution, theory of, 60–61, 62, 105, 169, 174, 186–87, 212–13
expeditions for AMNH, Brown's. *See* American Museum of Natural History, Brown's fieldwork for
expeditions for oil industry, Brown's. *See* oil industry consultant, Brown's employment as

Falconer, Hugh, 175, 194
Falkenbach, Charles, 155, 157*fig.*
Fantasia (motion picture), 290–91, 293
Farlow, James, 57
father of Barnum Brown. *See* Brown, William
Faunce, Wayne, 282, 283
Felton, Herbert, 66, 73, 75
Fernie Formation, Alberta, Canada, 276
Field, Marshall, 80
Field Museum, Chicago, 31, 80
fieldwork for AMNH, Brown's. *See* American Museum of Natural History, Brown's fieldwork for
Figgins, Jesse, 233, 234
fish, fossilized, 157, 236, 276
flatboat, used in Brown's Canadian expedition, 130, 132*fig.*, 133, 135, 137, 139, 140
Florida, 136
flowers, Brown's interest in, 8–9
Folsom, New Mexico, 233, 234–36, 237, 238, 240, 242–43, 260

footprints, dinosaur. *See* dinosaur tracks, discovery of
Foremost Formation, Alberta, Canada, 129
Fort Pierre Formation, South Dakota, 94, 99
Fort Union Formation, Montana, 101, 153, 260
France, 225
Frick, Childs, 181, 193, 246, 265, 272, 273, 280
Fruitland Formation, New Mexico, 103
Fuchs, Theodor, 27, 317n18

Gaffney, Gene, 33, 262
Gaindatherium, 194
Gall (Native American chief), 16
Gallup, New Mexico, 103
Gavialis, 193, 194
Gazella lydekkeri, 194
Gehrig, Lou, 280
Geochelone, 194
Geological Survey, Canadian, 128, 131, 139, 143, 144
Geological Survey, U.S., 23, 154
Geological Survey of India, 175, 194, 198
Gidley, James W., 61
Giraffokeryx, 193
Glasgow, Montana, 131–32
Glen Rose, Texas, 274, 276–77, 278*fig.*, 280
glyptodont, 65, 72, 73, 78, 265
Gobi Desert, Mongolia, 44, 69, 177, 246
Gomphotherium, 194
gorilla, 212–13
Gould, Stephen Jay, 175
government, U.S., Brown's work for, xi, 149, 154, 158, 159, 284, 285–87
Grand Canyon, Arizona, 105
Granger, Walter, 43–44, 53, 55, 61, 133, 177, 241, 242, 246, 247, 249, 266, 268, 271–72, 280, 282
Grant, Madison, 137
Great Falls, Montana, 251, 260, 280–81
Greece: political conflict in, 209; war between Turkey and, 172, 214. *See also* Samos, Greece
Gregory, William King, 212–13, 220, 229, 231, 232

INDEX 359

Greybull, Wyoming, 250, 253, 258, 273
Guadagni, G. B., 273
Guatemala, 288–89
Guyot, Arnold, 41

hadrosaur bones, discovery of: in Alberta, Canada, 131, 134, 135, 137, 138, 141, 146, 148, 149, 150; in Montana, 82, 109, 111, 113, 154
hadrosaur tracks, discovery of, 271
Haile Selassie (Emperor of Ethiopia), 161, 169
Hall of Advanced Mammals (AMNH), 78, 83, 194, 223
Hall of Fossil Mammals (AMNH), 47, 236
Hall of Fossil Reptiles (AMNH), 47
Hall of Mammals and Their Extinct Relatives (AMNH), 78, 136
Hall of Ornithischian Dinosaurs (AMNH), 93, 103, 111, 143, 243, 249, 269, 278
Hall of Saurischian Dinosaurs (AMNH), 244, 269, 277
Hall of Vertebrate Origins (AMNH), 93, 100, 194, 237, 240, 265
Hapalops, 78
Harbicht, Darwin, 241, 246, 248, 249, 250
Hard, Roy, 141
Harlowton, Montana, 251
Harrar, Ethiopia, 162, 166
Harrar Province, Ethiopia, 166, 169
Harvard University, 155
Hassan, Mohammed Abdullah ("Mad Mullah"), 163, 165, 166
Hatcher, John Bell, xiii, 31–32, 38, 63–67, 68*fig.,* 69–71, 73, 75, 79, 80
Havre, Montana, 125
Hell Creek Formation, Montana, xi, 85–92, 92–93*figs.,* 103, 106–9, 108*fig.,* 111, 113, 114, 115, 117, 124–25, 129, 153–54, 239, 260, 265, 273, 276, 281, 309
Hellman, Milo, 213, 220, 229
Hermann, Adam, 42, 310
Herring, George H., 160
Hesperopithecus, 228–32, 233, 236
Hesperosuchus, 237
Hexaprotodon, 182

Hipparion, 185, 211, 224
hippopotamus, 180, 182, 191, 193, 200, 206
Hodder, Frederick, 30
Holmes, William Henry, 233, 234
hominid fossils, 169, 189, 190–91, 192, 193, 220, 227–32
Homo erectus, 175, 229
Honduras, 289
Hopkins, Edwin M., 30
Hoplitosaurus, 242, 247, 248
Hoplophoneus, 29
Hornaday, William T., 84–86, 87, 88, 89, 90
horse, fossilized, 60–61, 78, 83, 180, 182, 206, 215, 218, 223, 242, 265, 279
Horseshoe Canyon Formation, Alberta, Canada, 129
Howe, Barker M., 247, 256–57, 262
Howe Quarry, Wyoming, 247, 249, 250, 251–54, 254*fig.,* 255*fig.,* 256–57, 258*fig.,* 260–63
Hrdlička, Aleš, 234, 235
Huffman, L. A., 86
Hughes, Charles Evans, 220, 221
human evolution, 169, 174, 175–77, 208, 212–13, 228–29
humans, prehistoric, 233–36, 237–39, 288, 300
Husband, Rachel, 236
Huxley, Thomas Henry, 40, 42, 60, 176, 243–44
Hypacrosaurus, 149, 151, 154
Hypohippus, 83, 83*fig.*
Hyracotherium, 47*fig.*
Hystrix, 223

Ice Age. *See* Pleistocene Epoch
Ictitherium, 223
iguanodont, 267
impact theory of mass extinctions, 114–16
incest case involving BB's father and sister, 13–14, 296–97
India: Brown's expedition to, 180–95, 181*fig.,* 183*fig.,* 192*fig.,* 215, 217, 218, 224; European paleontological expeditions to, 174–75, 181, 182, 194; geological history of, 115, 180; social unrest in, 182–83
India, fossils discovered in: dinosaurs, 177;

hominids, 189, 190–91, 192, 193, 220; mammals, 175, 180, 182, 183*fig.*, 185, 187, 189, 191, 192, 193, 194, 211, 212, 215; turtles, 191, 192, 193
Indiana, 99, 265
Indians. *See* Native Americans
Indosuchus, 177
iridium layer, 114, 115
Irrawaddy Series, Burma, 200, 201, 203, 205, 206
Issa tribe, in Abyssinia, 167, 168, 168*fig.*
Italy, war between Ethiopia and, 169

"Java Man," 175, 229
Javelina Formation, Texas, 278
Jonas, Louis Paul, 291
Jordan, Arthur, 88–89
Jordan, Montana, 87, 88, 89, 91, 102, 107, 309, 310
Judith River Formation, Montana, 86, 102, 125, 154, 260, 266, 278
Jurassic Period fossils: in AMNH collection, 79, 129; scientific papers on, 23
Jurassic Period fossils, discovery of: in Alberta, Canada, 276; in Arizona, 240; in Colorado, 22–23, 236; in Cuba, 157; in Montana, 95, 249; in South Dakota, 94; in Utah, 240, 241; in Wyoming, 23, 46–47, 48, 49, 55, 79, 80, 261

Kaisen, Peter: as Brown's assistant in Alberta, Canada, 129, 130, 131, 132, 134–35, 136–38, 140–42, 144–45, 146, 147, 148, 150; as Brown's assistant in Montana, 111, 117, 119, 121, 122, 124, 153, 154, 240, 241, 248; as Brown's assistant in New Mexico, 236; as Brown's assistant in Wyoming, 246, 249, 251–52, 253, 260; and laboratory examination of *T. rex* fossils, 105, 310
Kansas: Brown family's homestead in, 2, 5, 7–11; Brown's university education in, 19–21, 20*fig.*, 29, 30–31; Civil War in, 3–4; Sternberg's activity in, 139
Kansas-Nebraska Act, 2, 3
Kashmir, 184, 185, 188, 190
Kirtland Shale, New Mexico, 103

Knight, Charles R., 42
Knight, Wilbur C., 48, 52
Kootenai Formation: in Canada, 280; in Montana, 251, 281
Kritosaurus navajovius, 103

labyrinthodont, 239
Lakes, Arthur, 22, 123
Lakota Formation, Montana, 241
Lamarckian theory of evolution, 61
Lambe, Lawrence, 128, 129, 131, 134, 135, 138, 143
Lance Formation: in Montana, 309; in South Dakota, 81; in Wyoming, 83, 103, 129
Langston, Wann, 159, 299
Laramie, Wyoming, 46, 48, 49, 50, 53
Lawrence, Kansas, 3, 48
lawsuits filed against Brown, 158–59, 160, 171–72, 173, 179–80, 187, 190, 262, 298
lead deposits, 190
Lehigh University, 253
Leidyosuchus, 148
Leptoceratops, 135, 151
Lewin, Roger, 176–77
Lewis, G. Edward, 241, 251, 252, 253, 258, 266, 295
Lewis and Clark expedition, 87, 128, 131
lignite beds, 110, 131
Lincoln, Abraham, 3
litoptern, 78
Little Bighorn, battle of, 16, 30, 86
Los Angeles County Museum, 239
Louisiana Purchase, 128
Lucas, O. W., 23
Lucy skeleton *(Australopithecus afarensis)*, 169
Lull, Richard Swann, 88, 89, 93, 105, 121, 309
Lydekker, Richard, 175, 194

Machaeroprosopus, 265
"Mad Mullah." *See* Hassan, Mohammed Abdullah
Magellan, Ferdinand, 77
Major, Charles I. M., 208
malaria, Brown's bouts with, 204–6, 219–20

mammals, in evolutionary theory, 60–61, 62, 114, 174, 224, 319n20
mammals, extinct, exhibition of: at American Museum of Natural History, 47, 78, 83, 136, 223, 236; at World's Fair in Chicago, 250
mammals, fossilized, discovery of: in Alberta, Canada, 133; in Arkansas, 96, 104; in Burma, 200, 201, 203–4, 205, 206, 215; in Central Asia, 193; in Colorado, 62, 83, 236; in Cuba, 136, 155; in Ethiopia, 169; in Indiana, 99; in Indian subcontinent, 175, 180, 182, 183*fig.*, 185, 187, 189, 191, 192, 193, 194, 211, 212; in Mexico, 136; in Mississippi, 136; in Montana, 112, 233; in New Mexico, 43, 45, 233–36, 242–43; in Oklahoma, 158; in Patagonia, 65, 68, 69, 72, 73–74, 78; in Samos, Greece, 208, 215, 216, 217, 218, 222, 223, 224; in South Africa, 95; in South Dakota, 25–30; in Texas, 136, 264–65; in Wyoming, 41–42, 45, 47, 48, 49, 50, 52, 53, 56
mammoth, 99, 136, 250
Mandalay, Burma, 198, 204, 205–6
Manospondylus gigas, 123
marine invertebrates, 102, 169, 281
marine reptiles, 79, 94, 141
Marsh, Othniel Charles, 21–24, 26, 31, 32, 41, 42, 47, 48, 49, 52, 56, 57, 60, 63, 71, 79, 80, 90, 123, 134, 139
Martes lydekkeri, 194
mastodon, 185, 187, 189, 193, 206, 215, 265, 288
Matthew, William Diller, 132, 154, 229, 232, 319n20; Osborn's relations with, 54, 59–60, 99, 121, 122, 212, 213, 232; scientific work of, 54–55, 59–62, 99, 121, 194, 208–9, 232
Matthew, William Diller, Brown's relations with, 59, 62, 83, 111, 135, 171, 174, 206, 302; and BB's Burma expedition, 200, 203, 204, 206, 207; and BB's Canadian expedition, 142, 143, 146, 147, 148, 150, 155, 156, 158–59; and BB's daughter's visit to AMNH, 201–2; and BB's Indian expedition, 180, 182, 183, 185, 186, 187, 188, 189, 190, 191, 193, 194, 198; and BB's Samos expedition, 208, 211–12, 213, 215, 216, 219, 221, 222, 224

Mayan ruins, in Honduras, 289
Mayr, Ernst, 293
McIntosh, John S., 244–45
McIntyre, D. A. "Mac," 257–60
McLaughlin, Lilian. *See* Brown, Lilian
Meeker, Herman E., 97
Megalocnus, 136, 155
Meltzer, David, 233
Menelek II (Emperor of Ethiopia), 161
Menke, Harold W., 50, 52, 55, 56
Merychyus, 62
Merycochoerus, 62
Mesaverde Formation, Wyoming, 265, 266, 272, 273, 278, 301
Mesocnus, 155
Mesohippus, 239, 279
Mesozoic Era, 83, 114, 257, 279, 291
Messmore and Damon firm, 250
Metamynodon, 203, 206
meteorites, investigated by Brown, 249, 260, 281
methodology, paleontological. *See* techniques, paleontological
Metzler, Rosalind, 5
Mexico: Brown's fossil-collecting activity in, 136; U.S. war against, 2
Microcnus, 155
Microvenator, 251
Miles City, Montana, 85, 87, 88, 91, 309, 310
Miller, Paul, 105
mines, fossils discovered in, 11, 60, 96, 103, 233, 236, 266, 268–69, 270*fig.*
mining industry consultant, Brown's employment as, 154–55, 156
Miocene Epoch, 25, 29, 60, 62, 66*fig.*, 67, 83, 174, 176, 208, 211, 212, 215, 232
Mioclamus, 45
Mississippi, 136
Missouri Breaks, 86, 87, 89, 102, 114
Missouri Compromise, 3
Moberly, Ralph, 242
monetary value of fossils, 106, 109, 124, 148, 230, 282, 309
Mongolia, 44, 177, 213, 215, 218, 224, 300

Monoclonius, 86, 137, 143, 144, 148, 150, 151, 240
Montana: Brown's fossil-collecting activity in, xi, 86–94, 92–93*figs.*, 100–102, 106–9, 111–12, 113, 114, 117–21, 120*fig.,* 122*fig.,* 124–25, 129, 131–32, 153–54, 232–33, 239–44, 247, 248–49, 250–51, 280–81, 289–90, 309–11; Brown's wagon trip with his father through, 16, 17, 299; Brown's wife Marion as expedition member in, 100–102, 111–12; Brown's work for oil industry in, 275–76
Montana, fossils discovered in: crocodile, 100, 101, 106, 112; dinosaurs, xi, 86–94, 92–93*figs.*, 107–10, 108*fig.*, 111–13, 117–21, 120*fig.*, 122*fig.*, 124–25, 129, 131, 153–54, 232–33, 239, 241–42, 247, 248–49, 250, 251, 266, 273, 276, 280–81, 289, 309–11; mammals, 112, 233; plants, 112; turtle, 101, 112
Moon, Harry F., 160, 162, 163
Morgan, J. P., 218, 246
Moropus, 30
Morrison Formation, 80, 129, 232–33, 245, 273; in Wyoming, 46, 48, 80, 95, 260, 261
mosasaur, 29, 79, 88, 94, 100, 260, 291
mother of Barnum Brown. *See* Brown, Clara
motorboat, used in Brown's Canadian expedition, 141–42
Mudge, Benjamin Franklin, 21, 139
Muslims, in Abyssinia, 160, 163, 165, 166
Myanmar. *See* Burma
"Mystery Dinosaur," 266, 268, 269, 272
Mytilini Formation, Samos, Greece, 215

National Academy of Sciences, U.S., 23, 228
National Geographic (periodical), 150
National Park Service, U.S., 240
Native Americans, 2, 16, 77, 85, 86, 100, 101*fig.*, 103, 106, 239, 240, 250, 281. *See also* Paleoindians
Natural History (periodical), 161, 172, 199, 215, 217, 234, 256, 261
Nature (periodical), 228
Navajo Indians, 103, 238, 239

Nebraska, 2, 3, 15, 25, 54, 211–12; purported primate fossils discovered in, 227–32
Nesodon, 78
Nevada, 237, 238
New Mexico: Brown's fieldwork in, 43–45, 103, 232, 233, 234–36, 235*fig.*, 240, 242–43
New Mexico, fossils discovered in: dinosaurs, 103; mammals, 43, 45
New York, Brown's activity in. *See under* American Museum of Natural History; Columbia University
New York Academy of Sciences, 290
New York Zoological Society, 84
Nichols, Rachel, 244–45
Nielsen, T. G., 281
Niobrara Formation, South Dakota, 94, 99
Nodosaurus, 82
North Carolina, fossils discovered in, 60
notoungulate, 78

Office of Strategic Services (OSS), 285, 286, 287
oil industry consultant, Brown's employment as, 154–55, 156–58, 159, 275, 287, 288; in Abyssinia, 159–61, 169; in Alberta, Canada, 148, 149, 273, 275; in Burma, 198–99, 200*fig.*; in Cuba, 156–57; in Montana, 275
oil production, 160, 199
Ojo Alamo Formation, New Mexico, 103
Oklahoma, 158, 264
Oldman Formation, Alberta, Canada, 129, 137, 138, 142, 143–44, 145, 146, 150
Oligocene Epoch, 25, 60, 62, 176, 279
Olsen, George, 140, 142
Onas (natives of Patagonia), 77
Ophiacodon, 264
oreodont, 25, 26, 62, 83
Ornatotholus, 151
Ornitholestes, 291
ornithomimid, 112, 121, 134, 141, 145, 146, 148, 149, 150
ornithopod, 232, 241, 242, 247
Osage County, Kansas, 2, 4, 13, 14
Osage Oil Company, 155
Osborn, A. Perry, 283
Osborn, Henry Fairfield, 40–42, 47–48,

Osborn, Henry Fairfield *(continued)*
59–61, 62–63, 70, 80, 121–23, 212, 247–48, 264, 283, 319n20; Central Asiatic Expeditions supervised by, 177, 189, 193, 194, 212, 213, 215, 271; evolutionary theory of, 61, 169, 174, 175–77, 186, 213, 228–29; monograph on elephant evolution by, 264; and purported anthropoid fossils, 227–32; scientific papers on *Tyrannosaurus rex* by, 94, 105–6, 110, 112–13, 121–22, 311–12; and Scopes trial, 186–87; studies of Canadian fossils by, 128–29, 134, 135; two-volume monograph on Proboscidea by, 99, 136

Osborn, Henry Fairfield, Brown's relations with, 95, 125, 126, 158–59, 171, 172, 247–48, 264, 297, 299; and Columbia scholarship, 48–49, 57–58, 299; and popular articles, 215, 217; and scientific publications, 71, 112, 121, 122, 300; and *Tyrannosaurus rex* fossils, 94, 105–6, 107–10, 112, 117–23, 311–12

Osborn, Henry Fairfield, as supervisor of Brown's fossil-collecting activity: in Alberta, Canada, 125, 129, 135, 137, 138, 145, 148; in Arizona, 83; in Arkansas, 96; in Burma, 203, 205, 206; in Colorado, 62; in Indiana, 99; in Indian subcontinent, 181, 186, 187, 189–90, 191, 192–94, 195, 198; in Montana, 86, 87, 89–90, 93–94, 106–10, 111–12, 117–22, 153, 154; in New Mexico, 103; in Patagonia, xii, xiv, 62–64, 70–72; in Samos, Greece, 208, 211, 215, 216, 217, 218, 219, 220, 221, 222; in South Dakota, 79, 80–83, 94; in Wyoming, 49–57, 54*fig.*, 70, 79, 80, 122

Osborn, Loulu, 248
ostracod, 242
Ostrom, John H., 232, 242–45
Ottawa Museum, fossil-collecting expedition sponsored by, 140, 141, 142, 144–47
oviraptorosaur, 251
Oxford, New York, 97, 98, 125, 126, 135, 139, 146, 171, 294, 294*fig.*

Pachyama, 45
pachycephalosaur, 151

Pachycephalosaurus, 281
Pakistan, 175, 183
Palaeocastor, 316n18
Palaeopithecus, 185
Palaeoscincus, 82, 278
Paleocene Epoch, 43, 45
Paleoindians, 233–36, 237–39, 300
Paleolithic Period, xii, 233–36
Paleozoic Era, 200
Paleozoic Era fossils: in Alberta, Canada, 276; in Kansas (Brown's childhood collection), 10–11, 17, 19; in Montana, 154; in Texas, 279
Paluxy River, Texas, 274–75, 278*fig.*
Pantolambda, 45
Paratetralophodon, 194
Parkin, Leroy, 100, 101–2, 106
Parotosaurus, 253
Paskapoo Formation, Alberta, Canada, 128, 133, 137
Patagonia: Brown's expedition to, xii–xiv, 62–78; fossilized dinosaurs discovered in, 143; fossilized mammals discovered in, 65, 68, 69, 72, 73–74, 78
Peabody, George, 21–22
Peabody Museum of Natural History, Yale University, 22, 24
Pecarichoerus, 194
peccary, fossilized, 194, 227, 231, 232
Peck, W. H., 281
Peltosaurus, 249
Pennsylvanian Subperiod, 11
Pentaceratops, 278
Permian Period fossils, discovery of: in Kansas, 264; in Oklahoma, 158, 264; in South Africa, 95; in Texas, 264
Peterson, Olaf A., xiii, 38, 41, 63, 64, 66, 67, 69, 70, 75, 121
Petrified Forest, Arizona, 104–5, 265
petroglyphs, in Utah, 289
Phenacodus, 47*fig.*
Philadelphia Academy of Sciences, 21, 22
photography, Brown's practice of, 32, 257–60
phytosaur, 60, 83, 239, 274
Pierre–Fox Hills Group, Alberta, Canada, 128, 129

Pikermi, Greece, 219
Pike's Peak, Colorado, 102
Pilgrim, Guy, 174–75, 180, 181, 182, 193, 194, 204, 213
Pithecanthropus, 229
plants, fossilized, 83, 112, 155, 265, 269, 273
plate tectonics, 180, 209
Pleistocene Epoch, 65, 96, 98*fig.*, 103, 104, 136, 155, 169, 174, 200, 233, 236, 237, 264, 265, 288, 300
plesiosaur, 94, 99, 141, 146, 249, 289, 291
Pliocene Epoch, 169, 174, 198, 211, 228
points. *See* projectile points
Pondaung Formation, Burma, 198, 200, 201, 203, 207
Pondaungia, 204, 207
Popular Science (periodical), 250
pottery, Paleoindian, 238–39
Powell, George W., 160
Price, Clayton Sumner, 131
Price, L. I., 239
primate fossils, discovery of: in Burma, 204, 206, 207, 215; in Indian subcontinent, 174, 175, 180, 185, 186, 191, 192, 211, 212, 213, 215, 220; in Nebraska (later retracted), 227–32
primates, in evolutionary theory, 212–13, 228–29
Princeton University, xiii, 38, 41, 42, 61, 62, 63, 70
Proboscidea, 99, 136, 187, 215, 216, 217, 218
projectile points, xii, 233, 234, 235, 236, 238, 240, 242–43
Propaleohoplophorus, 73, 78
Prosaurolophus, 143, 151
Prosthennops, 231, 232
Prostrepsiceros, 223
Protanancus, 194
Protoceratops, xi, 300
Protogonia, 45
Protohippus, 83
Protosuchus, 240
Pryor, Montana, 95, 232, 242, 250
Pryor Gap, Montana, 100, 101*fig.*
Pryor Mountain, Montana, 246, 247, 250, 258
Psittacotherium, 45

pterodactyl, 94
pterosaur, 157, 291
Pueblo Bonito, New Mexico, 103, 239, 260
Puercan Age, 45
punctuated equilibria, theory of, 175
Punta Arenas, Patagonia, xiii, 64–65, 75, 77, 321n61

Quakers, 22

Radcliffe College, 293
radio broadcasts, Brown's participation in, 266, 268, 271, 272, 301
Rainger, Ronald, 40, 41, 60, 121–22
Ramoceros, 83
Rangoon, Burma, 197, 198, 204, 206
Red Deer River Valley, Alberta, Canada, 125, 128–36, 130*map*, 132*fig.*, 136–52, 138*fig.*, 144*fig.*, 149*fig.*, 153, 154
Reed, W. H., 23
Reed, William, 48, 49, 50, 52, 53, 56
Regal, Brian, 41, 61, 175–76, 228
religion, and Osborn's theory of evolution, 61, 186
Republican River Beds, in Nebraska, 211–12
rhinoceros, fossilized, 62, 78, 180, 185, 189, 193, 194, 215, 218, 223
Rhizomyides, 194
Riggs, Elmer S., 26, 28, 29, 35, 36*fig.*, 45, 80, 205
Ring, Uwe, 209, 215
Rockefeller, John D., 160, 169
Rock Springs, Wyoming, 265–68, 271, 272
Roosevelt, Franklin D., 280
Royal Tyrrell Museum, Alberta, Canada, 130, 151
Rutiodon, 60
Ryan, James, 266

saber-toothed tiger, 29, 62, 96, 104, 250
Saltpeare, Mr., 75–77, 321nn61,63
Samos, Greece: Brown's fossil-collecting activity in, 208–25, 212*map*, 214*fig.*; fossilized mammals discovered in, 208, 215, 216, 217, 218, 222, 223, 224
Samotherium, 216, 218, 222
Sams, J. P., 34, 35

San Juan Basin, New Mexico, 43
Santa Cruz Formation, Patagonia, 67, 72
Saurolophus, 141, 151
Sauropelta, 242, 243, 247, 248
sauropod, 47, 56, 80, 95, 143, 212, 242, 246–47, 250, 253, 261, 274, 275, 276–79
Savage, Donald, 207
Schaffer, Bobb, 285
Schlaikjer, Erich, 266, 273, 277, 279*fig.,* 300
Schwachheim, Carl, 234, 235*fig.*
Science (periodical), 231, 234
Scollard Formation, Alberta, Canada, 129
Scopes trial, 186–87, 231
Scott, Gayle, 236
Scott, William Berryman, 41–42, 62, 63, 70
seashells, fossilized, 102
Selenoportax, 194
Shea, George, 248, 252, 258, 273, 290
Sherwood, George, 227
shipwreck in Patagonia, Brown's experience of, 77, 205, 216
Sieber, Max, 309
Silver, Charles (BB's maternal grandfather), 2, 12
Simpson, George Gaylord, 232, 252, 285, 293
Simpson, Wallis, 287
Sinclair, Harry, 251, 257, 266
Sinclair Oil Corporation, 169, 250, 251, 257, 262, 271, 276, 277, 291
Sioux Indians, 16, 85, 86
Sitting Bull, 16
Sivacanthion, 194
Sivaceros, 194
Sivapithecus, 193, 212
Sivatherium, 185
Siwalik strata, in southern Asia, 175, 180, 181*fig.,* 182, 184–85, 189, 190, 191, 193, 194, 198, 200, 211, 212, 213, 215, 216, 217, 224, 300
Skouphos, Theodore, 209, 211, 215, 218, 221–22
sloth, 78, 136, 155, 250, 288
Smithsonian Institution, Washington, D.C., 83, 84, 88, 233, 234, 276
Snake Creek deposits, Nebraska, 227, 228, 230
Snell, Mary, 5

Snow, Francis, 24
soap holes, 67
Sohio Petroleum Company, 288
Solounias, Nikos, 209, 215
Somalia, 163, 165
Sorensen, Carl, 251–52, 253, 256, 260, 261
South Africa, plans for expedition to, 94–95
South Dakota: Brown's fossil-collecting activity in, 25–29, 63, 79, 80–83, 94, 99–100, 273; Lewis's photographs of badlands in, 252
South Dakota, fossils discovered in: dinosaurs, 79, 81–82, 94; mammals, 25–30; marine reptiles, 79, 94
Southern Methodist University, 277
Spalding, David, 140
Stammosaurus, 239
Standard Oil Trust, 160
Stanocephalosaurus, 253
States, Charles, 268
stegocephalian, 253
Stegoceras, 151
Stegodon, 191, 200, 201, 203, 205, 215
Stegolophodon, 182
Stegosaurus, 47, 95, 291
Sternberg, Charles H., 86, 139–40, 142, 146, 147, 150
Sternberg, George, 139, 140, 141, 144–45
Stevens, William Chase, 30
stock market crash of 1929, 237, 238, 246
Stone Age. *See* Paleolithic period
Stravinsky, Igor, 291
Strepsiportax, 194
Struthiomimus, 134, 141
Stucker, Gil, 33–34, 266, 272
Sturges, Jonathan, 40
Styracosaurus, 151, 267
Sweet, William O., 277
synapsid, 95

Tafari, Ras, 161, 162
Tanke, Darren, 130, 134, 141
Tanzania, 131
Taylor, Melissa (*née* Brown; BB's sister), 2, 6, 6*fig.,* 8, 13–14
techniques, paleontological: drainage, 155,

157*fig.*; excavation, 26–27, 55, 56, 91, 92, 92*fig.*, 116*fig.*, 310; jacketing, 27, 29, 56, 57, 215, 217, 301; laboratory, 105, 158, 310–11; quarry map, 255*fig.*, 261
Tecumseh, 2
Teleorhinus, 100
Tendaguru, Tanzania, 131
Tenontosaurus, 241, 242, 247, 251, 273
Tertiary Period, 110, 114, 128, 133, 174, 175, 177, 180, 319n20
Tertiary Period fossils, discovery of: in central Asia, 193; in Indian subcontinent, 180, 193; in Montana, 112; in South Dakota, 25; in Texas, 279
Tertiary Period fossils, scientific papers on, 61
Texas, 136, 264, 274, 275, 276–80, 278*fig.*
theropod, 82, 123, 146, 148, 151, 244, 245, 261, 274, 277
Thoatherium, 78
Thomson, Albert, 53, 61, 122, 229, 230, 231
Tierra del Fuego, 76, 77
titanothere, 26, 212, 272, 279
tortoise, 180, 194
Trachodon, 123
tracks, dinosaur. *See* dinosaur tracks, discovery of
Tragocerus, 194
Tragoportax, 194
Treasury Department, U.S., Brown's work for, 154, 158
Triassic Period fossils, in AMNH collection, 83
Triassic Period fossils, discovery of: in Arizona, 237, 240, 265; in North Carolina, 60; in South Africa, 95
Triceratops, 31–32, 34–36, 43, 79, 80, 81, 86, 90, 93, 102, 107, 110, 113, 121, 124, 239, 250, 276
Troodon, 281
Tsipouras, Mary White, 209, 211, 213, 221, 222
Tullock Formation, Montana, 114, 115, 153
Turkey, 172–73, 186, 214
turtle, fossilized, 101, 112, 136, 148, 149, 191, 192, 193, 236, 272
Two Medicine Formation, Montana, 154

typothere, 68
Tyrannosaurus rex fossils: Brown's discovery of, xi, xii, 90–91, 94, 105–6, 154, 310–11; and Brown's post-discovery fieldwork, 106–10, 108*fig.*, 112, 117–21, 120*fig.*, 122*fig.*, 124; Brown's writings on, 123, 124; laboratory techniques applied to, 105, 310–11; media coverage of, 110, 113; Osborn's scientific papers on, 94, 105–6, 110, 112–13, 121–22, 311–12; public exhibitions of, 113, 123–24, 292*fig.*, 293, 311; and sale of type specimen to Carnegie Museum, 124, 281–82, 309; and specimens found prior to Brown's discovery, 122–23
Tyrrell, Joseph B., 128

Union Pacific company, 266, 271, 272, 301
University of California at Berkeley, 115, 232
University of Kansas, 4, 19, 21, 24, 29, 30–31, 42–43, 50, 52, 205
University of Nebraska, 25–28, 38
University of Oklahoma, 239, 279
University of Pennsylvania, 237
University of Texas, 277
University of Utah, 239–40
University of Wyoming, 48, 53
Utah, 232, 237, 238, 239–40, 259, 260, 265, 289
Utterbach, William, 106

Vanderbilt, Cornelius, 60
Velociraptor, 244
Victoria Memorial Museum, Ottawa, Canada, 131
Vokes, Harold, 275

Wabaunsee Group, 11
Walcott, Charles D., 276
Walker, Elisha, 287
Walker Museum at University of Chicago, 264
Walter Herring Endowment Expedition, 237
Wasatchian Age, 45
Washington, D.C., 285–86
Weber, Rudolph, 42
Wells College, New York, 115*fig.*

Weston, Robert C., 128, 131
Weston County, Wyoming, 81, 105
whale, 69, 136
Whitemud Formation, Alberta, Canada, 129
Whitney, William C., 60
Willis, Bess, 117, 118*fig.*, 119
Willis, John, 311
Williston, Samuel Wendell, 19–21, 24–32, 34, 36*fig.*, 36–37, 38, 39, 48, 49, 50, 52, 63, 79, 99, 205, 297
Winkley, William, 281
Wisconsin, 2, 12
Wissler, Clarke, 237, 238
Works Progress Administration (WPA), 276, 277
World's Fair: in Chicago, 249–50; in New York, 291, 293, 301
World War I, 148, 150, 154, 236
World War II, 124, 262, 281, 284, 309
Wortman, Jacob, 38–39, 42, 43, 45–46, 48–49, 53, 55, 57, 63, 71, 80, 121
Wyoming: Brown's fossil-collecting activity in, 32, 34–36, 36*fig.*, 45–46, 49–56, 54–55*figs.*, 70, 95, 124, 129, 232, 240, 246–47, 249, 250, 251–54, 254*fig.*, 255*fig.*, 256–57, 258*fig.*, 260–63, 266–68, 271–72; Brown's wagon trip with his father through, 16
Wyoming, fossils discovered in: dinosaurs, 23, 31–32, 34–36, 47, 49, 50, 52, 53, 55, 56, 79, 80, 95, 106, 122, 124, 129, 246–47, 253, 261, 266–67, 268, 271, 272; mammals, 41–42, 45, 47, 48, 49, 50, 52, 53, 56

Yahgans (natives of Patagonia), 77
Yale University: Hatcher's work at, 63; Marsh's work at, 21–22, 24, 32, 41, 48, 60, 63, 123; Ostrom's work at, 243, 245; Peabody Museum of Natural History at, 22, 24; Williston's work at, 24
Yellowstone Park, 16, 17
Yuma, Arizona, 238

Zaoditou (Queen of Ethiopia), 161
Zeuglodon, 136
Zygolophodon, 194

Text:	11.25/13.5 Adobe Garamond
Display:	Perpetua, Adobe Garamond
Compositor:	BookMatters, Berkeley
Indexer:	Andrew Joron
Cartographer:	Bill Nelson
Printer and binder:	Thomson-Shore, Inc.